Gas Injection for Disposal and Enhanced Recovery

Scrivener Publishing
100 Cummings Center, Suite 541J
Beverly, MA 01915-6106

Publishers at Scrivener
Martin Scrivener(martin@scrivenerpublishing.com)
Phillip Carmical (pcarmical@scrivenerpublishing.com)

Gas Injection for Disposal and Enhanced Recovery

Edited by

Ying Wu
Sphere Technology Connection

John J. Carroll
Gas Liquids Engineering

Qi Li
Chinese Academy of Sciences

Scrivener Publishing

WILEY

Copyright © 2014 by Scrivener Publishing LLC. All rights reserved.

Co-published by John Wiley & Sons, Inc. Hoboken, New Jersey, and Scrivener Publishing LLC, Salem, Massachusetts.
Published simultaneously in Canada.

No part of this publication may be reproduced, stored in a retrieval system, or transmitted in any form or by any means, electronic, mechanical, photocopying, recording, scanning, or otherwise, except as permitted under Section 107 or 108 of the 1976 United States Copyright Act, without either the prior written permission of the Publisher, or authorization through payment of the appropriate per-copy fee to the Copyright Clearance Center, Inc., 222 Rosewood Drive, Danvers, MA 01923, (978) 750-8400, fax (978) 750-4470, or on the web at www.copyright.com. Requests to the Publisher for permission should be addressed to the Permissions Department, John Wiley & Sons, Inc., 111 River Street, Hoboken, NJ 07030, (201) 748-6011, fax (201) 748-6008, or online at http://www.wiley.com/go/permission.

Limit of Liability/Disclaimer of Warranty: While the publisher and author have used their best efforts in preparing this book, they make no representations or warranties with respect to the accuracy or completeness of the contents of this book and specifically disclaim any implied warranties of merchantability or fitness for a particular purpose. No warranty may be created or extended by sales representatives or written sales materials. The advice and strategies contained herein may not be suitable for your situation. You should consult with a professional where appropriate. Neither the publisher nor author shall be liable for any loss of profit or any other commercial damages, including but not limited to special, incidental, consequential, or other damages.

For general information on our other products and services or for technical support, please contact our Customer Care Department within the United States at (800) 762-2974, outside the United States at (317) 572-3993 or fax (317) 572-4002.

Wiley also publishes its books in a variety of electronic formats. Some content that appears in print may not be available in electronic formats. For more information about Wiley products, visit our web site at www.wiley.com.

For more information about Scrivener products please visit www.scrivenerpublishing.com.

Cover design by Kris Hackerott

Library of Congress Cataloging-in-Publication Data:

ISBN 978-1-118-93856-0

Printed in the United States of America

10 9 8 7 6 5 4 3 2 1

Contents

Preface   xvii

Section 1: Data and Correlations

1 Densities of Carbon Dioxide-Rich Mixtures Part I: Comparison with Pure CO_2   1
Erin L. Roberts and John J. Carroll
   1.1 Introduction   1
   1.2 Density   2
   1.3 Literature Review   2
      1.3.1 CO_2 + Methane   2
      1.3.2 CO_2 + Nitrogen   4
   1.4 Calculations   4
      1.4.1 Kay's Rule   6
      1.4.2 Modified Kay's Rule   12
      1.4.3 Prausnitz-Gunn   19
   1.5 Discussion   19
   1.6 Conclusion   27
      References   27

2 Densities of Carbon Dioxide-Rich Mixtures Part II: Comparison with Thermodynamic Models   29
Erin L. Roberts and John J. Carroll
   2.1 Introduction   29
   2.2 Literature Review   30
   2.3 Calculations   30
   2.4 Lee Kesler   31
   2.5 Benedict-Webb-Rubin (BWR)   37
   2.6 Peng-Robinson   43
   2.7 Soave-Redlich-Kwong   49
   2.8 AQUAlibrium   54

	2.9	Discussion	60
	2.10	Conclusion	62
	References	63

3 On Transferring New Constant Pressure Heat Capacity Computation Methods to Engineering Practice — 65
Sepideh Rajaeirad and John M. Shaw

3.1	Introduction	65
3.2	Materials and Methods	66
3.3	Results and Discussion	67
3.4	Conclusions	70
	References	70

4 Developing High Precision Heat Capacity Correlations for Solids, Liquids and Ideal Gases — 73
Jenny Boutros and John M. Shaw

4.1	Introduction	73
4.2	Databases and Methods	75
4.3	Results and Discussion	77
4.4	Conclusion	77
	References	77

5 Method for Generating Shale Gas Fluid Composition from Depleted Sample — 79
Henrik Sørensen, Karen S. Pedersen and Peter L. Christensen

5.1		Introduction	79
5.2		Theory of Chemical Equilibrium Applied to Reservoir Fluids	80
5.3		Reservoir Fluid Composition from a Non-Representative Sample	83
	5.3.1	Depleted Gas Condensate Samples	83
	5.3.2	Samples from Tight Reservoirs	86
5.4		Numerical Examples	87
	5.4.1	Depleted Gas Condensate Samples	87
	5.4.2	Samples from Tight Reservoirs	92
5.5		Discussion of the Results	94
5.6		Conclusions	96
5.7		Nomenclature	97

		Greek letters	97
		Sub and super indices	97
		References	98

6	Phase Equilibrium in the Systems Hydrogen Sulfide + Methanol and Carbon Dioxide + Methanol	99
	Marco A. Satyro and John J. Carroll	
	6.1 Introduction	100
	6.2 Literature Review	101
	6.2.1 Hydrogen Sulfide + Methanol	101
	6.2.2 Carbon Dioxide + Methanol	101
	6.3 Modelling With Equations Of State	102
	6.4 Summary	107
	6.5 Nomenclature	108
	Greek	109
	Subscripts	109
	References	109

7	Vapour-Liquid Equilibrium, Viscosity and Interfacial Tension Modelling of Aqueous Solutions of Ethylene Glycol or Triethylene Glycol in the Presence of Methane, Carbon Dioxide and Hydrogen Sulfide	111
	Shu Pan, Na Jia, Helmut Schroeder, Yuesheng Cheng, Kurt A.G. Schmidt and Heng-Joo Ng	
	7.1 Introduction	111
	7.2 Results and Discussion	112
	7.2.1 Experimental	112
	7.2.2 Vapour Liquid Equilibrium and Phase Density Modeling	113
	7.2.3 Liquid-Phase Viscosity Modeling	117
	7.2.4 Interfacial Tension Modeling	118
	7.2.5 Commercial Software Comparison	119
	7.3 Conclusions	122
	7.4 Nomenclature	122
	7.5 Acknowledgement	125
	References	124
	Appendix 7.A	125

Section 2: Process Engineering

8 Enhanced Gas Dehydration using Methanol Injection in an Acid Gas Compression System   129
M. Rafay Anwar, N. Wayne McKay and Jim R. Maddocks

    8.1 Introduction   129
    8.2 Methodology   130
        8.2.1 Modeling Software   130
        8.2.2 Simulation Setup   131
    8.3 CASE I: 100 % CO_2 132
        8.3.1 How Much to Dehydrate   132
        8.3.2 Dehydration using Air Coolers   135
        8.3.3 Methanol injection for hydrate suppression   136
        8.3.4 Methanol Injection for Achieving 2:1 Water Content   136
        8.3.5 DexPro™ for Achieving 2:1 Water Content   137
    8.4 CASE II: 50 Percent CO_2, 50 Percent H_2S 140
        8.4.1- How Much to Dehydrate?   140
        8.4.2 Dehydration using Air Coolers   141
        8.4.3 Methanol Injection for Hydrate Suppression   141
        8.4.4 Methanol Injection for Achieving 2:1 Water Content   141
        8.4.5 DexPro™ for Achieving 2:1 Water Content   142
    8.5 CASE III: Enhanced Oil Recovery Composition   142
        8.5.1 How Much to Dehydrate?   142
        8.5.2 Enhanced Oil Recovery using Methanol   146
    8.6 Conclusion   150
    8.7 Additional Notes   151
    References   151

9 Comparison of the Design of CO_2-capture Processes using Equilibrium and Rate Based Models   153
A.R.J. Arendsen, G.F. Versteeg, J. van der Lee, R. Cota and M.A. Satyro

    9.1 Introduction   155
    9.2 VMG Rate Base   155
    9.3 Rate Based Versus Equilibrium Based Models   157
        9.3.1 Physical Absorption   158
        9.3.2 Isothermal Absorption with Chemical Reactions   160

	9.4	Process Simulations	162
		9.4.1 Configuration	162
		9.4.2 Absorber	162
		9.4.3 Absorber and Regenerator	167
		9.4.4 Temperature Profile	171
	9.5	Conclusions	173
		References	**174**
10	**Post-Combustion Carbon Capture Using Aqueous Amines: A Mass-Transfer Study**		**177**
	Ray A. Tomcej		
	10.1	Introduction	178
	10.2	Mass Transfer Basics	179
	10.3	Factors Influencing Mass Transfer	182
		10.3.1 Concentration Driving Force	182
		10.3.2 Reaction Rate Constant	184
		10.3.3 Interfacial Area	186
	10.4	Examples	188
		10.4.1 Venturi/Spray Tower System	188
		10.4.2 Amine Contactor with Pumparound	189
	10.5	Summary	190
		References	**191**
11	**BASF Technology for CO_2 Capture and Regeneration**		**193**
	Sean Rigby, Gerd Modes, Stevan Jovanovic, John Wei,		
	Koji Tanaka, Peter Moser and Torsten Katz		
	11.1	Introduction	195
	11.2	Materials and Methods	197
		11.2.1 HiPACT™ Laboratory Screening [4]	197
		11.2.2 HiPACT™ Pilot Plant [4]	197
		11.2.3 HiPACT™ Demonstration Plant [5]	199
		11.2.4 HiPACTTM Case Study [4,5]	201
		11.2.5 OASE™ blue Laboratory Screening [6, 7, 8, 9]	203
		11.2.6 OASETM blue Miniplant [7, 9]	203
		11.2.7 OASE™ blue Pilot Plant: Niederaussem [7,8,10]	203
		11.2.8 OASE™ blue Case Study [1,2]	205
	11.3	Results	206
		11.3.1 HiPACT™ CO_2 Capture Technology for Natural Gas Treating	207

		11.3.2	HiPACT™Solvent Stability and Losses	208
		11.3.3	HiPACT™ Solvent CO_2 Absorption Capacity and Kinetics	209
		11.3.4	HiPACT™ Materials Compatibility	211
		11.3.5	HiPACT™ Energy Requirements	212
		11.3.6	HiPACT™ CO_2 Stripping Pressure	212
		11.3.7	HiPACT™ Economics	213
		11.3.8	OASE™ blue CO_2 Capture Technology for Flue Gas Treating	215
		11.3.9	OASE™ blue Solvent Stability and Losses	215
		11.3.10	OASE™ blue Process Materials Compatibility	218
		11.3.11	OASE™ blue Solvent Capacity, Kinetics, Energy Requirements, and CO_2 Stripping Pressure	219
		11.3.12	OASE™ blue Economics	220
		11.3.13	OASE™ blue Emissions	222
	11.4	Conclusions		223
	11.5	Acknowledgements and Disclaimer		225
	References			226
12	**Seven Deadly Sins of Filtration and Separation Systems in Gas Processing Operations**			**227**
	David Engel and Michael H. Sheilan			
	12.1	Gas Processing and Contamination Control		228
		12.1.1	Feed and Effluent Separation	229
		12.1.2	Unit Internal Separation	230
		12.1.3	Seven Sins of Separation Devices in Gas Processing Facilities	230
	12.2	The Seven Deadly Sins of Filtration and Separation Systems in Gas Processing Operations		231
		12.2.1	Sin 1. Unsuitable Technology for the Application	231
		12.2.2	Sin 2. Incorrect Compatibility (thermal, chemical, mechanical)	233
		12.2.3	Sin 3. Deficient Vessel Design	234
		12.2.4	Sin 4. Inappropriate Sealing Surfaces	235
		12.2.5	Sin 5. Wrong Internals & Media	236
		12.2.6	Sin 6. Lack of or Incorrect Maintenance Procedures	237
		12.2.7	Sin 7. Instrumentation Deficiencies	239
	12.3	Concluding Remarks		240

Section 3: Acid Gas Injection

13 Development of Management Information System of Global Acid Gas Injection Projects — 243
Qi Li, Guizhen Liu and Xuehao Liu

- 13.1 Background — 243
- 13.2 Architecture of AGI-MIS — 244
- 13.3 Data management — 246
- 13.4 Data mining and information visualization — 248
  - 13.4.1 Injection formation — 248
  - 13.4.2 Pipeline — 249
  - 13.4.3 Injection rate — 250
  - 13.4.4 Leakage events — 250
- 13.5 Interactive program — 251
- 13.6 Conclusions — 252
- 13.7 Acknowledgements — 252
- References — 253

14 Control and Prevention of Hydrate Formation and Accumulation in Acid Gas Injection Systems During Transient Pressure/Temperature Conditions — 255
Alberto A. Gutierrez and James C. Hunter

- 14.1 General Agi System Considerations — 255
- 14.2 Composition And Properties Of Treated Acid Gases — 256
- 14.3 Regulatory And Technical Restraints On Injection Pressures — 258
- 14.4 Phase Equilibria, Hydrate Formation Boundaries And Prevention Of Hydrate Formation In Agi Systems — 259
  - 14.4.1 Hydrate Formation Conditions in AGI Compression Facilities — 259
  - 14.4.2 Hydrate Controls in AGI Compression Facilities — 260
- 14.5 Formation, Remediation And Prevention Of Hydrate Formation During Unstable Injection Conditions – Three Case Studies — 263
  - 14.5.1 Case 1: CO_2 – rich TAG (90% CO_2, 10%H_2S) Injection into a 2,000 m Deep Clastic Reservoir — 263
  - 14.5.2 Case 2: CO_2-Rich TAG (75% CO2, 25% H2S) Injected Into a 3050 m Deep Carbonate Reservoir — 267

		14.5.3	Case 3: CO2-Rich TAG (82% CO_2, 18% H_2S) Injected Into a 2950 m Deep Carbonate/Clastic Reservoir	270

	14.6	Discussion And Conclusions	272
	References		273
15	**Review of Mechanical Properties Related Problems for Acid Gas Injection**		**275**
	Qi Li, Xuehao Liu, Lei Du and Xiaying Li		
	15.1	Introduction	276
	15.2	Impact Elements	276
		15.2.1 Well	277
		15.2.2 Reservoir	280
		15.2.3 Caprock	281
	15.3	Coupled Processes	285
	15.4	Failure Criteria	286
	15.5	Conclusions	286
	15.6	Acknowledgements	287
	References		287
16	**Comparison of CO_2 Storage Potential in Pyrolysed Coal Char of different Coal Ranks**		**293**
	Pavan Pramod Sripada, MM Khan, Shanmuganathan Ramasamy, VajraTeji Kanneganti, Japan Trivedi and Rajender Gupta		
	16.1	Introduction	294
	16.2	Apparatus, Methods, & Materials	295
		16.2.1 Sample Characterization	297
	16.3	Results And Discussion	298
		16.3.1 Repeatability of adsorption experiments	298
		16.3.2 Adsorption capacities of coal	299
		16.3.3 Adsorption capacities of coal chars	300
		16.3.4 Effect of temperature on blank test	301
	16.4	Conclusion	302
	References		302

Section 4: Carbon Dioxide Storage

17 Capture of CO_2 and Storage in Depleted Gas Reservoirs in Alberta as Gas Hydrate — 305
Duo Sun, Nagu Daraboina, John Ripmeester and Peter Englezos
  17.1 Experimental — 306
  17.2 Results And Discussion — 307
  17.3 Conclusions — 310
  Reference — 310

18 Geological Storage of CO_2 as Hydrate in a McMurray Depleted Gas Reservoir — 311
Olga Ye. Zatsepina, Hassan Hassanzadeh and Mehran Pooladi-Darvish
  18.1 Introduction — 312
  18.2 Fundamentals — 313
    18.2.1 Gas Flow — 313
    18.2.2 Hydrate Phase Equilibrium — 313
    18.2.3 Assumptions — 314
  18.3 Reservoir — 314
    18.3.1 Geological Model — 314
    18.3.2 Base Case — 316
  18.4 Sensitivity Studies — 322
    18.4.1 Effect of the Injection Rate — 322
    18.4.2 Effect of the number of wells — 324
    18.4.3 Effect of the initial saturation of water — 325
    18.4.4 Effect of the heat removal — 325
  18.5 Long-term storage — 326
  18.6 Summary and conclusions — 327
  18.7 Acknowledgements — 329
  References — 329

Section 5: Reservoir Engineering

19 A Modified Calculation Method for the Water Coning
Simulation Mode in Oil Reservoirs with Bottom Water Drive   331
Weiyao Zhu, Xiaohe Huang and Ming Yue
     19.1   Introduction   331
     19.2   Mathematical Model   332
     19.3   Solution   334
     19.4   Results and Discussion   335
     19.5   Conclusions   336
     19.6   Nomenclature   336
     References   337

20 Prediction Method on the Multi-scale Flow Patterns and
the Productivity of a Fracturing Well in Shale Gas Reservoir   339
Weiyao Zhu, Jia Deng and M.A. Qian
     20.1   Introduction   340
     20.2   Multi-scale flow state analyses of the shale gas reservoirs   340
     20.3   Multi-scale seepage non-linear model in shale
            gas reservoir   343
            20.3.1   Non-linear model considering on diffusion
                         and slippage effect   343
            20.3.2   Multi-scale seepage model considering of
                         diffusion, slippage and desorption effect   347
     20.4   Productivity prediction method of fracturing well   348
            20.4.1   Productivity prediction method of
                         vertical fracturing well   348
            20.4.2   Productivity method of horizontal well
                         with multi transverse cracks   349
     20.5   Production Forecasting   351
     20.6   Conclusions   354
     20.7   Acknowledgements   354
     References   355

21 Methane recovery from natural gas hydrate in porous
sediment using gaseous CO_2, liquid CO_2, and CO_2 emulsion   357
Sheng-li Li, Xiao-Hui Wang, Chang-Yu Sun,
Qing-Yuan and Guang-Jin Chen
     21.1   Introduction
     21.2   Experiments   359

		21.2.1	Apparatus and materials	359
		21.2.2	Procedure	360
	21.3	Results and Discussion		361
		21.3.1	The replacement percent of CH_4 with gaseous CO_2	362
		21.3.2	The replacement percent of CH_4 with liquid CO_2	364
		21.3.3	The replacement percent of CH_4 with CO_2-in-water emulsion	366
	21.4	Conclusion		368
	21.5	Acknowledgements		369
	References			369

Section 6: Hydrates

22 On the Role of Ice-Solution Interface in Heterogeneous Nucleation of Methane Clathrate Hydrates — 371
PaymanPirzadeh and Peter G. Kusalik

	22.1	Introduction	371
	22.2	Method Summary	373
	22.3	Results and Discussion	373
	22.4	Summary	378
	References		379

23 Evaluating and Testing of Gas Hydrate Anti-Agglomerants in (Natural Gas + Diesel Oil + Water) Dispersed System — 381
Chang-Yu Sun, Jun Chen, Ke-Le Yan, Sheng-Li Li, Bao-ZiPeng and Guang-Jin Chen

	23.1	Introduction		381
	23.2	Experimental Apparatus And Analysis		382
	23.3	Results And Discussion		382
		23.3.1	Measurement of water-droplet size in emulsion	382
		23.3.2	Morphology of hydrate slurry formed in emulsion	383
		23.3.3	Gas consumption in the hydrate formation process in emulsion	383
		23.3.4	Flow characteristic and morphology of hydrate slurry in a flow loop apparatus	383
	23.4	Conclusion		385

Section 7: Biology

24 "Is That a Bacterium in Your Trophosome, or Are You Just Happy to See Me?" - Hydrogen Sulfide, Chemosynthesis, and the Origin of Life   **387**
Neil Christopher Griffin
    24.1   Introducing the extremophiles   387
    24.2   Tempted by the guts of another   388
    24.3   Chemosynthesis 101   389
    24.4   Chemosynthetic bacteria and the origins of life   391
    References   392

Index   **399**

Preface

The Fourth International Acid Gas Injection Symposium (AGIS IV) was held in Calgary in September of 2013. The papers in this volume are a selection of the papers presented at the Symposium.

The main core of the Symposium remains the science and technology of the injection of acid gases (hydrogen sulfide, carbon dioxide, and their mixtures) for disposal or for enhanced recovery. This includes acid gas injection (AGI), carbon capture and sequestration (CCS), and enhanced oil recovery (EOR).

There was sub-theme of gas hydrates at AGIS IV, with many papers on this subject both from a pure hydrates stand point and as related to gas injection. Included in this volume are papers discussing the storage of CO_2 in the subsurface in the form of a gas hydrate, a relatively new technology for CCS.

In addition, there are several papers on the topic of carbon capture, including new solvents, theoretical analysis, and simulation tools.

This year there was a contribution from the biological sciences which shows that not all life forms on earth find H_2S toxic – some rely on it for their lives.

YW, JJC & QL

1
Densities of Carbon Dioxide-Rich Mixtures Part I: Comparison with Pure CO_2

Erin L. Roberts and John J. Carroll

Gas Liquids Engineering, Calgary, AB, Canada

Abstract

The design of a gas injection scheme requires knowledge of the physical properties of the injection stream. These are required for both the design of the surface equipment and the modeling flow in the reservoir. One of the important physical properties is the density of the stream. The physical properties of pure carbon dioxide have been measured over a very wide range of pressure and temperature and there are several reviews of these measurements. However, the stream injected in the field is rarely pure carbon dioxide. For acid gas injection, the common impurity is methane and for carbon capture and storage, the common impurity is nitrogen.

This paper reviews the literature for measurements of the density of carbon dioxide with methane containing less than 20 mol% methane and for mixtures of carbon dioxide with nitrogen again with less than 10 mol% nitrogen.

1.1 Introduction

The injection of carbon dioxide into subsurface reservoirs is one tool to combat increasing carbon dioxide in the atmosphere. Typically the CO_2 comes from the combustion of fossil fuels, but can also come from other industrial processes such as the production of natural gas.

The transport properties of the fluid to be injected, and the density in particular, are important in the design of these processes. For example, to estimate the pressure required to inject the stream requires the density in order to calculate the hydrostatic head of fluid in the well.

To inject the gas stream it must be compressed to sufficient pressure to achieve injection. It is also important to know the density of the fluid during compression. High speed compressors are not design to handle high density fluids.

The CO_2 to be injected is rarely in the pure form. If it is separated from eat natural gas then methane is a common impurity, whereas if it comes from flue gas then the major impurity is nitrogen. These mixtures tend to be rich in carbon dioxide with only a few per cent of impurities.

1.2 Density

Typically the density is expressed as the mass density in kg/m³ or the molar density in kmol/m³. However, depending upon the experimental technique used and the personal preference of the investigator, various other quantities can be used. For example, the specific volume, m³/kg, and molar volume, m³/kmol, are merely reciprocals of the density expression given above.

It is also common to express the density in terms of the compressibility factor or z-factor. The z-factor is defined as

$$\rho = \frac{M\,P}{z\,R\,T} \tag{1.1}$$

where:
ρ – density, kg/m³
M – molar mass, kg/kmol
z – compressibility factor, unitless
P – pressure, kPa
R – universal gas constant, 8 314 m³•Pa/kmol•K
T – absolute temperature, K

1.3 Literature Review

A review of the literature was undertaken to find all of the experimental data for the density (in its various forms) for mixtures of CO_2+CH_4 and $CO_2 + N_2$ regardless of the concentration of the various components. The results of that review are summarized in this section and the data of importance to this new study are highlighted.

1.3.1 CO_2 + Methane

Table 1.1 summarizes the experimental data for mixtures of carbon dioxide and methane. Many of the density data were taken in association with

Table 1.1 Summary of Experimental Measurements of the Density of Carbon Dioxide + Methane Mixtures

Temperature (°C)	Pressure (MPa)	Composition (mol% CO_2)	Comments	Ref.
38–238	up to 70	15 to 100	report only compressibility factors tables published in paper are smoothed values	1
−20, 0, 15	2.6 to 8.5	45 to 100	molar volumes of saturated vapor and liquid	2
0, 15	2.4 to 14.5	50 to 96	molar volumes in the gas phase	2
13 to 127	up to 35	98	see text for more discussion of this data set	3
30 to 60	0.7 to 13	0 to 100	mixtures contain a small amount of N_2 (less than 1%)	4
27, 47	0.2 to 10	10 to 100		5
−68 to 47	0.1 to 48	~50		6
30, 40, 50	less than 0.1	not specified	second virial coefficients very low pressure	7
28	6.9 to 7.7	96 to 100	molar volumes of saturated vapor and liquid	8
50 to 300	10 to 100	10 to 90		9
50 to 300	20 to 100	20 to 80	mixtures contain 10 to 80 mol% nitrogen	10
−48 to 77	2 to 45	10 to 90		11

vapor-liquid equilibrium measurements and thus are the density for the saturated phases.

The first significant measurements of the densities of CO_2 + methane mixtures were those of Reamer et al. [1]. They report compressibility factors for five compositions: pure CO_2, 79.65 mol% (91.48 wt%) CO_2, 59.44 mol% (80.09 wt%) CO_2, 39.50 mol% (64.17 wt%) CO_2, and 15.31 mol% (33.15 wt%) CO_2. The temperatures and pressure of this study are such that all of the data are for the gas phase. Although the composition is slightly outside of the range of interest in this study, the density for the 79.65% CO_2 will be examined in detail.

The paper of Magee and Ely [3] is particularly interesting to this study. They measured the density of a mixture of CO_2 (98 mol%) and methane (2 mol%) over a wide range of temperatures -46° to 127°C (-55° to 260°F) and pressures up to 34.5 MPa (up to 5000 psia). However most of their data are for temperatures less than 77°C (170°F); only one isochore[1] had measurements as high as 127°C (260°F). They state that the measured densities are accurate to ±0.1%. They also report a few points for the density of pure CO_2 and their measured values are almost all within ±0.1% of the calculated value from Span and Wager (1996) with the exception of a single point and there is a typographical error in the table presented by Magee and Ely [3].

1.3.2 CO_2 + Nitrogen

As with methane and ethane, there is a significant amount of data available for the density of carbon dioxide nitrogen mixtures. These experimental studies are summarized in Table 1.2.

1.4   Calculations

An attempt was made to compare the experimental data to the compressibilities of pure carbon dioxide using the principle of corresponding states with pure CO_2 as the reference fluid.

Four different methane mixtures were investigated, 2% methane from Magee and Ely [3], two mixtures of 10% methane from Hwang et al. [11] and Brugge et al. [5], and 20% methane from Reamer et al. [1]. The 10% methane mixture from Brugge et al. [5] had data taken entirely in the vapour phase.

One nitrogen mixture of 10% was investigated, with data from two papers by Brugge et al. [5, 12].

An additional data set by Arai et al. [2] containing mixtures ranging from 4.3% to 22% methane was used. However due to each mixture having few data points, all near the critical point, the data was not included in this analysis.

Several methods for estimating the mixture critical properties where employed.

---

[1] An isochore is a line of constant volume or equivalently a line of constant density.

Table 1.2 Summary of Experimental Measurements of the Density of Carbon Dioxide + Nitrogen Mixtures

Temperature (°C)	Pressure (MPa)	Composition (mol% CO_2)	Comments	Ref.
−20, 0, 15	2.4 to 14.5	43 to 100	molar volumes of saturated vapour and liquid	2
0, 15	2.3 to 15	50 to 100	molar volumes in the gas phase	2
30 to 60	0.6 to 12.6	0 to 99.98	compressibility factors mixtures contain 0 to 99 mol% methane	4
27, 47	0.2 to 10.6	10 to 90	also report cross virial coefficients	8
−68 to 47	0.1 to 48.4	55		9
28, 30	6.9 to 8.1	96 to 100	saturated vapor and liquid densities	8
50 to 300	10 to 100	10 to 90		9
50 to 300	20 to 100	20 to 80	mixtures contain 20 to 80 mol% methane	10
−66 to 300	7 to 78	40, 50		13
50 to 125	3 to 50	25, 50	tables published in paper are smoothed values	14
−48 to 177	1 to 70	10 to 90		12
30, 40, 50	0.6 to 13	25 to 74		15

Two objective functions were calculated for all methods to minimize the error. The absolute average difference, AAD, is defined as:

$$\text{AAD} = \frac{1}{\text{NP}} \Sigma \left| z_{exp} - z_{calc} \right| \quad (1.2)$$

where:
  NP – number of points
  z_{exp} – experimental z-factor
  z_{calc} – calculated z-factor

A similar equation could be used for the densities, however for densities the average absolute errors, AAE, were used.

$$\text{AAE} = \frac{1}{\text{NP}} \Sigma \frac{|\rho_{calc} - \rho_{exp}|}{\rho_{calc}} \times 100\% \qquad (1.3)$$

where: ρ_{exp} – experimental density
ρ_{calc} – calculated density

Two other error functions were also used in the analysis but not in the optimization. For the compressibility factors the average deviations, AD, were also calculated.

$$\text{AD} = \frac{1}{\text{NP}} \Sigma \left(z_{exp} - z_{calc} \right) \qquad (1.4)$$

For the density, the average errors were calculated.

$$\text{AE} = \frac{1}{\text{NP}} \Sigma \frac{\rho_{calc} - \rho_{exp}}{\rho_{calc}} \times 100\% \qquad (1.5)$$

1.4.1 Kay's Rule

As a first approximation the pseudo-critical temperatures and pressures mixture were calculated using Kay's rule, mole fraction-weighted averages of the pure component properties:

$$pT_c = \Sigma y_i T_{ci} \qquad (1.6)$$

where: pT_c – pseudo-critical temperature, K
pP_c – pseudo-critical pressure, kPa
y_i – mole fraction of component i, unitless

$$pP_c = \Sigma y_i P_{ci} \qquad (1.7)$$

where: T_{ci} – critical temperature of component i, K
P_{ci} – critical pressure of component i, kPa

The critical temperatures and pressures for carbon dioxide, methane, and nitrogen used in this study are summarized in Table 1.3.

The experimental compressibility factors were compared to those from pure CO_2 calculated from the pseudo-reduced pressures and

Table 1.3. Critical Temperature, Volume, Pressure and Compressibility for Carbon Dioxide, Methane and Nitrogen*

Component	Critical Temperature (K)	Critical Pressure (MPa)	Critical Volume (m³/kmol)	Critical Compressibility
Carbon Dioxide	304.13	7.38	0.094	0.267
Methane	190.56	4.59	0.099	0.286
Nitrogen	126.20	3.39	0.089	0.288

* Data for carbon dioxide was obtained from software EOS-SCx Ver.02w by Tsutomu Ohmori and the data for methane and nitrogen was obtained from Perry's Handbook.

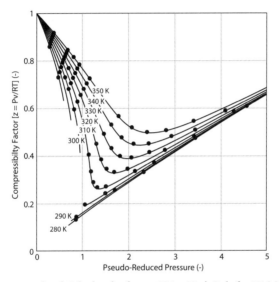

Figure 1.1 Experimental and Calculated z-factors Using Kay's Rule for 2% Methane Mixture [3].

pseudo-reduced temperatures based on Kay's Rule. For each mixture the results are shown in Figures 1.1 through 1.5. For the 2% methane, only the isotherms of 280 K through 350 K are shown, however all data was included in the error calculations. These plots show that this is a reasonable approach to calculating the z-factors for these mixtures although these can be improved. The AAD for the 2 mol%, 10 mol%, and 20 mol% mixtures are 0.002 75, 0.009 78 [11], 0.001 11 [5], and 0.007 22 respectively. The AAD for the 9% nitrogen mixture was 0.002 13.

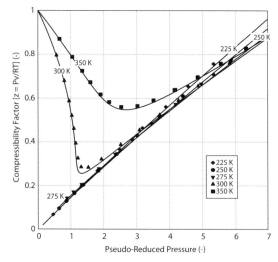

Figure 1.2 Experimental and Calculated z-factors Using Kay's Rule for 9.9% Methane Mixture 11].

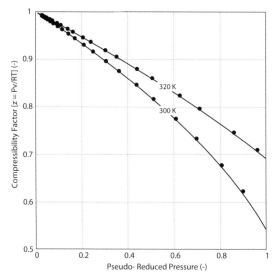

Figure 1.3 Experimental and Calculated z-factors Using Kay's Rule for 9.9% Methane Mixture [12].

Figures 1.6 through 1.10 show the experimental densities compared to the calculated densities using this approach. The predicted densities are reasonable but appear less accurate than the z-factors. The 2%, 10%, 20% methane and 9% nitrogen mixtures had AAEs of 0.633%, 2.44% [11], 0.141% [5], 0.951% and 0.423% respectively.

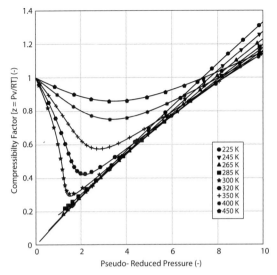

Figure 1.5 Experimental and Calculated z-factors Using Kay's Rule for 9.1% Nitrogen Mixture [5, 12].

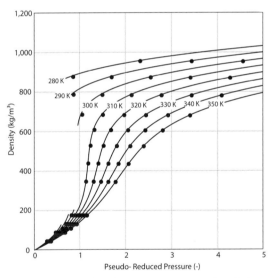

Figure 1.6 Experimental and Calculated Densities Using Kay's Rule for 2% Methane Mixture [3].

For the 2 mol% mixture, the maximum absolute difference was 0.017 79 occurring at a pseudo-reduced temperature of 1.027 (310 K) and a pseudo-reduced pressure of 1.19 (8.71 MPa). The maximum error in density was at the same pressure and temperature and was 5.30%. The 2 mol% mixture contained data taken at eight different isotherms, ranging in temperatures of 280 K

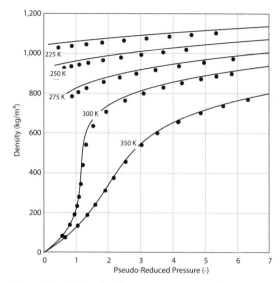

Figure 1.7 Experimental and Calculated Densities Using Kay's Rule for 9.9% Methane Mixture [11].

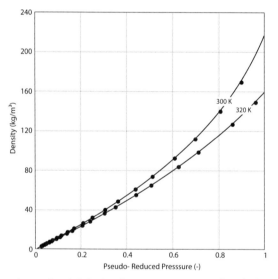

Figure 1.8 Experimental and Calculated Densities Using Kay's Rule for 9.9% Methane Mixture [12].

to 350 K (pseudo-reduced temperatures from 0.745 to 1.325). Each isotherm reached a maximum difference at a different pseudo-reduced pressure, with the higher isotherms have a maximum at a higher pseudo-reduced pressure. Isotherms below the critical temperature had negative maximum differences occurring at low pseudo-reduced pressures. From a pseudo-reduced pressure

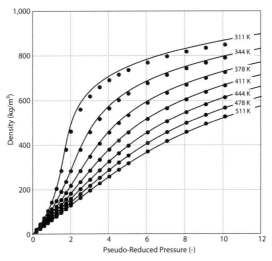

Figure 1.9 Experimental and Calculated Densities Using Kay's Rule for 20.4% Methane Mixture [1].

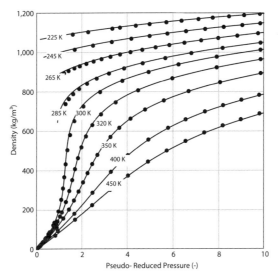

Figure 1.10 Experimental and Calculated Densities Using Kay's Rule for 9.1% Nitrogen Mixture [5, 12].

of about 3 to about 4.5 (the highest pseudo-reduced pressure), the difference in z-factors was less than about 0.002 for all isotherms.

The maximum absolute difference of z-factors for the Hwang *et al.* [11] mixture was 0.055 68 occurring at a pseudo-reduced temperature of 1.024 (300 K) and a pseudo-reduced pressure of 1.22 (8.67 MPa). For the

Brugge et al. [5] methane data, the maximum was 0.006 24 at 300 K but at a pseudo-reduced pressure of 0.898 (6.38 MPa). This was the highest pressure in the data set. The maximum error in densities occurred at the same pseudo-reduced temperature and pseudo-reduced pressure for both mixtures. The Hwang et al. [11] mixture had a maximum error of 17.1%, which is very close to the critical point for this mixture. As can be seen from Figure 1.7, densities change rapidly with changes in pressure for this isotherm near a pseudo-reduced pressure of 1. The Brugge et al. [5] mixture had a maximum error of 1.01%. The Hwang et al. [11] data set consisted of five isotherms ranging from 225 K to 350 K (pseudo-reduced temperatures of 0.768 to 1.195). The 225 K, 250 K, and 275 K isotherms all steadily increased from a difference in z-factors of about 0.002 to 0.01 across the measured pseudo- reduced pressure range of 0.45 to 6.32.

The 20 mol% mixture had a maximum absolute difference in z-factors of 0.034 56 and a maximum error in densities of 7.33% at a pseudo-reduced temperature of 1.106 (311 K) and a pseudo-reduced pressure of 1.77 (12.07 MPa). As with the other mixtures, each isotherm reached a maximum difference at a pressure near the critical point. At high pseudo-reduced pressures greater than around 6, the 311 K isotherm started to greatly increased in difference. The 478 K and 511 K isotherms also started to increase in difference in z-factors, however they increased in the negative direction.

The 9 mol% nitrogen mixture had an AAD in z-factors of 0.002 13, with a maximum of 0.017 21 occurring at a pseudo-reduced temperature of 1.041 (300 K) and a pseudo-reduced pressure of 1.36 (9.57 MPa). The maximum error in density was 5.30% occurring at the same pressure and temperature as for the z-factors.

1.4.2 Modified Kay's Rule

Although the original Kay's Rule gave reasonable estimates for the compressibilities, a modification was used here. The pseudo-critical temperatures and pressures were modified using a binary parameter. These mixing rules are discussed in this section.

1.4.2.1 Modified Pseudo-Critical Temperature

The modified pseudo-critical temperature is given by:

$$pT_c = \sum_i y_i T_{ci} + \sum_i \sum_j y_i y_j \tau_{ij} \tag{1.8}$$

where $\tau_{ij} = \tau_{ji}$ and $\tau_{ii} = 0$. For a binary mixture this becomes:

$$pT_c = y_1 T_{c1} + 2y_1 y_2 \tau_{12} + y_2 T_{c2} \quad (1.9)$$

New values of the z-factor were calculated using the modified pseudo-critical temperatures with different values of τ_{12} and the original critical pressure from Kay's Rule. An optimal τ_{12} for each mixture was found by minimizing the AAD in z-factors. The optimal τ_{12} value for the 2 mol% methane mixture was -14.09 K corresponding to an AAD in z-factors of 0.001 67 and an AAE in density of 0.343%. For the Hwang et al. [11] 10 mol% mixture the optimal τ_{12} value was -9.91 K giving an AAD in z-factors of 0.006 59 and an AAE in densities of 1.62%. The Brugge et al. [5] 10% mixture had an optimal value of τ_{12} of -4.22 giving an AAD in z-factors of 0.000 05 and an AAE in densities of 0.007%. The 20 mol% methane mixture had an optimal τ_{12} value of -8.47 K giving an AAD in z-factors of 0.004 26 and an AAE in densities of 0.508%.

It was observed that there was some variability among the data sets for the τ_{12} for the methane mixtures. An overall optimum τ_{12} of -9.75 K was estimated using a weighted average based on the number of data points in each data set.

The optimal τ_{12} value for the 9% nitrogen mixture was 0.98 K corresponding to an AAD of z-factors of 0.002 01 and an AAE in densities of 0.417%

1.4.2.2 Modified Pseudo-Critical Pressure

The pseudo-critical pressure is given by:

$$pP_c = \sum_i y_i P_{ci} + \sum_i \sum_j y_i y_j \pi_{ij} \quad (1.10)$$

where $\pi_{ij} = \pi_{ji}$ and $\pi_{ii} = 0$. For a binary mixture this becomes:

$$pP_c = y_1 P_{c1} + 2y_1 y_2 \pi_{12} + y_2 P_{c2} \quad (1.11)$$

New values of the z-factor were calculated using the modified pseudo-critical pressure with different values of π_{12} and the original critical temperature from Kay's Rule. An optimal π_{12} was found by minimizing the AAD in z-factors for each mixture. The 2 mol% methane mixture had an optimal π_{12} value of -0.177 MPa corresponding to an AAD in z-factors of 0.003 99 and an AAE in density of 0.900%. The optimal value of π_{12} for the Hwang et al. [11] 10 mol% mixture was -0.541 MPa giving an AAD of 0.008 02 in z-factors and an AAE in density of 1.98%. The optimal value for the Brugge et al. [5] 10% mixture was 0.371 giving an AAD in z-factors of 0.000 10 and a AAE in density of 0.894%. The 20 mol % mixture had an optimal π_{12} value of -0.195 MPa giving an AAD of 0.006 67 and an AAE in density of 0.894%.

As with τ_{12} there was variability with the π_{12} amongst the methane mixtures. The overall optimum π_{12} value for all methane mixtures was -0.183 MPa, which was obtained using a weighted average as with the τ_{12}.

The optimal π_{12} for nitrogen was -0.092 MPa resulting in an AAD of 0.001 88 and an AAE in density of 0.367%.

1.4.2.3 Combined

The optimal τ_{12} and π_{12} for each mixture were used in conjunction to calculate new z-factors and compared to the experimental z-factors as shown in Figures 1.11 through 1.15. These were optimized separately and may not be the optimum in a two-dimensional sense. By comparing Figures 1.1 through 1.5 to Figures 1.11 through 1.15 one can see a general improvement in the estimated compressibility factors for all of the mixtures considered here.

The AAD in z-factors were 0.001 43, 0.002 49 [11], 0.001 05 [5], 0.003 54, and 0.001 84 for the 2%, 10%, 20% methane and 9% nitrogen mixtures respectively. The experimental and calculated densities are shown in Figures 1.16 through 1.20. The AAE in densities were 0.282%, 0.716% [11], 0.139% [5], and 0.371% for the 2%, 10%, 20% methane and 9% nitrogen mixtures respectively.

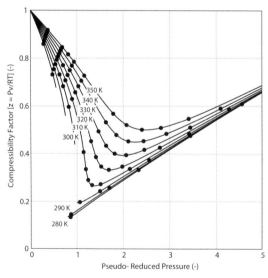

Figure 1.11 Experimental and Calculated z-factors Using Combined Modified Kay's Rule for 2% Methane Mixture (Magee and Ely, 1988).

For the combined modified pseudo-critical pressure and temperature method, the smallest AAD in z-factors were achieved for all mixtures except for the 10% methane mixture by [5], where the modified pseudo-critical pressure and modified pseudo-critical temperature methods on

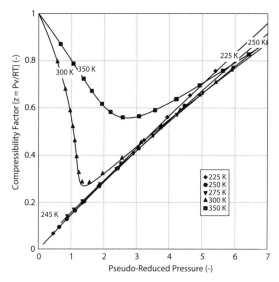

Figure 1.12 Experimental and Calculated z-factors Using Combined Modified Kay's Rule for 9.9% Methane Mixture [11].

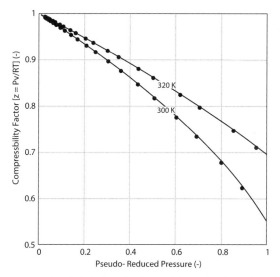

Figure 1.13 Experimental and Calculated z-factors Using Combined Modified Kay's Rule for 9.9% Methane Mixture [12].

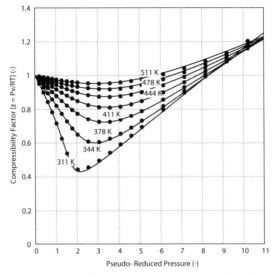

Figure 1.14 Experimental and Calculated z-factors Using Combined Modified Kay's Rule for 20.4% Methane Mixture [1].

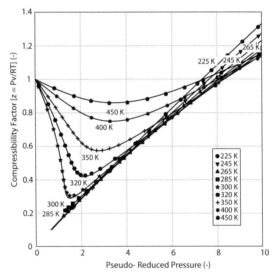

Figure 1.15 Experimental and Calculated z-factors Using Combined Modified Kay's Rule for 9.1% Nitrogen Mixture [5, 12].

the own achieved smaller differences in z-factors. However the combined method did achieve smaller difference than the original Kay's Rule for the 10% Brugge et al. [5] mixture.

The smallest AAE in densities for the 2%, 10% Hwang et al [11] and 20% mixture were observed in the combined method, however the 10% Brugee

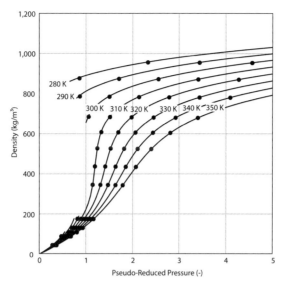

Figure 1.16 Experimental and Calculated Densities Using Combined Modified Kay's Rule for 2% Methane Mixture [3].

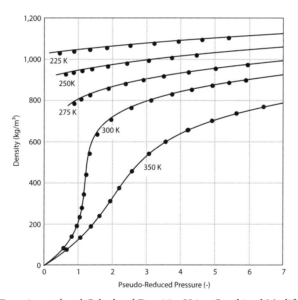

Figure 1.17 Experimental and Calculated Densities Using Combined Modified Kay's Rule for 9.9% Methane Mixture [11].

et al [5] methane mixture had the smallest error in densities in the modified pseudo-critical pressure and modified pseudo-critical temperature method. The 9% nitrogen mixture had the smallest AAE in the modified pseudo-critical pressure method.

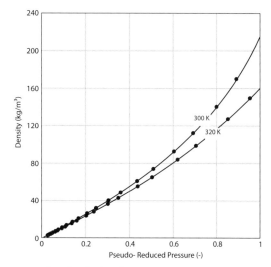

Figure 1.18 Experimental and Calculated Densities Using Combined Modified Kay's Rule for 9.9% Methane Mixture [12].

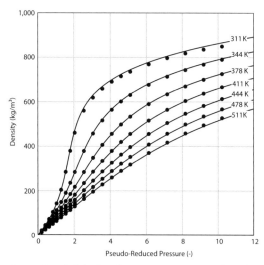

Figure 1.19 Experimental and Calculated Densities Using Combined Modified Kay's Rule for 20.4% Methane Mixture [1].

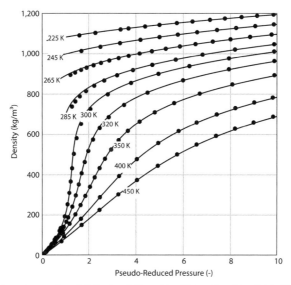

Figure 1.20 Experimental and Calculated Densities Using Combined Modified Kay's Rule for 9.1% Nitrogen Mixture [5, 12].

1.4.3 Prausnitz-Gunn

The original pseudo-critical temperature from Kay's rule was combined with the modified critical pressure estimate from Prausnitz and Gunn:

$$pP_c = \frac{R\left(\sum_i y_i z_{ci}\right) pT_c}{\sum_i y_i v_{ci}} \quad (1.12)$$

The experimental and calculated z-factors using this method can be seen in Figures 1.21 through 1.25, and the experimental and calculated densities can be seen in Figures 1.26 through 1.30. The AAD in z-factors were determined to be 0.002 56, 0.009 64 [11], 0.001 25 [5], 0.007 10, and 0.003 33 for the 2%, 10%, 20% methane and 9% nitrogen respectively. The AAE in density were 0.524%, 2.41% [11], 0.158% [5], 0.939%, and 0.607% for the 2%, 10%, 20% methane mixtures and the 9% nitrogen mixture.

1.5 Discussion

A summary of the AAD in z-factors, AD in z-factors, AAE in densities, and AE in densities can be found in Tables 1.4 through 1.8.

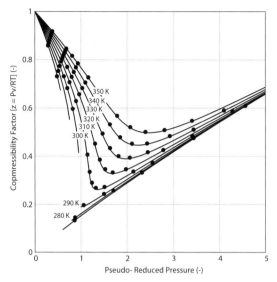

Figure 1.21 Experimental and Calculated z-factors Using Prausnitz-Gunn Equation for 2% Methane Mixture [3].

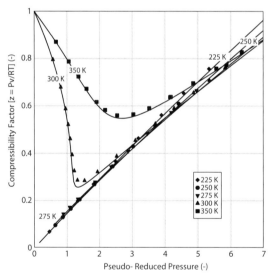

Figure 1.22 Experimental and Calculated z-factors Using Prausnitz-Gunn Equation for 9.9% Methane Mixture [11].

Densities of Carbon Dioxide-Rich Mixtures   21

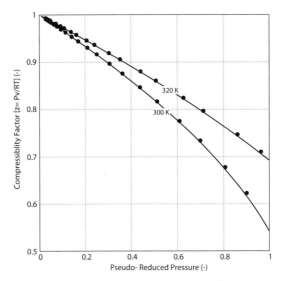

Figure 1.23 Experimental and Calculated z-factors Using Prausnitz-Gunn Equation for 9.9% Methane Mixture [12].

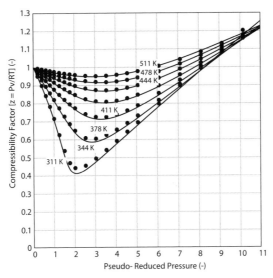

Figure 1.24 Experimental and Calculated z-factors Using Prausnitz-Gunn Equation for 20.4% Methane Mixture [1].

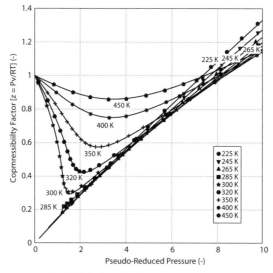

Figure 1.25 Experimental and Calculated z-factors Using Prausnitz-Gunn Equation for 9.1% Nitrogen Mixture [5, 12].

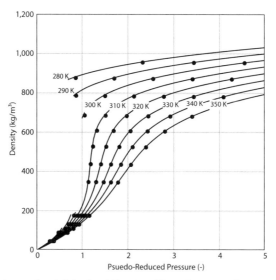

Figure 1.26 Experimental and Calculated Densities Using Prausnitz-Gunn Equation for 2% Methane Mixture [3].

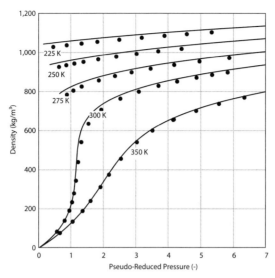

Figure 1.27 Experimental and Calculated Densities Using Prausnitz-Gunn Equation for 9.9% Methane Mixture [11].

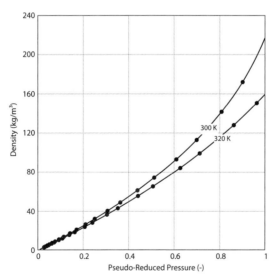

Figure 1.28 Experimental and Calculated Densities Using Prausnitz-Gunn Equation for 9.9% Methane Mixture [12].

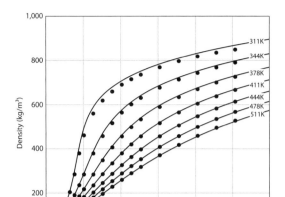

Figure 1.29 Experimental and Calculated Densities Using Prausnitz-Gunn Equation for 20.4% Methane Mixture [1].

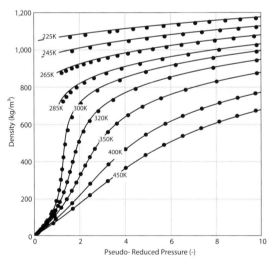

Figure 1.30 Experimental and Calculated Densities Using Prausnitz-Gunn Equation for 9.1% Nitrogen Mixture [5, 12].

For the 2%, 10% Hwang et al. [11], and 20% methane mixtures, the AAD in z-factors between Kay's Rule and Prausnitz-Gunn-method were within 2% of each other, with the z differences for Prausnitz- Gunn being slightly smaller than Kay's Rule. For the 10% Brugee et al. [5] and nitrogen

Table 1.4 Calculated Errors Using Kay's Rule

Mixture	AD	AAD	AE	AAE
2.0% Methane	0.00174	0.00257	0.528	0.633
10% Methane (Hwang et al.)	0.00978	0.00978	2.44	2.44
20.4% Methane	0.00689	0.00722	0.921	0.951
10% Methane (Brugge et al.)	0.00111	0.00111	0.141	0.141
9.1% Nitrogen	0.00007	0.00213	0.138	0.423

Table 1.5 Calculated Errors Using Modified Pseudo-Critical Temperature Method

Mixture	AD	AAD	AE	AAE
2.0% Methane	−0.00067	0.00170	−0.023	0.343
9.9% Methane (Hwang et al.)	0.00410	0.00660	1.14	1.62
20.4% Methane	0.00160	0.00430	0.200	0.508
9.9% Methane (Brugge et al.)	0.000004	0.00005	−0.003	0.007
9.1% Nitrogen	0.00041	0.00201	0.206	0.417

Table 1.6. Calculated Errors Using Modified Pseudo-Critical Pressure Method

Mixture	AD	AAD	AE	AAE
2.0% Methane	0.00325	0.00399	0.807	0.900
9.9% Methane (Hwang et al.)	0.00802	0.00802	1.98	1.98
20.4% Methane	0.00571	0.00667	0.807	0.895
9.9% Methane (Brugge et al.)	0.00004	0.00010	0.003	0.014
9.1% Nitrogen	−0.00037	0.00189	0.083	0.367

Table 1.7 Calculated Errors Using Combined Pseudo-Critical Method with Optimal τ_{12} and π_{12} Used

Mixture	τ_{12}	π_{12}	AD	AAD	AE	AAE
2.0% Methane	−14.09	−0.177	−0.00062	0.00143	−0.023	0.282
9.9% Methane (Hwang et al.)	−9.91	−0.541	0.00232	0.00249	0.688	0.716
20.4% Methane	−8.47	−0.195	0.00040	0.00354	0.088	0.417
9.9% Methane (Brugge et al.)	−4.22	0.371	−0.00105	0.00105	−0.139	0.139
9.1% Nitrogen	0.98	0.092	−0.00004	0.00184	0.151	0.371

Table 1.8 Calculated Errors Using Prausnitz-Gunn

Mixture	AD	AAD	AE	AAE
2.0% Methane	0.00179	0.00256	0.301	0.524
9.9% Methane (Hwang et al.)	0.00964	0.00964	2.41	2.41
20.4% Methane	0.00667	0.00710	0.900	0.939
9.9% Methane (Brugge et al.)	0.00125	0.00125	0.158	0.158
9.1% Nitrogen	0.00097	0.00333	0.249	0.607

mixture, Kay's rule differed from Prausnitz-Gunn by differences of 11.3% and 44.0% respectively, with Kay's Rule achieving the smaller differences.

For all mixtures, the Combined Modified Kay's rule resulted in smaller AAD in z-factors than Kay's Rule, and Prausnitz- Gunn method, with differences of 57.0%, 118% [11], 5.74% [5], 68.4%, and 14.5% less than Kay's Rule for the 2%, 10% and 20% methane mixtures and 9% nitrogen mixture respectively.

The optimum method for all mixtures, except the 10% Brugge et al. [5] mixture, was the Combined Modified Kay's Rule. This method achieved z-factors that deviated, on average, 0.281%, 0.717%, 0.417%, and 0.371% from the experimental values for the 2%, 10% [11], 20% methane and 9% nitrogen mixture respectively.

The optimum method for the 10% Brugge et al. [5] mixture was the modified pseudo-critical temperature method, with z-factors that deviated, on average, 0.005% from the experimental values. For all methane mixtures, the modified pseudo-critical temperature method achieved smaller differences than the modified pseudo-critical pressure method. For the nitrogen mixture, the opposite trend was observed.

1.6 Conclusion

This corresponding states approach shows promise for predicting compressibility factors and densities for carbon dioxide mixtures with small percentages of impurities. In fact, for many engineering applications the use of the simple Kay's Rule may be of sufficient accuracy.

For all mixtures, except for the 10% Hwang et al. [11] data, all methods predicted densities, on average, within 1% of the experimental values, and z-factors within 0.0008, on average. For the Hwang et al. [11] data, the densities were predicted within 2.5%, on average, and the z-factors were predicted, on average, within 0.0100. The Prausnitz-Gunn method predicted the densities and z-factors more accurately for all mixtures except for the 10% methane Brugge et al. [5] mixture.

For all mixtures, for the methods examined here, the Modified Kay's Rule achieved the smallest differences in z-factors and errors in densities between the experimental and calculated data. The Combined Modified Kay's Rule method gave the lowest errors for the 2%, 10% Hwang et al. [11], and 20% methane mixture. The Pseudo-Critical Pressure method gave the lowest errors for the 9% nitrogen mixture. Using the optimal methods, z-factors were estimated within 0.7%, on average, of the experimental values for all mixtures, as well as densities were estimated, on average, within 0.8% for all mixtures.

References

1. Reamer, H.H., Olds, R.H., Sage, B.H., and Lacey, W.N., "Phase Equilibrium in Hydrocarbon Systems. Methane-Carbon Dioxide System in the Gaseous Region", *Ind. Eng. Chem.*, 36, 88–90, (1944).
2. Arai, Y., Kaminishi, G.-I., and Saito, S., "The Experimental Determination of the P-V-T-X Relations for the Carbon Dioxide-Nitrogen and the Carbon Dioxide-Methane Systems", *J. Chem. Eng. Japan*, 4, 113–122, (1971).
3. Magee, J.W. and Ely, J.F., "Isochoric (p, v, T) Measurements on CO_2 and (0.98 CO_2 + 0.02 CH_4) from 225 to 400 K and Pressures to 35 MPa", *Internat. J. Thermophysics*, 9, 547–557, (1988).

4. McElroy, P.J., L.L. Kee, and C.A. Renner, "Excess Second Virial Coefficients for Binary Mixtures of Carbon Dioxide with Methane, Ethane, and Propane", *J. Chem. Eng. Data*, 35, 314–317, (1990).
5. Brugge, H.B., C.-A. Hwang, W.J. Rogers, J.C. Holste, K.R. Hall, W. Lemming, G.J. Esper, K.N. Marsh, and B.E. Gammon, "Experimental Cross Virial Coefficients for Binary Mixtures of carbon Dioxide with Nitrogen, Methane and Ethane at 300 and 320 K", *Physica A*, 156, 382–416, (1989).
6. Esper, G.J., D.M. Bailey, J.C. Holste, and K.R. Hall, "Volumetric Behavior of Near-Equimolar Mixtures for CO_2+CH_4 and CO_2+N_2", *Fluid Phase Equil.*, 49, 35–47, (1989).
7. McElroy, P.J., L. Leong, and C.A. Renner, "Excess Second Virial Coefficients for Binary Mixtures of Carbon Dioxide with Methane, Ethane, and Propane", *J. Chem. Eng. Data*, 35, 314–317, (1990).
8. Bian, B., Y. Wang, J. Shi, E. Zhao, and B.C.-Y. Lu, "Simultaneous Determination of Vapor-Liquid Equilibrium and Molar Volumes for Coexisting Phases up to the Critical Temperature with a Static Method", *Fluid Phase Equil.*, 90, 177–187, (1993).
9. Seitz, J.C., Blencoe, J.G., and Bodnar, R.J., "Volumetric Properties for $\{(1-x) CO_2 + x CH_4\}$, $\{(1-x) CO_2 + x N_2\}$, and $\{(1-x) CH_4 + x N_2\}$ at the Pressures (9.94, 19.94, 29.94, 39.94, 59.93, 79.93, and 99.93) MPa and Temperatures (323.15, 373.15, 473.15, and 573.15) K", *J. Chem. Thermo.*, 28, 521–538, (1996a).
10. Seitz, J.C., Blencoe, J.G., and Bodnar, R.J., "Volumetric Properties for $\{x_1 CO_2 + x_2 CH_4 + (1-x_1-x_2) N_2\}$ at the Pressures (19.94, 39.94, 59.93, and 99.93) MPa and Temperatures (323.15, 373.15, 473.15, and 573.15) K", *J. Chem. Thermo.*, 28, 521–538, (1996b).
11. Hwang, C-A., G.A. Iglesias-Silva., J.C. Holste, K.R. Hall, B.E. Gammon, K.N. Marsh, "Densities of Carbon Dioxide + Methane Mixtures from 225 K to 350 K at Pressures up to 35 MPa", *J. Chem. Eng. Data*, 42, 897–899, (1997)
12. Brugge, H.B., J.C. Holste, K.R. Hall, B.E. Gammon, K.N. Marsh, "Densities of Carbon Dioxide + Nitrogen from 225 K to 450 K at Pressures up to 70 MPa", *J. Chem. Eng. Data*, 42, 903–907, (1997).
13. Duarte- Garza, H.A., J.C. Holste, K.R. Hall, K.N. Marsh, and B.E. Gammon, "Isochoric pVT and Phase Equilbrium Measurements for Carbon Dioxide + Nitrogen", *J. Chem. Eng. Data*, 42, 897–899, (1997).
14. Haney, R.E.D and Bliss, H., "Compressibilities of Nitrogen- Carbon Dioxide Mixtures", *Ind. Eng. Chem.*, 36, 985–989, (1944).
15. Hacura, A., J-H. Yoon, F.G. Baglin, "Density Values of Carbon Dioxide and Nitrogen Mixtures from 500 to 2500 bar at 323 and 348 K", *J. Chem. Eng. Data*, 33, 152–154, (1988).

2

Densities of Carbon Dioxide-Rich Mixtures Part II: Comparison with Thermodynamic Models

Erin L. Roberts and John J. Carroll

Gas Liquids Engineering, Calgary, AB, Canada

The use of acid gas injection to store carbon dioxide requires knowledge of the physical properties, such as the density, of the stream. This is required to determine the necessary pressure for injection, among other considerations around the surface equipment and flow within the reservoir. The carbon dioxide used for injection is rarely pure with the common impurities of methane or nitrogen. Therefore, the effects of nitrogen and methane on the density of the carbon dioxide are important to understand.

This study compares five different models, Lee–Kesler, Benedict–Webb–Rubin (BWR), Peng–Robinson, and Soave–Redlich–Kwong (SRK) from VMGSim software, as well as AQUAlibrium software to determine the accuracy of the predictions compared to experimental data.

The Lee–Kesler model produced the most accurate predictions with an average error in density of 1.21% for all mixtures. The SRK model had the greatest errors with an average error of 5.10% for all mixtures.

2.1 Introduction

The injection of carbon dioxide into subsurface reservoirs is an effective means of removal of carbon dioxide from the atmosphere. However, the carbon dioxide injected is rarely pure. If its source is natural gas, then methane is a likely impurity. If the carbon dioxide came from flue gas, then nitrogen is the common impurity.

The density of the carbon dioxide stream is important for design considerations of the injection process, such as to determine the required pressure to achieve injection. Additionally, high speed compressors cannot be used with high density fluids.

Different Modelling software can be used to predict the density of such carbon dioxide mixtures. In this paper, five different models are analysed. Lee Kesler, Benedict-Webb-Rubin (BWR), Peng-Robinson, and Soave-Redlich-Kwong models were used with VMGSim, as well as the AQUAlibrium model, produced by FlowPhase Inc.

2.2 Literature Review

A review of the literature was performed in Carroll and Roberts (2013).

2.3 Calculations

Experimental compressibility factors as well as densities were compared to calculated values from computer based modelling programs. VMGSim was used with the Lee Kesler, BWR, Peng-Robinson, SRKproperty packages as well as AQUAlibrium software program.

Five different mixtures were evaluated using the five different methods. Four different methane mixtures, consisting of 2.0 mol% from Magee and Ely [1], two different 10 mol% mixtures from Hwang *et al.* [2] and Brugge *et al.* [3], and 20.4 mol% from Reamer *et al.* [4] were used. One nitrogen mixture, consisting of 9.1 mol% from Brugge *et al.* [3] was used. All mixtures consisted of data in both the vapour and liquid region, except for the Brugge *et al.* [3] 10mol% methane mixture, where only vapour was present.

Two objective functions were defined to compare the errors in the methods. The absolute average difference, AAD, was defined as:

$$AAD = \frac{1}{NP} \sum \left| z_{exp} - z_{calc} \right| \qquad (2.1)$$

where: NP – number of points
z_{exp} – experimental z-factor
z_{calc} – calculated z-factor

A different equation was used to compare the errors in the densities. The average absolute error, AAE, was used.

$$\text{AAE} = \frac{1}{NP}\sum \frac{|\rho_{calc} - \rho_{exp}|}{\rho_{calc}} \times 100\% \quad (2.2)$$

where: ρ_{exp} – experimental density

ρ_{calc} – calculated density

Two other error functions were used in the analysis to determine whether the z-factors and densities were being over-estimated or under-estimated by the programs. However, these functions were not used in the optimization. For the z-factors, the average difference AD, was calculated.

$$\text{AD} = \frac{1}{NP}\sum (z_{exp} - z_{calc}) \quad (2.3)$$

For the density, the average errors were calculated.

$$\text{AE} = \frac{1}{NP}\sum \frac{\rho_{calc} - \rho_{exp}}{\rho_{calc}} \times 100\% \quad (2.4)$$

2.4 Lee Kesler

Figures 2.1 through 2.4 show the experimental compressibility factors compared to the calculated compressibility factors using the Lee-Kesler model for the methane mixtures. The AAD in z factors for the 2%, 10%,

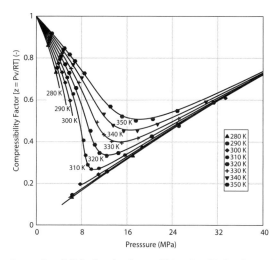

Figure 2.1 Experimental and Calculated z-factors Using Lee Kesler for 2% Methane [1].

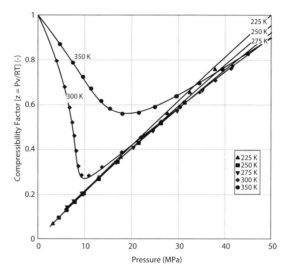

Figure 2.2 Experimental and Calculated z-factors Using Lee Kesler for 10% Methane [2].

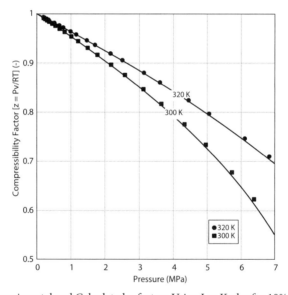

Figure 2.3 Experimental and Calculated z-factors Using Lee Kesler for 10% Methane [3].

and 20% methane mixtures were 0.005 01, 0.005 60 [2], 0.001 65 [3] and 0.006 06 respectively. The greatest difference for the 2% methane mixture was 0.032 52, occurring at 310 K and 8.35 MPa, similar to the greatest difference of the 10% Hwang *et al.* [2] mixture of 0.029 80, occurring at 300

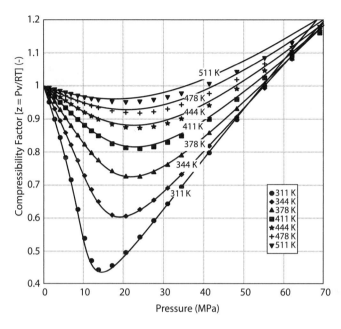

Figure 2.4 Experimental and Calculated z-factors Using Lee Kesler for 20% Methane [4].

K and 8.20MPa. The 10% Brugge *et al.* [3] data had a maximum difference in z-factors of 0.009 98 occurring at 300K and 6.38 MPa. The 20% methane mixture had data taken at much greater temperatures. The greatest difference of 0.026 30 in z-factors for the 20% mixture occurred at the highest measured temperature and pressure of 511 K and 70.0MPa.

Figure 2.5 shows the experimental and calculated Lee-Kesler z-factors for nitrogen. There were nine different points in the data where the software calculated a vapour-liquid mix, instead of solely a liquid or a vapour. These points were omitted from the data. The AAD was 0.013 72, with the greatest difference in z-factors of 0.040 28 occurring at a temperature of 225 K and pressure of 68.8MPa.

Figures 2.6 through 2.9 show the experimental densities compared to the calculated densities using the Lee-Kesler model. The AAE in densities for the 2%, 10% and 20% mixtures were 1.06%, 1.27% [2], 0.389% [3] and 0.670% respectively. The greatest error in densities for the 2% and 10% [2] methane mixtures were 8.00% and 7.56%occurring at the same temperature and pressure as for the z-factors. The 10% Brugge *et al.* [3] had a much smaller maximum error in density of 1.14% occuring at the

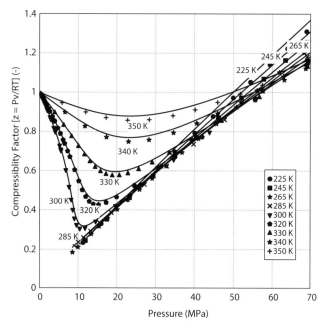

Figure 2.5 Experimental and Calculated z-factors Using Lee Kesler for 9% Nitrogen [3, 5].

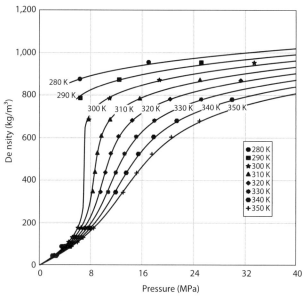

Figure 2.6 Experimental and Calculated Densities Using Lee Kesler for 2% Methane [1].

Densities of Carbon Dioxide-Rich Mixtures   35

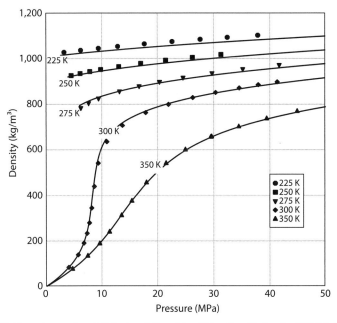

Figure 2.7 Experimental and Calculated Densities Using Lee Kesler for 10% Methane [2].

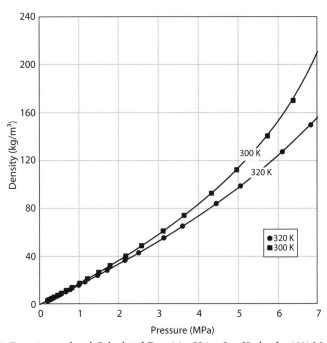

Figure 2.8 Experimental and Calculated Densities Using Lee Kesler for 10% Methane [3].

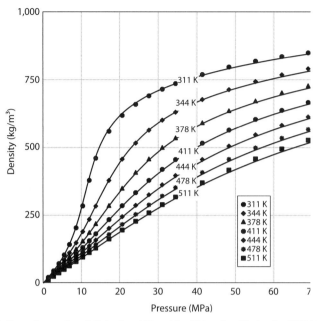

Figure 2.9 Experimental and Calculated Densities Using Lee Kesler for 20% Methane [4].

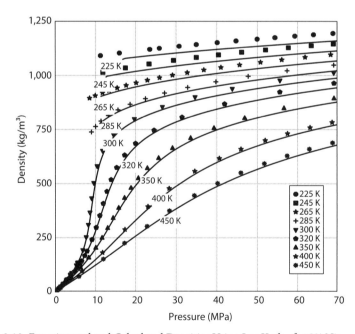

Figure 2.10 Experimental and Calculated Densities Using Lee Kesler for 9% Nitrogen [3, 5].

same temperature and pressure as the maximum difference in z-factors. However for the 20% mixture, the greatest density error of 3.52% occurred at a different temperature and pressure of 311 K and 10.3 MPa.

Figure 2.10 shows the experimental and calculated densities for the 9% nitrogen mixture. The AAE was 1.90 % with the maximum error of 9.30%occurring at a temperature of 300 K and 8.61MPa. For the 225 K, 245 K, 265 K and 285K isotherms, the experimental liquid densities extended to lower pressures than the calculated liquid densities, indicating that the software was measuring a vapour-liquid mix instead of a liquid as determined experimentally. The calculated points where there were both vapour and liquid present were omitted.

2.5 Benedict-Webb- Rubin (BWR)

Figures 2.11 through 2.14 show the experimental z-factors compared to the z-factors calculated using the BWR model for the methane mixtures. The AAD in z-factors were 0.013 96, 0.012 04 [2], 0.003 37 [3] and 0.009 56 for the 2, 10, and 20% methane mixtures. The greatest difference in z-factors

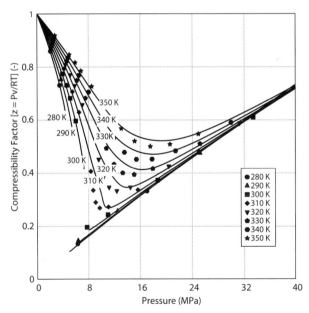

Figure 2.11 Experimental and Calculated z-factors Using BWR for 2% Methane [1].

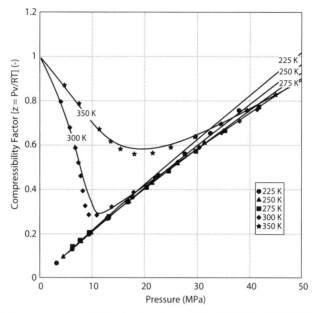

Figure 2.12 Experimental and Calculated z-factors Using BWR for 10% Methane [2].

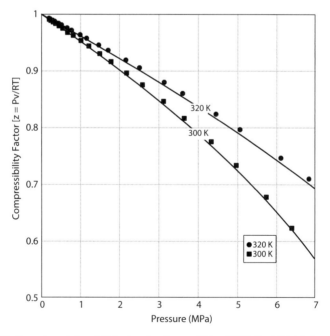

Figure 2.13 Experimental and Calculated z-factors Using BWR for 10% Methane[3].

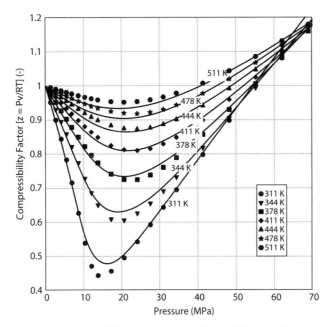

Figure 2.14 Experimental and Calculated z-factors Using BWR for 20% Methane [4].

for the 2% methane mixture was 0.101 55, occurring at a temperature of 310 K and a pressure of 9.08 MPa. The 10% Hwang *et al.* [2] methane mixture had a maximum difference of 0.094 80 at a temperature of 300 K and a pressure of 8.67 MPa. The 10% Brugge *et al.* [3] data had a much smaller maximum difference in z-factors of 0.009 30 at a 320 K and 6.11 MPa. The 20% methane mixture had a maximum difference in z-factors of 0.045 50 at a temperature of 311K and a pressure of 13.8 MPa.

Figure 2.15 shows the experimental and calculated z-factors for the 9% nitrogen mixture using BWR. The average absolute difference in z-factors was 0.012 42 and the maximum difference was 0.071 11 at 300 K and 9.57 MPa.

One data point for the 2% mixture, three points for the Hwang *et al.*, [2] 10% methane and two points for nitrogen mixture were omitted due to the software calculating a vapour-liquid mix.

Figures 2.16 through 2.19 shows the experimental densities compared to the calculated densities for the four methane mixtures. The AAE in densities for the 2%, 10% and 20% mixtures were 3.40%, 2.89% [2], 0.325% [3] and 1.20% respectively. The maximum error occurred at the same pressure and temperature for the 2%, and 20% mixtures resulting in differences of

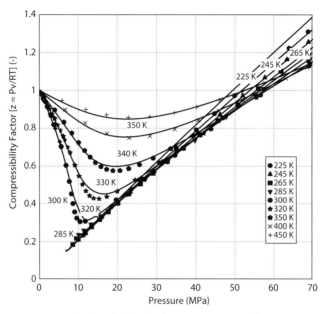

Figure 2.15 Experimental and Calculated z-factors Using BWR for 9% Nitrogen [3, 5].

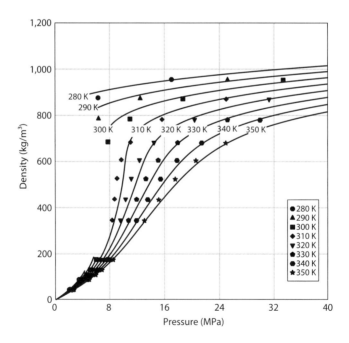

Figure 2.16 Experimental and Calculated Densities Using BWR for 2% Methane [1].

Densities of Carbon Dioxide-Rich Mixtures 41

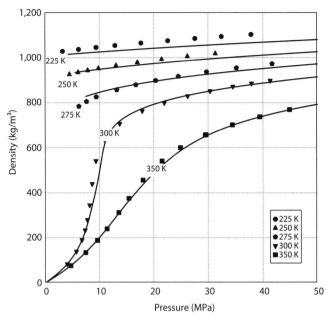

Figure 2.17 Experimental and Calculated Densities Using BWR for 10% Methane [2].

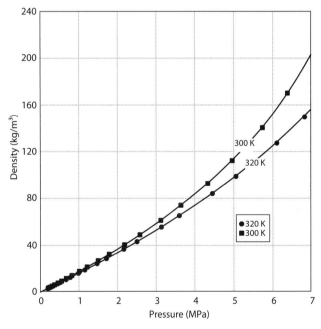

Figure 2.18 Experimental and Calculated Densities Using BWR for 10% methane [3].

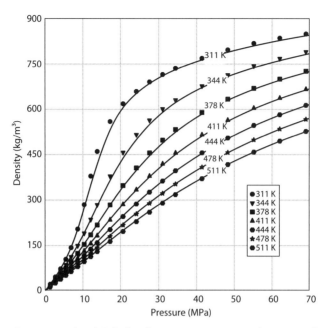

Figure 2.19 Experimental and Calculated Densities Using BWR for 20% Methane [4].

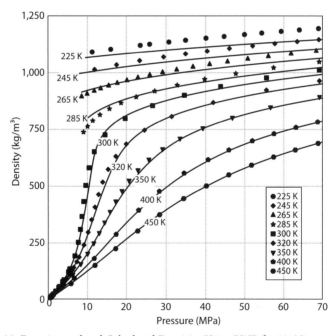

Figure 2.20 Experimental and Calculated Densities Using BWR for 9% Nitrogen [3, 5].

35.0% and 10.2% respectively. The maximum for the 10% Hwang *et al.* [2] mixture occurred at a different pressure and temperature of 9.36MPa and 300 K with an error of 30.7%. The 10% Brugge *et al.* [3] data had a maximum error in density of 0.809% occurring at 320 K and 6.83 MPa. The calculated points where there were both vapour and liquid present were omitted.

Figure 2.20 shows the experimental and calculated densities for the 9% nitrogen mixture. The AAE was 1.93% and the maximum error was 21.9% at a pressure of 9.57MPa and a temperature of 300 K.

2.6 Peng-Robinson

Figures 2.21 through 2.24 show the experimental z- factors compared to the calculated z-factors using the Peng-Robinson model for the methane mixtures. The AAD in z-factors were 0.014 47, 0.015 97 [2], 0.007 85 [3] and 0.013 98 for the 2%, 10%, 20% methane mixtures respectively.

For the 2% mixture, the maximum difference in z-factors was 0.035 50 occurring at 245K and 34.5MPa. The maximum for the 10% Hwang *et al.* [2] mixture was 0.048 94 occurring at 225 K and 37.9MPa. The Brugge *et al.* [3] mixture had a maximum difference in z-factors of 0.018 38 at 320 K

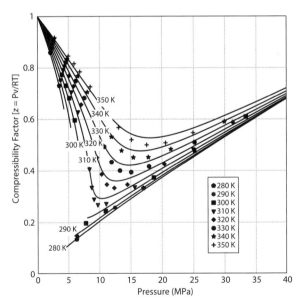

Figure 2.21 Experimental and Calculated z-factors Using Peng- Robinson for 2% Methane [1].

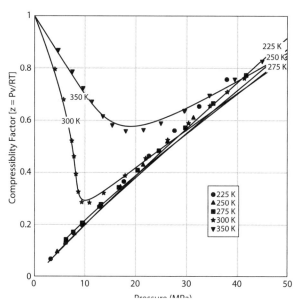

Figure 2.22 Experimental and Calculated z-factors Using Peng-Robinson for 10% Methane [2].

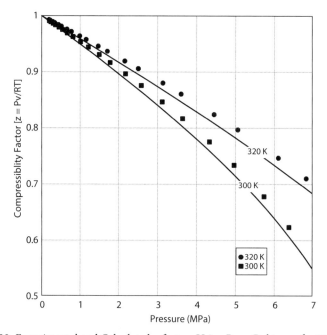

Figure 2.23 Experimental and Calculated z-factors Using Peng-Robinson for 10% Methane [3].

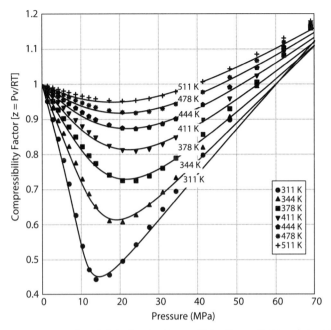

Figure 2.24 Experimental and Calculated z-factors Using Peng- Robinson for 20% Methane [4].

and 6.83 MPa. For the 20% mixture, the maximum was 0.093 33occurring at 311 K and 70.0MPa.

Figure 2.25 shows the experimental z-factors compared to the calculated z-factors for the 9% nitrogen mixture. The AAD in z-factors was 0.020 52 with a maximum difference of 0.082 43 occurring at 225 K and 68.8MPa.

Figures 2.26 through 2.29 show the experimental densities compared to the densities calculated using Peng- Robinson. The AAE in densities for the 2%, 10% and 20% mixtures were 3.19%, 3.33% [2], 0.696% [3] and 0.720% respectively. The maximum error for the 2% methane mixture was 10.3% occurring at a temperature of 300K and a pressure of 7.75MPa. The maximum error in densities for the 10% Hwang *et al.* [2] methane mixture was 6.46%, occurring at the same temperature and pressure as the maximum z-factor difference. The Brugge *et al.* [3] methane mixture had a maximum error in density of 2.25% at 300 K and 6.38 MPa. For the 20% methane mixture, the maximum was 7.76% occurring at the same pressure and temperature as the z-factor difference.

Figure 2.30 shows the experimental densities compared the calculated densities using Peng- Robinson. The AAE was 2.70% and the maximum error was 6.67% occurring at 265 K and 68.3MPa.

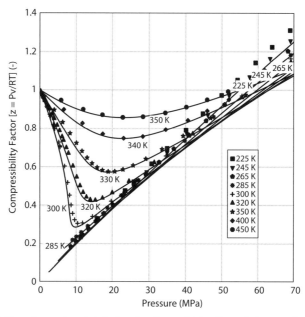

Figure 2.25 Experimental and Calculated z-factors Using Peng-Robinson for 9% Nitrogen [3, 5].

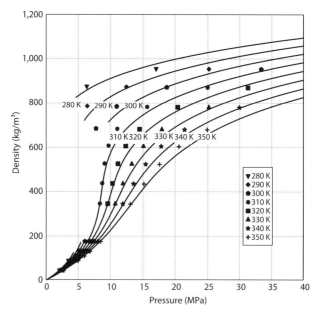

Figure 2.26 Experimental and Calculated Densities Using Peng-Robinson for 2% Methane [1].

Densities of Carbon Dioxide-Rich Mixtures 47

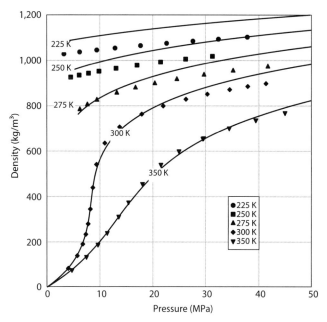

Figure 2.27 Experimental and Calculated Densities Using Peng-Robinson for 10% Methane [2].

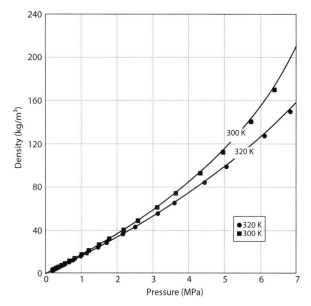

Figure 2.28 Experimental and Calculated Densities Using Peng-Robinson for 10% Methane [3].

48 Gas Injection for Disposal and Enhanced Recovery

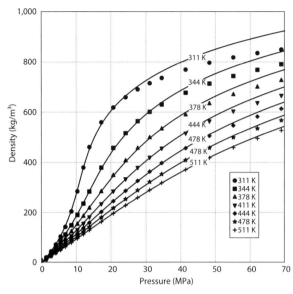

Figure 2.29 Experimental and Calculated Densities Using Peng-Robinson for 20% Methane [4].

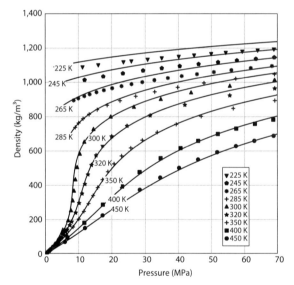

Figure 2.30 Experimental and Calculated Densities Using Peng-Robinson for 9% Nitrogen [3, 5].

2.7 Soave-Redlich-Kwong

Figures 2.31 through 2.35 show the experimental z-factors compared to the calculated z-factors for the methane mixtures using the Soave-Redlich-Kwong model. For the 2% methane mixture, the AAD was 0.029 45 with a maximum of 0.068 09 occurring at 350 K and 20.6 MPa. The 10% Hwang *et al.* [2] mixture had an AAD of 0.030 59 with a maximum of 0.063 19occurring at 350K and 25.0 MPa. The 10% Brugge *et al.* [3] mixture had a maximum difference of 0.010 27 at a temperature of 300K and a pressure of 6.38 MPa and an AAD in z-factors of 0.001 85. The 20% methane mixture had an AAD of 0.031 92with a maximum of 0.06 018 at 411K and 34.5 MPa.

Figure 2.36 shows the experimental z-factors compared to the calculated z-factors using Soave-Redlich-Kwong for the nitrogen mixture. The average absolute difference in z-factors was 0.032 41 with a maximum of 0.07 025 occurring at 450 K and 40.2 MPa.

Figures 2.37 through 2.39 show the experimental densities compared to the densities calculated using Soave-Redlich-Kwong. For the 2% methane mixture, the AAE in density was 7.53% with a maximum of 22.0% occurring at 300K and 7.75 MPa. For the 10% Hwang *et al.* [2] methane mixture, the AAE was 7.17% with a maximum of 14.5% at 300K and 10.92MPa. For

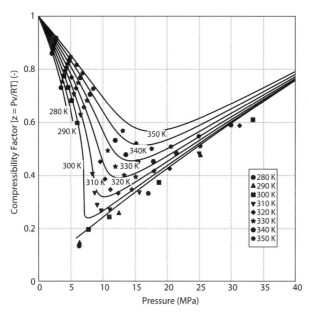

Figure 2.31 Experimental and Calculated z-factors Using Soave-Redlich-Kwong for 2% Methane [1].

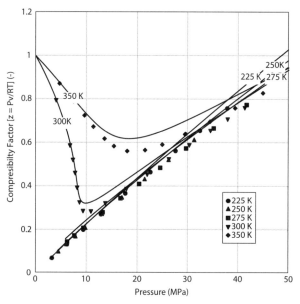

Figure 2.32 Experimental and Calculated z-factors Using Soave-Redlich-Kwong for 10% Methane [2].

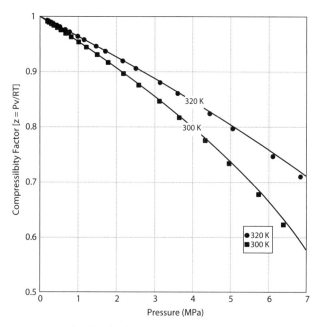

Figure 2.33 Experimental and Calculated z-factors Using Soave-Redlich-Kwong for 10% Methane [3].

DENSITIES OF CARBON DIOXIDE-RICH MIXTURES   51

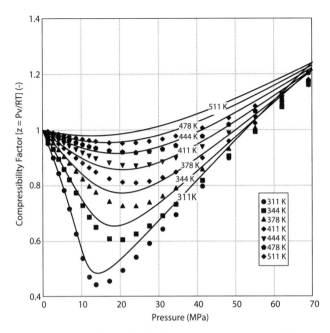

Figure 2.34 Experimental and Calculated z-factors Using Soave-Redlich-Kwong for 20% Methane [4].

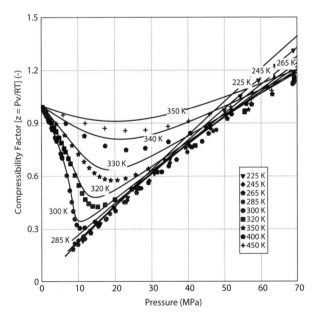

Figure 2.35 Experimental and Calculated Densities Using Soave-Redlich-Kwong for 9% Nitrogen [3, 5].

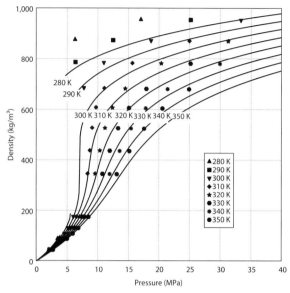

Figure 2.36 Experimental and Calculated Densities Using Soave-Redlich-Kwong for 2% Methane [1].

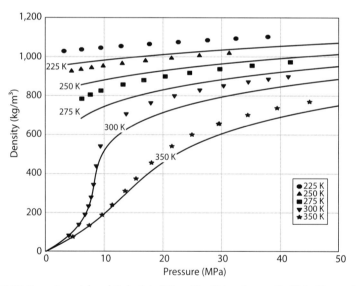

Figure 2.37 Experimental and Calculated Densities Using Soave-Redlich-Kwong for 10% Methane [2].

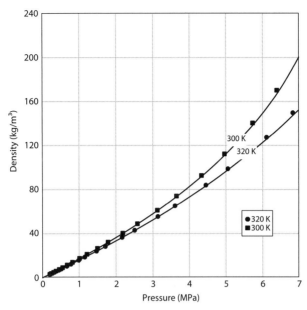

Figure 2.38 Experimental and Calculated Densities Using Soave-Redlich-Kwong for 10% Methane [3].

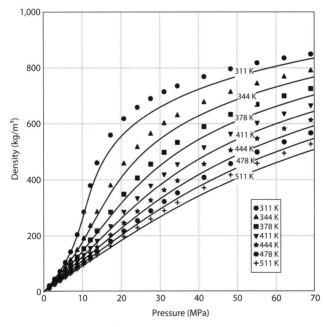

Figure 2.39 Experimental and Calculated Densities Using Soave-Redlich-Kwong for 20% Methane [4].

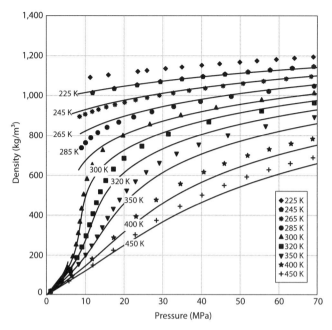

Figure 2.40 Experimental and Calculated Densities Using Soave-Redlich-Kwong for 9% Nitrogen [3, 5].

the 10% Brugge *et al.* [3] data, the AAE was 0.720% with a maximum of 2.12% at 300 K and 6.38 MPa. The 20% methane mixture had an AAE of 3.77% with a maximum of 10.2% at 311K and 17.2 MPa,

Figure 2.40 shows the experimental densities compared to the calculated densities for the nitrogen mixture. The AAE in densities was 5.19% with a maximum error of 13.8% occurring at 300K and 11.8 MPa.

2.8 AQUAlibrium

Figures 2.41 through 2.44 show the methane experimental z-factors and the z-factors calculated by the AQUAlibrium software. The AAD in z-factors for the 2%, 10% and 20% mixtures were 0.016 46, 0.014 67 [2], 0.008 04 [3] and 0.011 38 respectively. The maximum difference for the 2% mixture was 0.057 69 occuring at 34.5 MPa and 245 K. The maximum difference for the 10% Hwang *et al.* [2] mixture was 0.042 26 at 37.9 MPa and 225 K. For the Brugge *et al.* [3] mixture, the maximum difference was 0.019 22 at 320K and 6.82 MPa. For the 20% mixture, the maximum of 0.027 64 occurred at 70.0 MPa and 311 K.

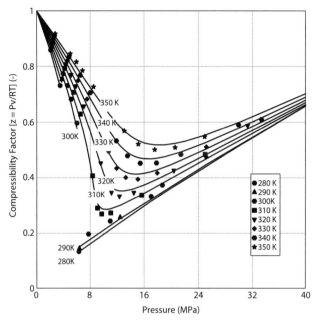

Figure 2.41 Experimental and Calculated z-factors Using AQUAlibrium software for 2% Methane [1].

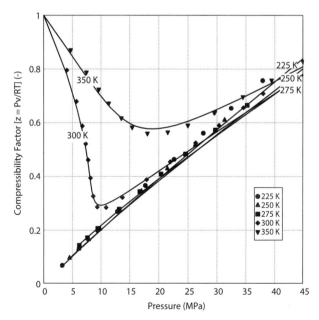

Figure 2.42 Experimental and Calculated z-factors Using AQUAlibrium software for 10% Methane [2].

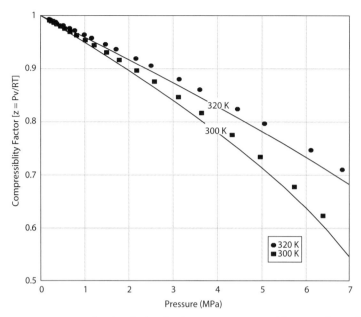

Figure 2.43 Experimental and Calculated z-factors Using AQUAlibrium software for 10% Methane [3].

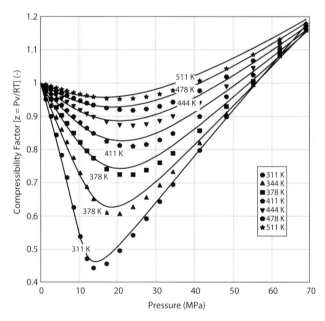

Figure 2.44 Experimental and Calculated z-factors Using AQUAlibrium software for 20% Methane [4].

Figure 2.45 shows the experimental and calculated z-factors for the nitrogen mixture. The AAD in z-factors was 0.034 64 and the maximum was 0.140 00 occurring at a pressure of 68.8 MPa and 225 K. At high pressures the experimental z-factors deviated greatly from calculated values.

Figures 2.46 through 2.49 show the experimental and calculated densities for the three methane mixtures as determined by the AQUAlibrium software. The AAE for the 2%, 10% and 20% mixtures were 3.51%, 3.05% [2] and 0.721% [3], and 1.38% respectively. The maximum error of 8.89% for the 2% mixture and 5.59 for the 10% Hwang *et al.* [2] mixture occurred at the same pressure and temperatures as the z-factors. The Brugge *et al.* [3] 10% mixture had a maximum error in density of 2.48% occuring at 300 K and 6.38 MPa. The 20% mixture had a maximum of 4.75% at 17.2 MPa and 311 K.

Figure 2.50 shows the experimental and calculated densities for the nitrogen mixture. The AAE was 4.47% with a maximum error of 10.8% at a pressure of 68.8 MPa and 225K. The two lowest isotherms, 225 K and 245 K, consistently had calculated densities 8-10% off the experimental values.

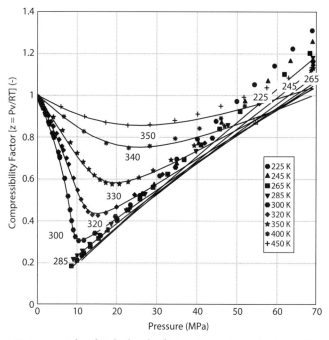

Figure 2.45 Experimental and Calculated z-factors Using AQUAlibrium software for 9.1% Nitrogen [3, 5].

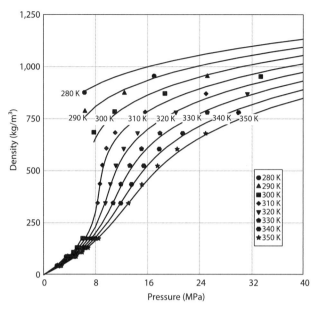

Figure 2.46 Experimental and Calculated Densities Using AQUAlibrium software for 2% Methane [1].

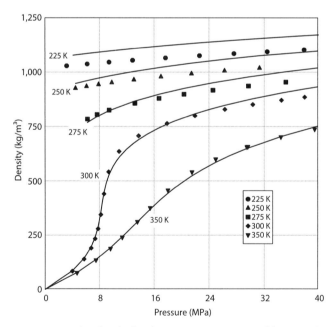

Figure 2.47 Experimental and Calculated Densities Using AQUAlibrium software for 10% Methane [2].

DENSITIES OF CARBON DIOXIDE-RICH MIXTURES   59

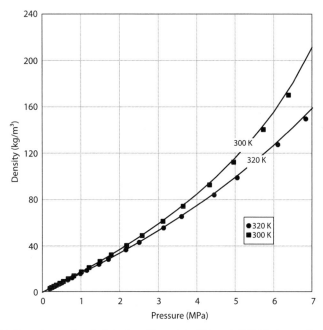

Figure 2.48 Experimental and Calculated Densities Using AQUAlibrium software for 10% Methane [3].

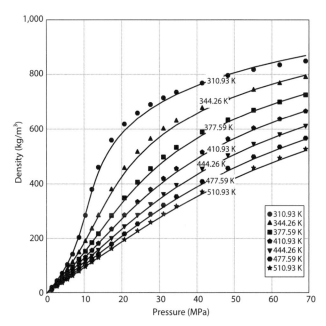

Figure 2.49 Experimental and Calculated Densities Using AQUAlibrium software for 20% Methane [4].

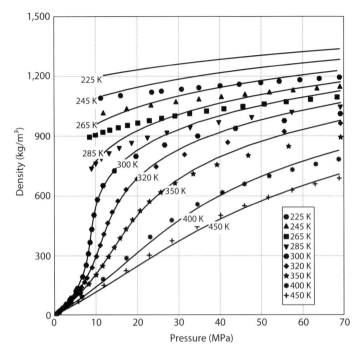

Figure 2.50 Experimental and Calculated Densities Using AQUAlibrium software for 9.1% Nitrogen [3, 5].

2.9 Discussion

A summary of the AAD and AD for z-factors, as well as AAE and AE for densities can be found in Tables 2.1–2.4.

The Lee Kesler model consistently produced the smallest errors and differences for all mixtures. For the Brugge et al. [3] 10% methane data and the 9% Nitrogen mixture, the Lee Kesler and the BWR model both produced similar results. The BWR model gave a smaller AAE for the Brugge et al. [3] 10% methane data, as well as produced a smaller AAD for the 9% Nitrogen mixture. In all other cases, Lee Kesler gave the smallest differences in z-factors and errors in density.

The BWR model had the second best overall average absolute error for all mixtures. However, this model had the highest maximum errors, with points around the critical point estimated up to 35% from the experimental values.

Both the Peng-Robinson and AQUAlibrium models achieved similar results to each other with the Peng-Robinson producing slightly smaller

Table 2.1 Absolute Average Difference in z-factors for the Methane and Nitrogen Mixtures Using the Different Models

	Lee Kesler	BWR	Peng-Robinson	SRK	AQUAlibrium
2% Methane	0.00501	0.01396	0.01447	0.02945	0.01646
10% Methane (Hwang et al.)	0.00560	0.01204	0.01597	0.03059	0.01467
10% Methane (Bruggeet al.)	0.00165	0.00337	0.00785	0.00184	0.00804
20% Methane	0.00606	0.00956	0.01398	0.03192	0.01138
9% Nitrogen	0.01372	0.01242	0.02052	0.03241	0.03464
Average	0.00808	0.01113	0.01605	0.02906	0.02037

Table 2.2 Average Difference in z-factors for the Methane and Nitrogen Mixtures Using the Difference Models

	Lee Kesler	BWR	Peng-Robinson	SRK	AQUAlibrium
2% Methane	0.00018	−0.00658	0.00201	−0.02945	0.00967
10% Methane (Hwang et al.)	0.00022	−0.00920	0.01222	−0.03059	0.01080
10% Methane (Bruggeet al.)	0.00165	0.00337	0.00785	0.00184	0.00804
20% Methane	−0.00415	0.00155	0.01227	−0.03187	−0.00462
9% Nitrogen	−0.01372	−0.00804	0.01590	−0.03241	0.03369
Average	−0.00566	−0.00433	0.01129	−0.02874	0.01393

Table 2.3 Average Absolute % Error in Densities for the Methane and Nitrogen Mixtures Using the Different Models

	Lee Kesler	BWR	Peng-Robinson	SRK	AQUAlibrium
2% Methane	1.06	3.40	3.19	7.53	3.51
10% Methane (Hwang et al.)	1.27	2.89	3.33	7.17	3.05
10% Methane (Bruggeet al.)	0.389	0.325	0.696	0.720	0.721
20% Methane	0.670	1.20	1.50	3.77	1.38
9% Nitrogen	1.90	1.93	2.70	5.19	4.47
Average	1.21	1.98	2.37	5.10	2.97

Table 2.4 Average % Error in Densities for the Methane and Nitrogen Mixtures Using the Different Models

	Lee Kesler	BWR	Peng-Robinson	SRK	AQUAlibrium
2% Methane	−0.064	−1.58	−0.249	−7.53	1.59
10% Methane (Hwang et al.)	0.147	−2.27	2.44	−7.17	2.12
10% Methane (Bruggeet al.)	−0.258	−0.061	0.486	−0.720	0.517
20% Methane	−0.359	0.022	1.25	−3.77	−0.620
9% Nitrogen	−1.90	−1.24	1.79	−5.19	3.38
Average	−0.765	−0.973	1.24	−5.09	1.56

errors. The error in both models increased as the pressure increased, with the maximum errrors for the densities occurring at the highest pressures.

The SRK model had the highest overall error in density, following a similar trend as the Peng-Robinson and AQUAlibrium models with error increasing with increasing pressure.

The Lee Kesler, BWR and SRK models all had negative AE and AD's for most mixtures, indicating that more often, the predicted density was less than the experimental density, and that the predicted z-factor was greater than the experimental z-factor. The Peng-Robinson and AQUAlibrium models had positive AE and AD's for most mixtures.

Average AAE's and AAD's were calculated based on a weighted average of the number of data points in each mixture. The AAE for the Lee Kesler, BWR, Peng-Robinson, SRK, and AQUAlibirium models were 1.21%, 1.98%, 2.37%, 5.10%, and 2.97% respectively. The AAD for the Lee Kesler, BWR, Peng-Robinson, SRK and AQUAlibrium models were 0.00808, 0.01113, 0.01605, 0.02906, 0.02037.

2.10 Conclusion

The five different models produced varying results for predicting carbon dioxide mixtures with small percentages of nitrogen and methane. The Lee Kesler model was found to achieve the smallest errors for all mixtures used, while the SRK model produced the greatest errors.

The Lee- Kelser and BWR models had the greatest errors near the critical point, while the SRK, Peng- Robinson and AQUAlibrium model typically increased in error as the pressure increased.

The Lee-Kesler model predicted within 1.21%, on average for all mixtures, of the experimental densities and within 0.00808, on average for all mixtures, of the z-factor. All data points were predicted within 8% of the experimental densities. The SRK model produced the least accurate predictions with errors of 5.10%, on average for all mixtures. AQUAlibrium, Peng-Robinson, and BWR predicted within 2-3%, on average for all mixtures.

References

1. Magee, J.W. and Ely, J.F., "Isochoric (p, v, T) Measurements on CO_2 and (0.98 CO_2 + 0.02 CH_4) from 225 to 400 K and Pressures to 35 MPa", *Internat. J. Thermophysics*, 9, 547-557, (1988).
2. Hwang, C-A., G.A. Iglesias-Silva., J.C. Holste, K.R. Hall, B.E. Gammon, K.N. Marsh, "Densities of Carbon Dioxide + Methane Mixtures from 225 K to 350 K at Pressures up to 35 MPa", *J. Chem. Eng. Data*, 42, 897-899, (1997)
3. Brugge, H.B., C.-A. Hwang, W.J. Rogers, J.C. Holste, K.R. Hall, W. Lemming, G.J. Esper, K.N. Marsh, and B.E. Gammon, "Experimental Cross Virial Coefficients for Binary Mixtures of carbon Dioxide with Nitrogen, Methane and Ethane at 300 and 320 K", *Physica A*, 156, 382-416, (1989).
4. Reamer, H.H., Olds, R.H., Sage, B.H., and Lacey, W.N., "Phase Equilibrium in Hydrocarbon Systems.Methane-Carbon Dioxide System in the Gaseous Region", *Ind. Eng. Chem.*, 36, 88-90, (1944).
5. Brugge, H.B., J.C. Holste, K.R. Hall, B.E. Gammon, K.N. Marsh, "Densities of Carbon Dioxide + Nitrogen from 225 K to 450 K at Pressures up to 70 MPa", *J. Chem. Eng. Data*, 42, 903-907, (1997).
6. AQUAlibrium 3.1.FlowPhase Inc. [software]. Calgary, Canada.
7. VMGSim v7.0.67. Virtual Materials Group Inc. [software]. Calgary, Canada.

3

On Transferring New Constant Pressure Heat Capacity Computation Methods to Engineering Practice

Sepideh Rajaeirad and John M. Shaw

Department of Chemical and Materials Engineering, University of Alberta, Edmonton, AB, Canada

Abstract

In the present work, a variety of available methods for isobaric liquid heat capacity estimation are evaluated relative to experimental data for divers hydrocarbon mixtures for which elemental analysis, refinery, and molecule-based characterization are experimentally accessible. The study focuses on identification of systemic deviations (over and underprediction, and skew) between specific calculation approaches and experimental data, so that implementation in process simulators can be optimized [1]. A comparison with experimental result showed that the Dadgostar–Shaw correlation for liquids provides accurate heat capacity values for a broad range of fluids at saturation and is, for example, preferred over the widely used Lee–Kesler correlation for liquids based on both accuracy and range of application. The API method for estimating heteroatom content was also shown to systematically underestimate heteroatom content.

3.1 Introduction

A series of predictive correlations for constant pressure heat capacity for crystalline organic solids, liquids and ideal gases were recently proposed [2–5]. Their primary application is to predict the thermal behavior of ill-defined hydrocarbons, where elemental analysis is one of the few certain composition properties available. Examples include, boiling cuts, or composition classes such as asphaltenes, or maltenes, where no molecular models or only primitive and imprecise molecular models are available for estimating heat capacity. The precision and accuracy of these correlations has warranted further evaluation, including applications arising with light hydrocarbons, where for example indirect calculation approaches for the heat capacity of liquids (ideal gas heat capacity + an equation of state based departure functions), or direct correlations (such as the Lee-Kesler correlation [8]) are typically employed. As hydrocarbon fluids can be characterized on a molecular or refinery basis, these element based correlations add two direct calculation options (elemental composition can be defined using an API calculation method [7] or from data) and up to four indirect calculation options for liquid phase Cp calculation per equation of state. Each of these approaches, summarized in Figure 3.1, along with the conventional approaches for predicting liquid heat capacity have advantages and disadvantages, and possess different input data requirements. Identification of hierarchies and the best niches for diverse combinations of methods is challenging, as is conveying this complexity to users. In the present work, a variety of available methods for isobaric liquid heat capacity estimation are evaluated relative to experimental data for divers hydrocarbon mixtures for which elemental analysis, refinery and molecule based characterization are experimentally accessible. The study focuses on identification of systemic deviations (over and under prediction, and skew) between specific calculation approaches and experimental data, so that implementation in process simulators [7] can be optimized.

3.2 Materials and Methods

Four liquid mixtures comprising: n-alkanes only (Mixture 1), a mixture of aromatic and n-alkane constituents (Mixture 2), a mixture of naphthenic and n-alkane consituents (Mixture 3), and a mixture of naphthenic and aromatic constituents (Mixture 4) were prepared. The compositions of these mixtures are listed in Table 3.1. Experimental isobaric heat capacity data, obtained using a differential scanning calarimeter, TG-DSC 111, were then compared with element based and conventional correlations over the temperature range 293 K to a temperature below the boiling point of the mixture (0.9 T_b) to avoid artifacts in the measurements introduced by sample vaporization. The possible impact of sample vaporiztion was further reduced by placing 90 mm^3 of sample in the 100 mm^3 sample cells. Heat flow profiles were calibrated and analyzed, and the heat capacity values obtained using established procedures [6]. The computation matrix

Table 3.1 Mixture Compositions.

Mixture 1		Mixture 2		Mixture 3		Mixture 4	
Composition	Wt %	Composition	Wt %	Composition	Wt %	Composition	Wt %
Nonane	33.3	1,2,4-TMB*	33.6	Trans-decalin	33.2	Trans-decalin	40.0
Decane	33.3	Decane	33.4	Decane	33.4	Durene	20.0
Undecane	33.3	Undecane	33.1	Undecane	33.3	1,2,4-TMB*	40.0

* 1,2,4-Trimethylbenzene

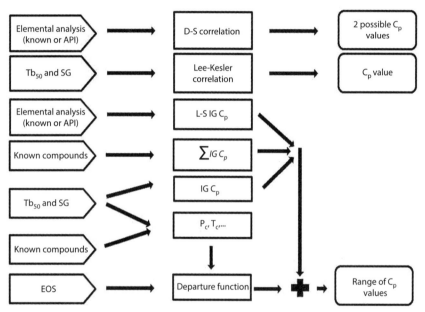

Figure 3.1 Computational matrix for the possible direct and indirect approaches for calculating isobaric liquid heat capacity.

based on two direct and four indirect calculation approaches for liquid heat capacity is illustrated in Figure 3.1.

3.3 Results and Discussion

The experimental and computed results are exemplified for the aromatic + n-alkane mixture (Mixture 2), comprising equal weight percents of 1,2,4-trimethylbenzene, decane, and unadecane. The measured and computed isobaric liquid-phase heat capacity values for this mixture are reported in Figure 3.2 and the mean absolute percentage error (MAPE)

Table 3.2 Deviation (MAPE) of liquid phase constant pressure heat capacity computational approaches from experimental data for mixtures 1-4.

					MAPE			
	L-S IG C_p (API based α)+APR based DF (known compounds)	D-S C_p (API based α)	L-K C_p	IG C_p + APR based DF (known compounds)	L-S IG C_p +APR based DF (petroleum cut)	L-S IG C_p +APR based DF (known compounds)	IG C_p + APR based DF (petroleum cut)	D-S C_p (known α)
Mixture 1	1.91	6.98	3.67	0.50	2.46	1.91	1.49	0.70
Mixture 2	8.57	6.31	3.35	0.49	1.25	0.76	2.07	2.69
Mixture 3	4.47	3.71	10.89	3.35	2.62	2.89	2.53	4.10
Mixture 4	7.86	16.89	20.1	0.92	2.22	2.24	1.27	13.24

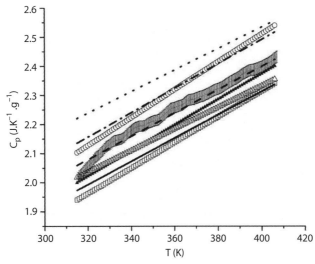

Figure 3.2 Isobaric liquid heat capacity of 1,2,4-Trimethylbenze, Decane, and Undecane mixture (Mixture 2) calculated by various methods: —, Experimental data with the error of 0.02 J.K^{-1}.g^{-1} shown as a shaded area; - - , IGC$_p$+APR base departure function (known compound); ▲, IGC$_p$+APR base departure function (petroleum cut); □, Lastovka-Shaw IGC$_p$+APR base departure function (known compound); ■, Lostovka-Shaw IGC$_p$+APR base departure function (petroleum cut); Δ, Dadgostar-Shaw C$_p$ (Actual α); ..., Dadgostar_Shaw C$_p$ (API base α); o, Lostovka-Shaw IGC$_p$ (API base α) +APR base departure function; _._, Lee-Kesler C$_p$

between the experimental and computed values for all four mixtures are shown in Table 3.2. For the purpose of illustration, the Lee-Kesler and Dadgostar-Shaw correlations (direct correlation), and two ideal gas heat capacity correlations (Lastovka-Shaw and the Benson method) combined with the Advanced Peng-Robinson equation of state, as implemented in VMGsim, (both molecular and refinery hydrocarbon speciation) provide four indirect liquid-phase constant pressure heat capacity calculation methods. The range of values obtained and the trends with temperature were unexpected. Elemental compositions defined on the basis of computed values (American Petroleum Institute [7]) yield heat capacities that are systematically too large. This arises because α, the number of atoms per unit mass and the basis of the heat capacty calculation, is over estimated for a broad range of compunds and compound classes as shown in Figure 3.3. More subtly, while the indirect calculations based on ideal gas heat capacity + departure functions fit the experimental heat capacity at near ambient conditions, the trend with temperature is overstated leading to systematic underprediction at low temperature and over prediction at high temperature.

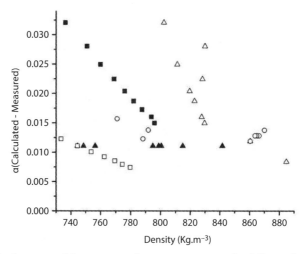

Figure 3.3 The deviation of the API procedure in estimating α for different families: ■, n-alkyne; ▲, $C_{10}H_{20}$ isomers; Δ, Aromatic; □, n-alkene; ○, Naphthenic

3.4 Conclusions

The diverse calculation approaches and methods available to practitioners for calculating liquid-phase constant pressure heat capacities, leads to a broad range of estimates at fixed temperature and saturated pressure. 10 % to 20 % variations among estimated values are common. Indirect calculation methods yield excellent estimates for liquid phase heat capacity at near ambient temperature but the trends with temperature are skewed relative to the experimental data yielding under estimates systematically at low temperature and over estimates systematically at higher temperatures. Direct calculation with the Dadgostar-Shaw correlation and known elemental composition is preferred over the Lee-Kesler method, [8] based both on the range of application and accuracy [4]. The element based correlations such as the Lastovka-Shaw and Dadgostar-Shaw correlations should only be used with experimental elemental analysis, as the API calculation method over estimates the number of atoms per mass and hence over estimates heat capacity systematically.

References

1. V. M. G. Inc, "New Ideal Gas Heat Capacity Estimation Methods for Oil Fractions Lastovka-Shaw IG Cp Estimation Method Dadgostar-Shaw IG Cp Estimation Method Implementation in VMGSim," pp. 1–8, 2011.

2. V. Lastovka, N. Sallamie, J. M. Shaw, Fluid Phase Equilibria, 268, 51-60, 2008
3. V. Lastovka and J. M. Shaw, "Predictive correlations for ideal gas heat capacities of pure hydrocarbons and   petroleum fractions." submitted 2013.
4. N. Dadgostar and J. M. Shaw, "A predictive correlation for the constant-pressure specific heat capacity of pure and   ill-defined liquid hydrocarbons" Fluid Phase Equilibria, 313 (2011) 211-226
5. N. Dadgostar and J. M. Shaw, "A predictive correlation for the constant-pressure specific heat capacity of pure and   ill-defined liquid hydrocarbons –including the critical point," Fluid Phase Equilibria, in press (2013).
6. M. Fulem, M. Becerra, M.D.A. Hasan, B. Zhao, and J. M. Shaw, "Phase behavior od Maya crude oil base don calorimetry and rheometry" Fluid Phase Equilibria, 272 (2008) 32–41
7. Technical data book-petroleum refining. American Petroleum Institute, 1997, p. 7D1_7D3.
8. B. I. Lee and M. G. Kesler, Hydrocarbon Processing, pp. 153–158, March 1976.

4

Developing High Precision Heat Capacity Correlations for Solids, Liquids and Ideal Gases

Jenny Boutros and John M. Shaw

Department of Chemical and Materials Engineering University of Alberta, Edmonton, AB, Canada

Abstract

Accurate and predictive correlations for heat capacity of organic solids, liquids, and ideal gases requiring temperature and elemental composition as inputs, have been developed by Lastovka and Shaw [4, 5] and Dadgostar and Shaw [1] in recent years. These correlations are based on a similarity variable concept, α, [2, 3] and are applicable to pure organic compounds as well as well-defined or ill-defined mixtures like Athabasca bitumen. The predictive correlations yield average relative deviations in test sets below 10% suggesting that these correlations are suitable bases for developing high-precision fluid family-specific correlations. In the present work, high-precision predictive correlations for specific mixed fluids and families of compounds are obtained by tuning the first term of the universal correlations, while retaining their overall reliability. Higher accuracy correlations for liquid paraffins, liquid naphthenes, and molten polymers have been obtained. This work is being extended to other families of compounds and phase states.

4.1 Introduction

Corresponding state or group contribution techniques require critical properties or molecular structure for estimating heat capacities. General empirical correlations, like the Lee-Kesler correlations for liquid heat capacity, possess variable accuracy and a limited range of applications. [1, 6] Hence the importance of predictive correlations for heat capacity like the ones introduced by Dadgostar and Shaw, and Lastovka and Shaw. These latter correlations, requiring only the value of an elemental composition based similarity variable and temperature as inputs, can play key roles, particularly for calculations related to ill-defined fluids such as petroleum cuts and bitumen where elemental analysis is one of the few readily available and reliable properties.

The liquid, solid and ideal gases predictive heat capacity correlations at constant pressure possess 6 [1], 7 [4], and 12 [5] universal coefficients respectively, and are valid for pure organic compounds and ill-defined mixtures such as heavy oil and bitumen over a wide range of temperatures. The three aforementioned correlations are functions of temperature and a similarity variable, α, which is proportional to the number of atoms per unit mass and hence a direct function of elemental analysis: [2, 3]

$$\alpha = \frac{N}{M} = \frac{\sum_{i=1}^{n} v_i}{\sum_{i=1}^{n} v_i M_i} = \frac{\sum_{i=1}^{n} x_i}{\sum_{i=1}^{n} x_i M_i} = \frac{\sum_{i=1}^{n} \frac{w_i}{M_i}}{\sum_{i=1}^{n} w_i} \qquad (4.1)$$

v_i is the stoichiometric coefficient for element i in a compound consisting of N atoms
n is the number of elements in a compound
M is the molar mass of the compound and
M_i is the molar mass of chemical element i (g.mol^{-1})
x_i is the mole fraction of element i in a compound
w_i is the mass fraction of element i.

The concept of the similarity variable is rooted in quantum mechanics and its validity for predicting constant pressure heat capacities for solids, liquids and ideal gases has been addressed. [1–3] Molecules sorb energy through skeletal and atomistic vibrations. Simple molecules possess three fundamental translation and three twisting or bending modes of vibration and 3N-6 fundamental atomistic vibration modes. Atomistic vibration dominates at high temperature and for large molecules. Specific heat capacity is therefore largely a function of the number of atoms per unit mass, α. To a first approximation, vibration frequency and saturation temperature differences among different types of atoms may be ignored. The details of molecular structure are also of secondary importance compared to elemental analysis and may be ignored. [1–3] These correlations have yielded average relative deviations in test sets with 6.8% [4], 3.5% [1], and 2.9% [5] for organic solids, liquids and ideal gases respectively, and they are clearly good bases for the development of high-accuracy fluid-specific correlations. In this work, the "Einstein term" that is shared by all three of the universal correlations, [1, 4, 5] is fit specific fluid families. It is expected that this single and common modification, across all three phase states, will yield high precision heat capacities estimates for all three phase states simultaneously.

4.2 Databases and Methods

Data and methods used for liquid heat capacity correlation modification illustrate the approach as the a_1 term is shared by all three heat capacity correlations (solid, liquid and ideal gas). Data for liquid heat capacity correlation modification were obtained from the same database used to fit and test the Dadgostar-Shaw correlation. [1] The Test and Training sets were grouped together, and then divided into 5 different chemical families: paraffins, naphthenes, molten polymers, aromatic and unsaturated cyclics, and compounds containing heteroatoms S, N, and O. The first term of (4.2) was then modified:

$$Cp = a_1 + (a_{21}a + a_{22}a^2)T + (a_{31}a + a_{32}a^2)T^2$$

$$a_1 = (a_{11}a + a_{12}a^2) \times 3R(\frac{\theta}{T})^2 \frac{\exp(\theta/T)}{[\exp(\theta/T)-1]^2} \qquad (4.2)$$

$$T > 200K$$

$$a_1 = (a_{11}a = a_{12}a^2) \times 24.5$$

The a_1 values were adjusted at an interval of ±10 % with a 1% percent increment until the least relative average error was obtained. As the data in the database possess experimental uncertainties of less than 1 %, the average absolute relative deviations (AARD) exceed the experimental uncertainty (Figure 4.1).

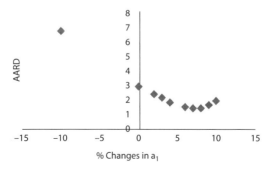

Figure 4.1 Absolute average relative deviation (AARD) distribution for liquid heat capacity of n-alkanes versus percent adjustments in the a_1 term of the Dadgostar-Shaw correlation.

Table 4.1 Average Absolute Relative Deviation (AARD) values for the Lee-Kesler, Dadgostar-Shaw, and the a1 Adjusted Coefficient correlations for the constant pressure heat capacity of paraffinic and naphthenic liquids, and molten polymers.

Correlations	Paraffins			Naphthenes			Molten Polymers		
	Lee-Kesler	Dadgostar-Shaw	Adjusted coefficient	Lee-Kesler	Dadgostar-Shaw	Adjusted coefficient	Lee-Kesler	Dadgostar-Shaw	Adjusted coefficient
AARD.100	2.37	2.81	1.36	1.98	3.84	1.72	N/A	2.01	1.95

4.3 Results and Discussion

Increasing the a_1 term by +7% for the liquid paraffins led to a reduction of the average absolute relative error from 2.81% to 1.36% as shown in Table 4.1. As for the naphthenes, decreasing the a_1 term by 9% lead to a reduction of the average absolute relative error from 3.84% to 1.72% as also shown in Table 4.1. Increasing the a_1 term by +3% also showed a reduction of the average absolute relative error from 2.01% to 1.95%. For the remaining families: aromatics and unsaturated cyclics, and compounds containing the heteroatoms S, N, and O, no improvement was realized, on average, by adjusting the a_1 value. Clearly other types of modification must be considered.

For liquids, illustrated in this overview, the results suggest the need to accommodate aromaticity and heteroatom content explicitly. From quantum mechanics, the type of atoms affects atomistic vibration (frequency, and mode saturation profile with temperature). Hence, the inclusion of the mass fraction of specific atoms as well as carbon types present in molecules or mixtures as correlated inputs. This type of speciation is either known, for pure compounds, or is in expensive to obtain. Elemental analysis and NMR analysis, for aromatic carbon content, are routine and available measurements in industry.

4.4 Conclusion

For paraffins, naphthenes, and molten polymers, the similarity variable concept can be used to produce accurate, compound-family-specific and predictive correlations for heat capacities, without the introduction of element or functional group specific parameters. Development of high accuracy correlations for aromatic and heteroatom containing compounds and mixtures will require the types and mass fractions of atoms present to be enumerated. These will comprise correlated groups in the correlation and permit extension of the similarity concept to bio-diesel (primarily liquids and gases), and pharmaceutics (primarily solids). It is unclear whether, modification can be restricted to the a_1 term, the term common to correlations for solids, liquids and ideal gases.

References

1. Dadgostar, N., & Shaw, J. M. (2012). A predictive correlation for the constant-pressure specific heat capacity of pure and ill-defined liquid hydrocarbons. *Fluid Phase Equilibria, 313*, 211–226.

2. Laštovka, V., Fulem, M., Becerra, M., & Shaw, J. M. (2008). A similarity variable for estimating the heat capacity of solid organic compounds. *Fluid Phase Equilibria, 268*(1-2), 134–141.
3. Laštovka, V., Sallamie, N., & Shaw, J. M. (2008). A similarity variable for estimating the heat capacity of solid organic compounds. *Fluid Phase Equilibria, 268*(1-2), 51–60.
4. V. Lastovka and J. M. Shaw, "Predictive correlations for ideal gas heat capacities of pure hydrocarbons and petroleum fractions." Fluid Phase Equilibria, in press, August 2013.
5. V. lastovka, J. M. Shaw. "Predictive correlation for the Cp of organic solids Based on Elemental Composition." *Journal of Chemical Engineering Data*, no. 52 (2007): 1160-1164.
6. B. I. Lee, M. G. Kesler Hydrocarbon Processing, pp. 153 -158, March 1976.

5

Method for Generating Shale Gas Fluid Composition from Depleted Sample

Henrik Sørensen, Karen S. Pedersen and Peter L. Christensen

Calsep A/S, Lyngby, Denmark

Abstract

Shale reservoirs are extremely tight and a high-pressure drawdown in the near wellbore area often makes the fluid split into two phases before reaching the well. Pressure buildup is most often insufficient to get the fluid back into single phase. For gas condensates, this means that only the gas phase is produced, while the liquid is left behind. Reservoir simulations require the original reservoir fluid composition and it is, therefore, a problem when a representative fluid composition cannot be sampled.

This chapter presents a theoretical method for recreating the original composition of a gas condensate based on an analysis to C_{36+} for a depleted sample. The sample can originate from a reservoir where the pressure has dropped below the saturation pressure, or from a tight reservoir where the fluid splits into two phases before reaching the well. The method is based on the theory of chemical reaction equilibria.

5.1 Introduction

The reservoir fluid composition must be known to evaluate the resources in a petroleum reservoir. Well established sampling techniques exist for how to get representative samples from undersaturated reservoirs of a reasonable permeability. For oil type of fluids bottom hole sampling is the preferred technique, while separator samples are preferable for gas condensates. No matter the type of sample it is essential that the reservoir fluid is single phase at bottom hole conditions when the sample is taken. That means the bottom hole pressure must be higher than the saturation pressure of the fluid.

Reservoirs exist from which a representative sample cannot be taken. A depleted gas condensate reservoir will contain free liquid condensed from the gas. Independent of sampling technique the condensed liquid will only be contained in the sample in negligible concentration.

Tight gas condensate reservoirs including shale gas reservoirs present a similar problem. Even though the fluid is single phase at reservoir conditions, the draw down in the near well bore area may be so high that the fluid will split into two phases downhole and only the gas phase will be sampled.

This paper outlines procedures for how to numerically recreate the original reservoir fluid composition from a non-representative sample.

5.2 Theory of Chemical Equilibrium Applied to Reservoir Fluids

A pattern has been observed in the compositions of oil and gas condensate reservoir fluids. For the C_{7+} fractions the logarithm of the mole fraction (z) versus carbon number (C_n) is seen to follow an approximately straight line [1]:

$$\ln(z_{C_n}) = A + B \times C_n \tag{5.1}$$

In this equation A and B are constants.

The theory of chemical reaction equilibria can be used to evaluate the theoretical molar ratio between the normal paraffins C_nH_{2n+2} and $C_{n+1}H_{2(n+1)+2}$ in a reservoir fluid composition. If the former component were to form from the pure elements, the reaction equilibrium would be

$$nC + (n+1)H_2 \leftrightarrow C_nH_{2n+2} \tag{5.2}$$

The equilibrium constant for this reaction is defined as

$$K_{C_n} = \frac{[C_nH_{2n+2}]}{[C]^n [H_2]^{n+1}} \tag{5.3}$$

and related to the Gibbs free energy of formation as

$$-RT \ln K_{C_n} = \sum v_i G_i^0 \qquad (5.4)$$

The v_i's are stoichiometric coefficients for the reactants and product in (5.2) and the term C_n is used for $C_n H_{2n+2}$. G_i^c is zero for pure elements, which allows (5.4) to be reduced to

$$-RT \ln K_{C_n} = \Delta G_{C_n}^0 \qquad (5.5)$$

where ΔG_C^0 is Gibbs free energy of formation for C_n. The ratio between the equilibrium constants for the reactions leading to C_n and C_{n+1} becomes

$$-RT \ln \frac{K_{C_n}}{K_{C_{n+1}}} = \Delta G_{C_n}^0 - \Delta G_{C_{n+1}}^0 \qquad (5.6)$$

which using (5.3) can be rewritten to

$$-RT \ln \frac{[C_n][C][H_2]}{[C_{n+1}]} = \Delta G_{C_n}^0 - \Delta G_{C_{n+1}}^0 \qquad (5.7)$$

or

$$-RT(\ln[C_n] - \ln[C_{n-1}]) = \Delta G_{C_n}^0 - \Delta G_{C_{n+1}}^0 + RT\ln[C] + RT\ln[H_2] \qquad (5.8)$$

Table 5.1 shows the Gibbs free energies of formation for the normal paraffins from nC_7 to nC_{20} in gas and liquid forms. Also shown is how much ΔG of formation increases from one C_n to the next one. For both gas and liquid states the increase is seen to be almost constant independent of carbon number. That means the term Äin (5.8) is a constant and so are the terms $RT\ln[C]$ and $RT\ln[H_2]$. With a constant difference between $\ln[C_n]$ and $\ln[C_{n-1}]$ (5.1) is reproduced.

Table 5.1 Gibbs energies of formation of normal paraffins at 25°C [5].

Component	ΔG formation gas J/mol	ΔG formation liquid J/mol	Difference in ΔG gas between C_n and C_{n-1} J/mol	Difference in ΔG liquid between C_n and C_{n-1} J/mol
nC7	8,033	1,004		
nC8	16,401	6,360	8,368	5,356
nC9	24,811	11,757	8,410	5,397
nC10	33,221	17,280	8,410	5,523
nC11	41,631	22,719	8,410	5,439
nC12	50,041	28,075	8,410	5,356
nC13	58,450	33,556	8,410	5,481
nC14	66,818	38,869	8,368	5,314
nC15	75,228	44,350	8,410	5,481
nC16	83,764	49,999	8,535	5,648
nC17	92,090	55,187	8,326	5,188
nC18	100,458	60,919	8,368	5,732
nC19	108,951	66,275	8,494	5,356
nC20	117,319	71,630	8,368	5,356

The data in Table 5.1 are for pure substances at 25°C. There will be an additional contribution to ΔG from transferring the hydrocarbons from pure form into a hydrocarbon mixture at a different temperature, but this contribution is an order of magnitude lower than ΔG of formation. Also the C_{7+} fractions will contain other components than n-paraffins. For heavy oils dominated by aromatics (5.1) may not apply or may not apply until after ~ C_{15} [2]. Most reservoir fluids are however dominated by paraffins and paraffinic side branches on aromatic and naphthenic molecules and for those reservoir fluids it can be concluded that the observed dependence of C_{7+} mole fractions versus carbon number expressed in (5.1) has a foundation in the theory of chemical reaction equilibria.

5.3 Reservoir Fluid Composition from a Non-Representative Sample

5.3.1 Depleted Gas Condensate Samples

In the laboratory the production from a gas condensate reservoir can be emulated experimentally by carrying out a Constant Volume Depletion (CVD) experiment as sketched in Figure 5.1. The simulation starts with a saturated reservoir fluid. When material is removed from a reservoir with a saturated gas condensate fluid, liquid will start to precipitate. This is in the CVD experiment emulated by initially expanding the volume to make the pressure decrease, and afterwards removing the excess volume of gas from the cell. This corresponds to the situation in a gas condensate reservoir, from which only the gas phase is produced while the liquid is left behind. As more material is removed, the pressure further decreases and still less liquid is contained in the produced gas. Simulation of an N-stage CVD experiment as sketched in Figure 5.1 can be used to get a rough idea about the variation with time in the composition of the produced fluid, or more correctly the variation in composition with decreasing reservoir pressure. It does however require that the initial reservoir fluid composition is known. That will only be the case if a fluid sample was taken before the reservoir pressure fell below the saturation pressure of the initial reservoir fluid.

If a depleted sample is taken, it will have the composition of the gas phase at the actual pressure. This composition is saturated with liquid at the current cell (reservoir) pressure.

At the time the depleted gas is sampled, measured data about the composition of the already produced fluid may be scarce or lacking and no data exists for the liquid left behind in the reservoir. Unless the gas sample is

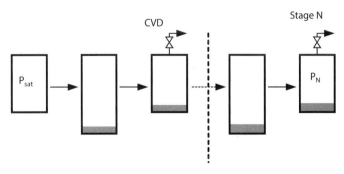

Figure 5.1 Schematic illustration of Constant Volume Depletion (CVD) experiment.

heavily depleted it does however provide enough information to estimate the original reservoir fluid composition.

1 mole of original reservoir fluid will consist of

- X_1 moles of already produced gas.
- X_2 moles of the fluid composition currently sampled (depleted gas).
- X_3 moles of the liquid remaining in the reservoir (dewpoint liquid).

If the already produced gas had the same composition as that currently produced, the problem would be reduced to finding the composition of the reservoir liquid at the sampling time and the molar ratio $(X_1+X_2)/X_3$.

The gas phase in the reservoir is saturated. The liquid composition in the reservoir at the time a depleted gas is sampled can therefore be found as the liquid in equilibrium with the gas at its saturation point at the reservoir temperature.

While (5.1) is assumed to hold for the original reservoir fluid composition, the below relation is assumed to apply in general for hydrocarbon fluids being processed

$$\ln(z_{C_n}) = A + B \times M_{C_n} + D \times M_{C_n}^2 \qquad (5.9)$$

M_C is the molecular weight of carbon number fraction C_n. D will be negative for a depleted sample. D will be zero for an original reservoir fluid composition, and D will be positive for a fluid composition that contains more liquid than the original reservoir fluid composition. Molecular weight is used in (5.9) instead of carbon number as in (5.1). Assuming the molecular weights of the C_{7+} fractions increase linearly with carbon number (5.1) and (5.9) will be fully compliant for D=0. Use of molecular weight instead of carbon number adds more importance to the curvature of the heavy end because $M_{C_n}^2 / M_{C_n}$ develops faster with carbon number than C_n^2/C_n.

An approximate ratio $(X_1+X_2)/X_3$ can be found by adding dewpoint liquid to the currently produced fluid until D becomes zero as is sketched in Figure 5.2. For each considered mixing ratio A, B and D are found by a least squares fit to the expression in (5.9). The data fitted to is mole%'s versus molecular weight of the fractions from C_7 to the highest carbon number fraction analyzed for. The mole%'s are for the combined fluid. X_1, X_2 and X_3 must sum to 1.0, which is used to determine (X_1+X_2) and X_3. The term V^{liq} is used for the molar volume of the dewpoint liquid at the current pressure.

X_1+X_2 moles of sampled fluid is combined with X_3 moles of dewpoint liquid to get an approximate original reservoir fluid composition. This is the starting point in Figure 5.3 (left hand cell). The molar volume (V^{res})

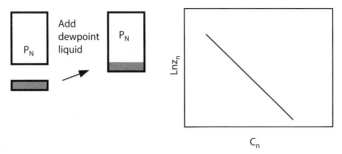

Figure 5.2 Liquid enrichment process to get a fluid composition compliant with (5.1) and with (5.9) with D=0.

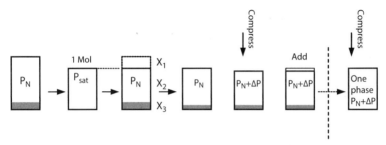

Figure 5.3 Procedure to recreate the reservoir fluid composition from a fluid sampled from a depleted gas condensate reservoir.

of this fluid at its saturation point at the reservoir temperature is a first estimate of the molar volume of the original reservoir fluid. At the current (sampling) pressure (P_N) the volumetric space represented by V^{res} is filled out by X_2 moles of gas and X_3 moles of liquid. The molar amount X_2 can be determined as the number of moles of depleted gas required to fill out the volume $V^{res} - X_3 \times V^{liq}$.

With X_2 determined the composition of the already produced gas can be found by carrying out an inverse CVD simulation starting with X_2 moles of depleted gas and X_3 moles of dewpoint liquid. The inverse CVD simulation is conducted as a series of stepwise increases of the pressure by 0.1 bar and at each pressure stage adding enough volume of the current gas phase to increase the total volume to V^{res}. The simulation is stopped when all liquid is dissolved in the gas phase. At this point the total molar amount should equal 1.0. If that is not the case, a new V^{res} is calculated using the fluid from the inverse CVD simulation and a new inverse CVD simulation is carried out using this molar volume.

It is afterwards to be checked that D in (5.9) is close to zero for the final composition. If that is not the case X_3 can be reestimated by determining the ratio $(X_1+X_2)/X_3$, which will give a D of zero in (5.9) using the correct composition for X_1. In the test calculations performed this additional round was not found to be needed.

5.3.2 Samples from Tight Reservoirs

If the production comes from a single phase tight gas condensate reservoir there may be a considerable pressure drop in the formation close to the well, a so-called draw down as schematically illustrated in Figure 5.4. If the pressure in the near well bore area falls below the saturation point of the reservoir fluid, the fluid will split into two phases. The separator sample will not contain any of the formed liquid in detectable quantity, but will consist of the gas flowing into the well.

As is sketched in the upper part of Figure 5.5, the fluid sampled from a tight reservoir not yet in production with a phase split occurring in the near wellbore area has undergone a different series of equilibrium stages than a fluid sampled from a depleted reservoir as illustrated in Figure 5.1. In a depleted reservoir the gas and liquid remain in contact and can at all time exchange components. That is not the case in a near well bore area with a continuously decreasing pressure. The gas will be transported towards the well and leave the condensed liquid at each pressure level behind.

The original reservoir fluid composition can numerically be recreated using a similar procedure as the one applied for a gas produced from a depleted reservoir. The procedure consists in a continuous series of additions of equilibrium liquid to the gas followed by a compression to the new saturation point. The procedure to recreate the reservoir fluid composition

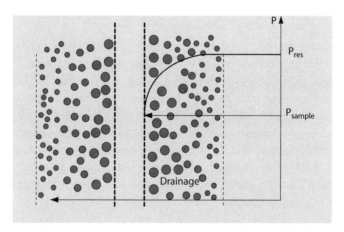

Figure 5.4 Drawdown causing a phase split in the near wellbore area.

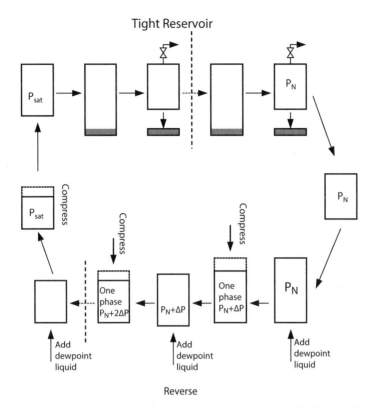

Figure 5.5 The upper part sketches the depletion process a gas sample from a tight reservoir undergoes. The lower part shows the procedure for how to recreate the original reservoir fluid composition from such sample.

from a gas sample from a tight reservoir is schematically illustrated in the lower part of Figure 5.5.

5.4 Numerical Examples

5.4.1 Depleted Gas Condensate Samples

Table 5.2 shows the weight% and mol% compositions of a gas condensate fluid sampled in the single phase region. The composition has been analyzed to C_{36+}. The fluid is characterized for the SRK-Peneloux equation Soave [3] and Peneloux et al. [4] using the characterization method of Pedersen at al. (1992). C_7-C_{35} are kept unlumped, while the C_{36+} fraction is lumped into one component.

Table 5.2 Weight% and mol% compositions of gas condensate reservoir fluid. The C_{7+} molecular weight is 145 and the C_{7+} density 0.762 g/cm3.

Component	Reservoir Fluid		Fluid depleted to 266 bar	Original Reservoir fluid from fluid at 266 bar	Fluid depleted to 226 bar	Original Reservoir fluid from fluid at 226 bar
	Weight%	Mol%	Mole%	Mole%	Mol%	Mole%
N_2	0.130	0.1301	0.1314	0.1300	0.1339	0.1310
CO_2	4.710	2.9994	3.0091	3.0000	3.0299	3.0002
H_2S	0.980	0.8059	0.8023	0.8060	0.7971	0.8050
C_1	42.660	74.5291	75.0142	74.4907	75.9553	74.7637
C_2	8.550	7.9697	7.9730	7.9701	7.9831	7.9705
C_3	5.890	3.ff7434	3.7260	3.7500	3.6902	3.7402
iC_4	1.700	0.8197	0.8127	0.8200	0.7975	0.8161
nC_4	3.510	1.6924	1.6739	1.6900	1.6343	1.6801
iC_5	1.820	0.7070	0.6956	0.7080	0.6706	0.7010
nC_5	2.100	0.8157	0.8011	0.8170	0.7690	0.8080
C_6	3.080	1.0016	0.9763	1.0000	0.9186	0.9881
C_7	3.514	0.9973	0.9634	1.0000	0.8859	0.9791
C_8	3.081	0.8159	0.7827	0.8190	0.7057	0.7980
C_9	2.741	0.6279	0.5969	0.6300	0.5239	0.6110
C_{10}	2.387	0.4808	0.4527	0.4830	0.3861	0.4660
C_{11}	2.060	0.4154	0.3869	0.4180	0.3194	0.4000
C_{12}	1.774	0.3088	0.2838	0.3110	0.2251	0.2950
C_{13}	1.517	0.2402	0.2174	0.2420	0.1648	0.2280

C_{14}	1.296	0.1826	0.1621	0.1840	0.1163	0.1720
C_{15}	1.105	0.1520	0.1317	0.1540	0.0883	0.1420
C_{16}	0.937	0.1189	0.1001	0.1200	0.0623	0.1100
C_{17}	0.786	0.0907	0.0739	0.0920	0.0425	0.0830
C_{18}	0.655	0.0759	0.0597	0.0772	0.0316	0.0687
C_{19}	0.540	0.0606	0.0460	0.0618	0.0225	0.0543
C_{20}	0.444	0.0467	0.0340	0.0477	0.0154	0.0416
C_{21}	0.370	0.0369	0.0253	0.0378	0.0103	0.0325
C_{22}	0.305	0.0296	0.0191	0.0304	0.0071	0.0260
C_{23}	0.250	0.0218	0.0132	0.0224	0.0045	0.0191
C_{24}	0.205	0.0178	0.0101	0.0184	0.0031	0.0154
C_{25}	0.168	0.0141	0.0073	0.0146	0.0021	0.0123
C_{26}	0.137	0.0116	0.0055	0.0120	0.0014	0.0098
C_{27}	0.112	0.0084	0.0036	0.0088	0.0008	0.0068
C_{28}	0.092	0.0066	0.0026	0.0069	0.0005	0.0051
C_{29}	0.075	0.0056	0.0020	0.0059	0.0004	0.0049
C_{30}	0.061	0.0041	0.0013	0.0043	0.0007	0.0152
C_{31}	0.050	0.0033	0.0009	0.0035		
C_{32}	0.040	0.0027	0.0007	0.0029		
C_{33}	0.033	0.0022	0.0005	0.0023		
C_{34}	0.026	0.0017	0.0003	0.0018		
C_{35}	0.021	0.0013	0.0002	0.0014		
C_{36+}	0.089	0.0047	0.0005	0.0050		

To get representative depleted fluid compositions two 40 stage CVD simulations were carried at the reservoir temperature of 129°C starting with the saturated original reservoir fluid at 306 bar. The end pressures were respectively 266 bar and 226 bar. The depleted compositions are shown in Table 5.2 and would be those sampled at 266 bar and 226 bar. A gas chromatographic (GC) analysis allows a weight% composition to be measured with 3 decimals corresponding to ~ 4 decimals for the mole%'s of the C_{30+} components. A compositional analysis to C_{36+} as that given for the reservoir fluid composition in Table 5.2 is only possible if all carbon number fractions to C_{35} as well as C_{36+} are present in concentrations of at least 0.001 weight%. It can be seen from Table 5.2 that the composition of the depleted fluid at 226 bar is only given to C_{30+}. The heavier components are not present in sufficient concentration to be quantitatively separated in a standard GC analysis.

Figure 5.6 shows a plot of mol% versus carbon number for the reservoir fluid and for each of the depleted fluids. While the reservoir fluid is seen to follow an approximately straight line consistent with (5.1), more bended curves are seen for the two depleted fluids.

For each of the three fluids the optimum values of the constants A, B and D in (5.9) are found by a data fit to the C_{7+} mole%'s versus carbon numbers in Table 5.2 ending with the last defined fraction (C_{35} or C_{29}). The optimum constants are shown in Table 5.3. The numerical value of D must be higher than 10^{-6} to be significant. It is seen that D for the reservoir fluid is numerically lower than 10^{-6}, which confirms that (5.1) is a good approximation for

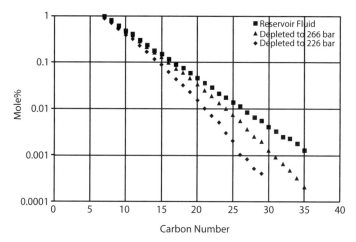

Figure 5.6 Mole% (in logarithmic scale) versus carbon number for the reservoir fluid in Table 5.2 and for the gas phases from CVD simulations to respectively 266 bar and 226 bar.

the reservoir fluid. D is negative for the two depleted fluids, which is consistent with the downwards bend in the ln(mol%) versus carbon number plots for those two fluids in Figure 5.6.

The procedure for recreating the original reservoir fluid from a depleted sample as outlined in the preceding section and sketched in Figures 5.2 and 5.3 led to the two estimated original reservoir fluid compositions shown in Table 5.2.

To test how well the recreated fluid compositions match the actual reservoir fluid composition CVD simulations were carried out on the reservoir fluid and on the two recreated "original reservoir fluid compositions". The simulated liquid dropout curves are plotted in Figure 5.7. Also shown are the liquid dropouts for the depleted fluids. This is to give an idea about how much "liquid loss" the recreation process must compensate for. It is seen that it was possible to almost perfectly recreate the original reservoir fluid composition from the depleted sample at 266 bar, which pressure is 40 bar lower than the saturation pressure of the original reservoir fluid. The liquid dropout from the fluid composition recreated based on the depleted fluid at 226 bar is slightly lower than that of the actual original reservoir

Table 5.3 Constants in (5.9).

Fluid	A	B	D
Reservoir fluid in Table 5.2	−3.0402	-1.6743×10^{-2}	-2.9279×10^{-7}
Fluid in Table 5.2 depleted to 266 bar	−3.3815	-1.2665×10^{-2}	-1.4862×10^{-5}
Fluid in Table 5.2 depleted to 226 bar	−3.3192	-1.2905×10^{-2}	-2.5184×10^{-5}

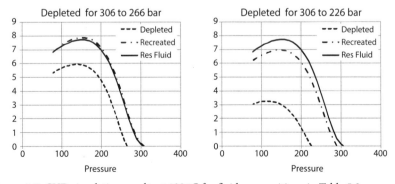

Figure 5.7 CVD simulation results at 129 °C for fluid compositions in Table 5.2.

fluid. For this fluid it must however be taken into consideration that the reservoir fluid composition was recreated from a depleted sample that had a max. liquid dropout of the order of 3 %, while the liquid dropout of the recreated fluid is more than twice as high and of the same order of magnitude as the liquid dropout from the true reservoir fluid.

As already mentioned the depleted fluid at 226 bar only has component mole%'s given to C_{30+}. The low concentration of the heavier components is the main reason for the discrepancy seen between the liquid dropout curves of the original reservoir fluid and of the recreated fluid.

5.4.2 Samples from Tight Reservoirs

Four series of "tight gas reservoir separation processes" were carried out to get the approximate fluid compositions that would be sampled from a tight reservoir with the fluid composition in Table 5.2 and a reservoir pressure of 306 bar or higher. It is assumed that the inlet pressure to the well because of a drawdown has fallen to respectively 286, 266, 246 and 226 bar in the four cases considered. The separation process is simulated as 40 consecutive flash calculations. At each stage the fluid is flashed at a pressure that is ΔP lower than the pressure of the previous stage and the liquid from the flash discarded. The process is schematically outlined in the upper part of Figure 5.5. The depleted fluid from the last stage would have approximately the composition of the fluid sampled from the tight reservoir if the pressure at the inlet to the well had dropped to the end pressure of the flash process.

Figure 5.8 shows plots of mol% versus carbon number for the reservoir fluids and for each of the four depleted fluids. As already established the reservoir fluid is seen to follow an approximately straight line compliant with (5.1). Bended curves are seen for the four depleted fluids.

A comparison of the molar distribution seen for the depleted fluids in Figure 5.6 and Figure 5.8 reveals that the depletion process in a tight reservoir strips off more C_{20+} components than the CVD process. On the other hand the CVD process strips off more C_{10}-C_{20} components than in the tight reservoir. Only for the tight reservoir sample taken at 286 bar would it be possible to get a compositional analysis to C_{36+}. The sample taken at 266 bar could be analyzed to C_{30+} while the fluid composition of the samples at the two lowest pressures at 246 bar and 226 bar would have to stop at C_{25+}.

To recreate the original reservoir fluid composition a series of consecutive saturation pressure calculations followed by addition of equilibrium liquid to the saturated gas is carried out for the depleted fluid as sketched

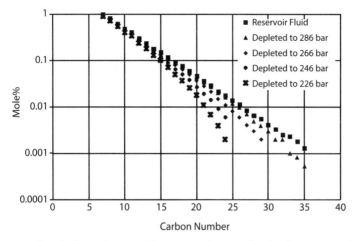

Figure 5.8 Mole% (in logarithmic scale) versus carbon number for the reservoir fluid in Table 5.2 and for the gas phases that would be sampled from a tight gas condensate reservoir at respectively 286, 266, 246 and 226 bar.

in Figure 5.5. The amount of liquid added in each stage is adjusted to correspond to an increase in saturation pressure of ~1.0 bar. The addition of liquid is stopped when D in (5.9) attains a non-negative value.

CVD simulations were carried out on the reservoir fluid and on the four recreated "original reservoir fluid compositions" to test how well the recreated fluid compositions match the actual reservoir fluid composition. The simulated liquid dropout curves are plotted in Figure 5.9. Also shown are the liquid dropouts for the depleted fluids. It is seen that it was possible to almost perfectly recreate the original reservoir fluid composition from the depleted sample at 286 bar, which pressure is 20 bar lower than the saturation pressure of the original reservoir fluid. The liquid dropout from the fluid composition recreated based on the depleted fluid at 266 bar is almost correct, but the simulated saturation point is somewhat too low. This is related to the fact that the separation process in the reservoir has stripped off the heaviest components. With those components lacking it is impossible to create a correct picture of the concentrations of the heaviest components in the liquid left back in the reservoir. The match of the dewpoint pressure is still poorer for the fluids depleted to lower pressures, but interestingly the max. liquid dropout is still almost correct. This has to do with most of the C_{10}-C_{20} components in the reservoir fluid being contained in the sampled gas. Knowing about the saturation point deficiency one might

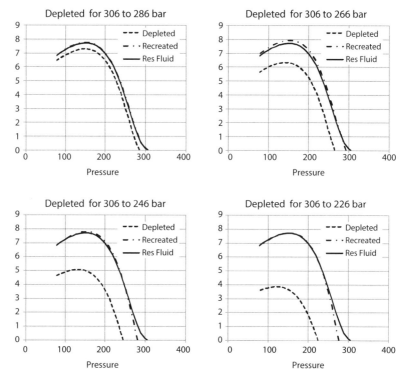

Figure 5.9 CVD simulation results at 129 °C for reservoir fluid composition in Table 5.2 and for fluid compositions sampled from tight reservoir.

add a tail to the CVD liquid dropout curves for the depleted fluids and add C_{30+} components in sufficient concentration to give such tail.

The procedure outlined for tight gas condensate reservoirs is assuming the sample is taken from a reservoir not yet in production. If the sample is taken from a tight gas condensate reservoir, which has produced for some time, the procedure outlined for depleted gas condensate reservoirs may be more appropriate.

5.5 Discussion of the Results

The outlined procedures are based on the assumption that there is a linear relation between the logarithm of the mole fraction of a C_{7+} component and its carbon number. This relationship is observed experimentally and confirmed by the theory of chemical reaction equilibria. The question addressed in the paper is to what extent this knowledge about the molar distribution of the C_{7+} fractions can be used to recreate the molar composition of the original reservoir fluid composition from the molar composition of a depleted fluid sample.

Two different processes are considered. One is for a reservoir depleted as a result of production and the other one is for a tight reservoir, which is not yet in production. The examples show that it is possible to recreate the original fluid composition almost correctly from a sample taken from a gas condensate reservoir in which the pressure is 40 bar lower than the saturation point of the original reservoir fluid. It is also shown that the match of the original reservoir fluid somewhat deteriorates if the down hole pressure is around 80 bar lower than the saturation point of the original fluid. It is however noteworthy that even at this pressure the recreation procedure generates a fluid with approximately the right liquid content. The liquid content of the original reservoir fluid is important when estimating gas and condensate in place. A significant discrepancy is only observed around the saturation pressure. This is due to the heaviest components virtually disappearing from the produced fluid at lower pressures. It is seen from the shown examples that a successful recreation of the original reservoir fluid is dependent on a compositional analysis providing information about the molar content of hydrocarbons to at least C_{30}. Components which cannot be detected in the compositional analysis are not available in the recreation procedure.

During production from a tight reservoir the drawdown is typically quite high, and the liquid drop out close to the saturation pressure will not be correct if simulated from a recreated fluid composition. This may however not present a major problem in reservoir simulations as long as the total liquid content is right.

The paper deals with gas condensate fluids. Oil compositions except heavy aromatic oils with an API lower than 25 will also follow (5.1), but the same recreation procedure cannot be used for oils as for gas condensates. If a reservoir oil splits into two phases, production may originate from either the gas phase, from the oil phase or from both phases. The liquid enrichment process carried out for gas condensates to get a D of zero in (5.9) requires that the gas phase is the dominant one and that is obviously not the case for oils. If the production consists of the oil present at reservoir conditions, the composition of the gas phase in equilibrium with the oil at reservoir conditions can be found from a saturation point calculation. Information may exist about the original saturation point or original GOR and that information can be used to find the recombination ratio, which will recreate the original reservoir fluid composition.

Most likely both reservoir oil and gas contribute to the production for which scenario a PT flash of the produced fluid at reservoir conditions will provide the equilibrium phase compositions present in the reservoir. Assuming the composition of the gas and oil phases in the reservoir have not changed significantly from the equilibrium compositions at the saturation point of the original reservoir fluid, the reservoir fluid composition can be generated by mathematical combination of the two fluid compositions

$$z_i = y_i \beta + x_i (1 - \beta) \; i = 1, 2 \ldots, N \tag{5.10}$$

z_i is the mole fraction of component i in the original reservoir fluid and y_i and x_i are respectively the mole fractions of component i in the gas and liquid phases. $\beta/(1-\beta)$ is the molar recombination ratio and N is number of components. β may be found from knowledge about initial GOR or initial saturation point.

If a reservoir produces from the two phase region, the produced fluid will be different from the original reservoir fluid or such fluid it is impossible to recreate the original reservoir fluid composition by recombining the sampled separator gas and separator oil. Recombination uses a fixed β in (5.10) independent of component and the recombined fluid is bound to be a linear composition of the two separator fluid compositions. One β will exist, which will give the produced fluid, but no β exists, which will give the original reservoir fluid composition.

5.6 Conclusions

The paper outlines the theory behind the observed linear trend between the logarithm of the mole fraction of a C_{7+} fraction and its carbon number seen for reservoir fluids. This linear trend exists as long as the reservoir is producing in single phase. Once the reservoir fluid splits into two phases, the produced fluid will no longer follow the same trend. For gas condensates the linear trend can be used to recreate the original reservoir fluid from a depleted sample. Different procedures are to be used for a sample from a reservoir depleted as a result of production and for a sample taken from a tight gas condensate reservoir not yet in production. The procedure will only be successful if the sampled fluid has been analyzed to at least C_{30+}. For reservoir oils either the original saturation point or the original GOR must be known to recreate the original reservoir fluid composition.

5.7 Nomenclature

A	Constant in molar distribution functions ((5.1) and (5.9))
B	Constant in molar distribution functions ((5.1) and (5.9))
CVD	Constant Volume Depletion PVT experiment
C_n	Carbon number fraction n
D	Constant in molar distribution function for depleted fluids (5.9)
G	Gibbs free energy
GC	Gas chromatography
GOR	Gas/oil ratio
K	Equilibrium Constant
M	Molecular weight
N	Number of components
P	Pressure
R	Gas Constant
T	Temperature
V	Molar volume
X_1	Moles of already produced gas.
X_2	Moles fluid of the composition sampled currently (depleted gas)
X_3	Moles of the liquid remaining in the reservoir (dewpoint liquid).
x	Component mole fraction in liquid phase
y	Component mole fraction in gas phase
z	Component mole fraction

Greek letters

β Vapor phase mole fraction

Sub and super indices

i	Component index
liq	Liquid
res	Reservoir

References

1. Pedersen, K.S., Blilie, A.L., and Meisingset, K.K., PVT calculations on petroleum reservoir fluids using measured and estimated compositional data for the plus fraction, *I&EC Research* 31, 1378–1384, 1992.
2. Krejbjerg, K. and Pedersen, K.S., "Controlling VLLE Equilibrium with a cubic EOS in heavy oil modeling", 57th Annual Technical Meeting of the Petroleum Society (Canadian International Petroleum Conference), Calgary, Canada, June 13–15, 2006.
3. Soave, S., "Equilibrium constant from a modified Redlich-Kwong equation of state", *Chem Eng. Sci.* 27, 1972, 1197–1203.
4. Peneloux, A., Rauzy, E. and Fréze, R. A., "A consistent correction for Redlich-Kwong-Soave volumes, Fluid Phase Equilibria 8, 1982, 7–23.
5. Journal of Physical and Chemical reference data, vol 11, supp. 2, 1982.

6

Phase Equilibrium in the Systems Hydrogen Sulfide + Methanol and Carbon Dioxide + Methanol

Marco A. Satyro[1] and John J. Carroll[2]

[1]*Virtual Materials Group, Calgary, AB, Canada*
[2]*Gas Liquids Engineering, Calgary, AB, Canada*

Abstract

Methanol is commonly used to prevent hydrate formation in the natural gas industry, including acid gas injection. Although the primary mechanism for hydrate inhibition is the same as in sweet gas systems and acid gas systems, the phase equilibrium is not. Acid gases have a significant solubility in methanol, whereas hydrocarbons, particularly liquid phase hydrocarbons, will also show significant solubility in methanol and water solutions as the concentration of methanol in the aqueous solution increases.

Phase equilibrium in the systems H_2S + water and CO_2 + water is fairly well understood and will not be re-examined here. However, there are two significant aspects of those systems that are important to this discussion: (1) CO_2 and H_2S exhibit liquid phase immiscibility with water and (2) both CO_2 and H_2S can form hydrates. On the other hand, both CO_2 and H_2S are completely miscible with methanol.

The purpose of this chapter is to start the investigation of methanol equilibrium in acid gas systems by examining the phase equilibrium in the binary systems H_2S + methanol and CO_2 + methanol. The first part of this chapter presents a literature review and the second part presents some modeling results. Equations of state (EoS), which are commonly used in the process industries, have trouble with predicting the behavior in these systems. This includes the prediction of false liquid phase immiscibility. That is, the EoS predicts two liquid phases which, as noted above, do not occur in nature. Thus, one must have a well-constructed EoS in order to model these systems.

The simultaneous modeling of water, acid gases, and liquid hydrocarbon equilibrium will be examined in a future paper.

6.1 Introduction

Gas hydrates are ice-like solids that are composed of water and other small molecules. The water molecules form cages inside of which the other molecule can reside. The other molecule is called a "hydrate former". The presence of the hydrate former stabilizes the water cage structure and thus a solid can precipitate. Both carbon dioxide and hydrogen sulfide are hydrate formers and thus acid gases are hydrate formers.

Hydrates can be a significant problem in an acid gas injection scheme. Hydrates are notorious for plugging pipelines (even wells) and for damaging process equipment. The main tool for dealing with hydrates in an acid gas injection system is to reduce the water content of the fluid to such appoint where a hydrate does not form for the range of temperatures and pressures experienced in the process.

However, sometimes dehydration is insufficient and some other technique is used. The other common tool used to combat hydrate formation is methanol. Methanol is a so-called thermodynamic inhibitor. The addition of methanol to a mixture reduces the hydrate temperature for a given pressure since it changes the activity of water in the liquid phase and therefore the conditions for thermodynamic equilibrium where acid gases stabilize the hydrate structure. Noticeable inhibition happens only when an aqueous phase is available for the dissolution of the inhibitor. Inhibitors are not limited to liquids; salts can also work as inhibitors. From a thermodynamic point of view inhibitors will affect the hydrate formation temperature in systems where no aqueous phase is present (for example hydrate in equilibrium with a gas phase only) but these effects are usually small.

To completely understand the effect of methanol on the hydrates in acid gas systems one must understand the fluid phase equilibria. The focus of this paper will be the equilibrium between hydrogen sulfide + methanol and carbon dioxide + methanol, however there will be some discussion of water.

The binary systems hydrogen sulfide + water and carbon dioxide + water exhibit liquid-liquid immiscibility (LLE). However, methanol +H_2S and methanol + CO_2 do not. The methanol + acid gas systems are completely miscible.

It is important to state that H_2S and methanol and CO_2 and methanol are completely miscible. This is unlike water which shows liquid-liquid equilibrium with these two components. If you are relying on a simulation that predicts a liquid methanol phase in equilibrium with liquefied acid gas then you have a problem since Nature will not collaborate with you. Unfortunately simple models based on cubic equations of state and quadratic mixing rules predict false phase splitting under some conditions.

6.2 Literature Review

They have been several experimental investigations of the equilibrium for the binary systems of interest in this study. The key studies are reviewed in this section.

Many of the experimental data are at higher temperatures than would be useful for this study of hydrates in acid gas mixtures. Such hydrates form only at temperatures below about 35°C. However, the higher temperature data are useful for building and testing models.

6.2.1 Hydrogen Sulfide + Methanol

The key experimental measurements for hydrogen sulfide + methanol are those of Leu et al. [1]. These cover four isotherms from 25°C to 175°C and pressures up to about 11 MPa. This included two measurements for the binary critical points. Although much of their data are for temperature greater that the hydrate region, they provide a good set for building and testing models.

Leu et al. [1] also reviewed other available data up to that date. Lower temperature data from Yorizane et al. [2] for isotherms at 248.055 K, 258.053 K and 273.1 K and pressures up to about 1 MPa is available but no vapour compositions were measured and therefore only useful for the validation of bubble pressure data. Short et al, [3] report two low pressure bubble pressure points at 263.1 K and 298.14 K at 0.101325 MPa.

6.2.2 Carbon Dioxide + Methanol

Hong and Kobayashi [4] measured the vapor-liquid equilibrium (VLE) for the binary system carbon dioxide + methanol for six isotherms from -43° to + 57°C and pressures from 0.7 MPa to just under 11 MPa. These are an excellent set for the current study as they cover the range of temperatures that are important for hydrate formation.

Leu et al. [5] measured the VLE for this system from 50° to 204°C and pressure from 0.4 to about 16 MPa. As with the data from this lab for H_2S + methanol, these data are at higher temperatures than are of interest from a hydrate point of view, but they are useful for building and testing models.

As part of a larger study, Yoon et al. [6] measured VLE for CO_2 + methanol at 40°C from 0.7 MPa up to the critical pressure, 8.2 MPa. Again, these data are at higher temperatures than would be of interest in the study of hydrates.

Liu et al. [7] measured the binary critical locus for this system and confirm that it is a continuous locus from the critical point of pure CO_2 to the critical point of pure methanol. Based solely on the work of Liu et al. [7], it is estimated that the locus has a maximum value of about 16.5 MPa at a temperature of about 140°C and 65 mol% methanol.

It is also important to note that when developing thermodynamic models for process simulation compromises are sought between data and accuracy. For example, from a purely thermodynamic point of view binary VLE data is preferred since it can be checked for thermodynamic consistency, but sometimes this data does not correspond to conditions where the model will actually be used. A good example of this is provided by the model parameters used by VMG's Advanced Peng-Robinson for Natural Gas (APRNG), a thermodynamic package widely used for the simulation of natural gas processes.

When the package was originally designed interaction parameters for H_2S and CO_2 were determined based on multicomponent data representing vapor-liquid equilibrium and vapor-liquid-liquid equilibrium (VLLE) for mixtures closely associated with light hydrocarbons, gas condensate, water and methanol commonly encountered in field operations as reported by Chen and Ng [8] and Ng and Chen [9] and therefore represent a compromise between VLE and VLLE modeling even though CO_2 + methanol and H_2S + methanol binaries do not exhibit VLLE.

6.3 Modelling With Equations Of State

It is well-known that the traditional equations of state used to model VLE in the natural gas processing, such as the Soave [10] and Peng and Robinson [11] equations, cannot be directly applied to systems containing polar substances such as water and alcohols. However with some recent improvements the predictions from models based on these equations are significantly improved.

The first requirement of an EoS is that it accurately predicts the vapor pressure of the pure components. This can be achieved for methanol by a relatively simple correction to the temperature dependence of the attractive term. The general form of a cubic equation of state is:

$$P = \frac{RT}{v - b_1} - \frac{a}{v^2 - b_2 v - b_3} \qquad (6.1)$$

The a and b_i's are unique to each equation of state although they are typically derived from the inflection of the critical isotherm at the critical point. Their values are fixed and determine the structure of the equation of state. For example if b_2 is equal to -1 and b_3 is equal to zero we have the familiar SRK equation of state. If b_2 is equal -2 and b_3 is equal to b^2 we have the PR equation of state. Note that we assume that a_c and b as obtained from the criticality conditions as applied to the critical point of a pure substance given by (6.2).

$$\left(\frac{\partial P}{\partial v}\right)_{T_c} = \left(\frac{\partial^2 P}{\partial v^2}\right)_{T_c} = 0 \qquad (6.2)$$

In this form:

$$a = a_c \alpha(T) \qquad (6.2a)$$

is the attractive term, which is modified to improve the prediction of the vapor pressure for the pure component. Or more precisely, the $\alpha(T)$ function is used to correlate vapor pressures. Many empirical forms exist but one that is particularly effective for the correlation of vapour pressures is the Mathias and Copeman [12] form defined using (6.3) and (6.4).

$$\tau = 1 - \sqrt{\frac{T}{T_c}} \qquad (6.3)$$

$$\alpha(T_r) = \left(1 + A\tau + B\tau^2 + C\tau^3\right)^2 \qquad (6.4)$$

For supercritical conditions (6.4) is extrapolated through the matching of the first and second derivatives calculated at the critical point using (6.5).

$$\alpha(T_r) = \exp\left(D\left[1 - T_r^E\right]\right) \qquad (6.5)$$

In (6.4) the pure component constants are determined through a least squares fit to vapor pressures. This procedure also provides a reasonable model for liquid heat capacities and enthalpies of vaporization as shown by Satyro and van der Lee [13].

Frequently the attractive parameter in (6.1) is modeled using a simple quadratic mixing rule shown in (6.6) and the repulsive parameter b is modeled using a simple linear mixing rule as shown in (6.7).

$$a = \sum_{i=1}^{nc}\sum_{j=1}^{nc} x_i x_j \sqrt{a_i a_j}\left(1 - k_{ij}\right) \qquad (6.6)$$

$$b = \sum_{i=1}^{nc} x_i b_i \qquad (6.7)$$

The parameter k_{ij} is the interaction parameter and it is used to fit thermodynamic equilibrium data in essence capturing all the inadequacies of the thermodynamic model into a simple adjustable parameter. The interaction parameter is sometimes required to vary with temperature.

A model can be quickly constructed using Virtual Materials Group Thermo Explorer regression package. The package works by minimizing the error between calculated and experimental bubble pressures and dew temperatures by changing the value of the interaction parameter. Results for two datasets at 298.15 K and 348.15 K from Leu et al. [1] are shown in Figure 6.1.

The errors are summarized in Figure 6.2.

The methanol and carbon dioxide binary is more challenging to model. The data from Hong and Kobayashi [4] was used for the development of the model and required the use of a temperature dependent interaction parameter. Even with this LLE behavior was predicted at low temperatures. The regression results are shown in Figures 6.3 and 6.4 and the predicted LLE region is shown in Figure 6.5.

For methanol and carbon dioxide mixtures a more flexible mixing rule is required. In this case we use VMG's implementation of Huron and Vidal's mixing rule (2013) that combines a Gibbs Excess energy model with a cubic equation of state by matching the Gibbs free energies estimated by the equation of state and by the activity coefficient model at infinite pressure. The results are shown in Figures 6.6–6.8.

Phase Equilibrium in the Systems 105

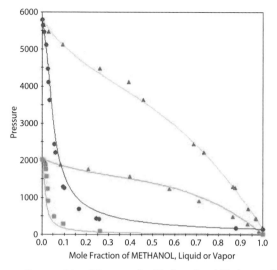

Figure 6.1: Pressure-Composition Diagrams for Methanol and Hydrogen Sulfide at 298.15 and 348.15 K Calculated Using VMG's Thermo Explorer (2009). The Interaction Parameter Was Determined to be 0.072.

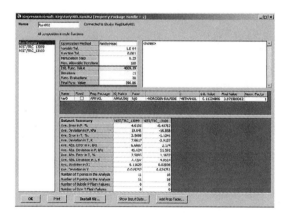

Figure 6.2: Error Summary for Methanol and Hydrogen Sulfide Binary.

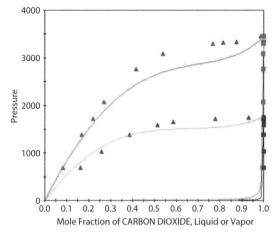

Figure 6.3: PXY Diagrams for Methanol and Carbon Dioxide at 250 and 290 K Calculated Using VMG's Thermo Explorer (2009) and APRNG model. The Interaction Parameter Was Determined to be Temperature Dependent $k_{12} = -27.0 + \dfrac{1095.6}{T} + 4.10 \ln T$ where T is in K.

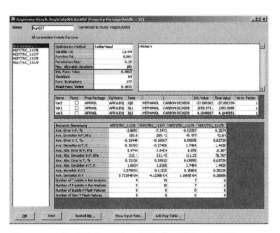

Figure 6.4: Error Summary for Methanol and Carbon Dioxide Binary

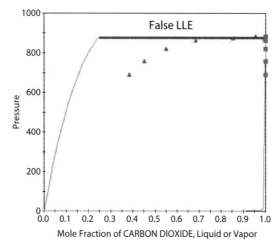

Figure 6.5: Predicted False Liquid-Liquid Equilibrium (LLE) at 230 K – APRNG with Quadratic Mixing Rules

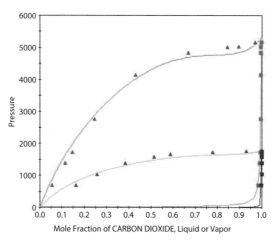

Figure 6.6: PXY Diagrams for Methanol and Carbon Dioxide at 250 and 290 K Calculated Using VMG's Thermo Explorer (2009) and GEPR Model. The Interaction Parameters Were Determined to be $b_{ij} = 523.94$, $b_{ji} = 221.22$ and $\alpha_{ij} = \alpha_{ji} = 0.75$.

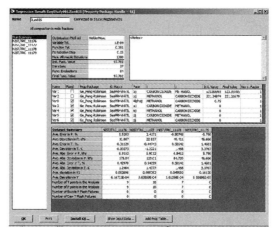

Figure 6.7: Error Summary for Methanol and Carbon Dioxide Binary

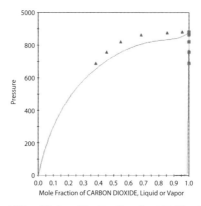

Figure 6.8: Liquids-Liquid Equilibrium Not Predicted at 230 K – GEPR using Huron-Vidal Mixing Rules

6.4 Nomenclature

Roman

a   attractive term in the cubic equation of state
A   parameter in the temperature dependence of the attractive term
b   co-volume in the cubic equation of state
B   parameter in the temperature dependence of the attractive term
C   parameter in the temperature dependence of the attractive term
D   parameter in the temperature dependence of the attractive term
E   parameter in the temperature dependence of the attractive term

k_{ij}	binary interaction parameter
nc	number of components
P	pressure
R	universal gas constant
T	absolute temperature
v	molar volume
x	mole fraction

Greek

$α(T)$	temperature dependence of the attractive term
τ	modified reduced temperature

Subscripts

c	critical
i	component i
j	component j
r	reduced

References

1. Leu, A.-D., Carroll, J.J., and Robinson, D.B., "The Equilibrium Phase Properties of the Methanol-Hydrogen Sulfide Binary System", *Fluid Phase Equil.*, **72**, 163–172, (1992).
2. Yorizane, M., Sadamoto, S., Masuoka, H., and Eto, Y.; "Solubility of Gases in Methanol at High Pressures"; *Kogyo Kagaku Zasshi*, 72, 2174–2176, (1969). – in Japanese.
3. Short, I., Sahgal, A., and Hayduk, W.; "Solubility of Ammonia and Hydrogen Sulfide in Several Polar Solvents", *J. Chem. Eng. Data*, **28**, 63-66, (1983).
4. Hong, J.H. and Kobayashi, R., "Vapor-Liquid Equilibrium Studies for the Carbon Dioxide-Methanol System", *Fluid Phase Equil.*, **41**, 269–276, (1988).
5. Leu, A.-D., Chung, S.Y.-K., and Robinson, D.B., "The Equilibrium Phase Properties of (Carbon Dioxide + Methanol)", *J. Chem. Thermodynamics*, **23**, 979–985, (1991).
6. Yoon, J.-H., Lee, H.-S., and Lee, H., "High-Pressure Vapor-Liquid Equilibria for Carbon Dioxide + Methanol, Carbon Dioxide + Ethanol, and Carbon Dioxide + Methanol + Ethanol", *J. Chem. Eng. Data*, **38**, 53–55, (1993).

7. Liu, J., Qin, Z., Wang, G., Hou, 6., and Wang, J., "Critical Properties of Binary and Ternary Mixtures of Hexane + Methano, Hexane + Carbon Dioxide, Methanol + Carbon Dioxide, and Hexane + Carbon Dioxide + Methanol", *J. Chem. Eng Data*, **48**, 1610–1613, (2003).
8. Chen, C.J. and Ng, H.-J.; "The Solubility of Methanol or Glycol in Water-Hydrocarbon Systems"; *GPA Research Report RR-117*, Gas Processors Assoc., Tulsa, OK, (1988).
9. Ng, H.-J. and Chen, C.J.; "Vapour - Liquid and Vapour- Liquid – Liquid Equilibria for H2S, CO2, Selected Light Hydrocarbons and a Gas Condensate in Aqueous Methanol or Ethylene Glycol Solutions"; *GPA Research Report RR-149*, Gas Processors Assoc., Tulsa, OK, (1995).
10. Soave, G., "Equilibrium Constants from a Modified Redlich-Kwong Equation of State", *Chem. Eng. Sci.*, **27**, 1197–1203, (1972).
11. Peng, D-Y. and D.B. Robinson, "A New Two-Constant Equation of State", *Ind. Eng. Chem. Fund.*, **15**, 59–64, (1976).
12. Mathias, P.M. and Copeman, T.W., "Extension of the Peng-Robinson Equation-of-State to Complex Mixtures: Evaluation of the Various Forms of the Local Composition Concept", *Fluid Phase Equilib.*, **13**, 91–108, (1983).
13. Satyro, M.A. and van der Lee, J.; "The Performance of State of the Art Industrial Thermodynamic Models for the Correlation and Prediction of Acid Gas Solubility in Water"; Wu/Carroll (ed.), *Acid Gas Injection and Related Technologies*, 21–36, Scrivener Publishing LLC, (2009).
14. Virtual Materials Group, Inc, Thermo Explorer User's Manual Version 1.0; Calgary, Alberta, Canada, (2009).
15. Virtual Materials Group, Inc. VMGSim User's Manual Version 7.0; Calgary, Alberta, Canada, (2013).

7

Vapour-Liquid Equilibrium, Viscosity and Interfacial Tension Modelling of Aqueous Solutions of Ethylene Glycol or Triethylene Glycol in the Presence of Methane, Carbon Dioxide and Hydrogen Sulfide

Shu Pan[1], Na Jia[1], Helmut Schroeder[1], Yuesheng Cheng[1], Kurt A.G. Schmidt[2] and Heng-Joo Ng[1]

[1]*Schlumberger, DBR Technology Center, Edmonton, AB, Canada*
[2]*Schlumberger, Abingdon Technology Center, Abingdon, United Kingdom*

Abstract

The Vapor-Liquid Equilibrium, equilibrium phase densities, liquid phase viscosity, and interfacial tension of methane and its mixtures with carbon dioxide and hydrogen sulfide in aqueous solutions of ethylene glycol and triethylene glycol were measured over a range of temperatures from 272 to 333 K. Pressures varied between 3447 and 13789.5 kPa. The phase densities and solubility data were predicted using the simplified PC-SAFT equation of state. The liquid phase viscosity was predicted with the Expanded Fluid viscosity model. Interfacial tensions were predicted with the Parachor method. The experimental and calculated results obtained in this investigation were also compared with the results obtained with a commercial software package.

7.1 Introduction

Presently, there is insufficient data to accurately design gas dehydration facilities that operate at pressures above 10000kPa.New data for high-pressure glycol dehydration contactors is needed to ensure the correct design of these columns. In addition, the solubility of gases in glycol solutions at high-pressures is of interest to properly design recompression facilities. To overcome the deficiency of data at these conditions, the Gas Processors Association sponsored DBR Technology Center, a division of Schlumberger Canada Limited, [1] to experimentally determine the phase behavior and transport properties of high pressure systems in the presence of ethylene glycol (MEG) and triethylene glycol (TEG). The Vapour-Liquid Equilibrium (VLE), phase densities, interfacial tension (IFT), and the liquid phase viscosity were measured at conditions relevant for the design of dehydration facilities that operate at pressures of 10000 to 15000kPa. To assist the tie of the new high-pressure data back to the existing data set, measurements were also made at pressures of 3447kPa.

7.2 Results and Discussion

7.2.1 Experimental

Experimental results, apparatus, and procedures are detailed in Ng et al. [1] and only a brief description of the methodology is discussed here. The VLE experiments were carried out in a DBR Technology Center visual PVT Cell. The composition of the equilibrium phases were determined by the analysis of subsamples of each phase with a HP5890 gas chromatograph equipped with multicolumn configuration and both a Flame Ionization Detector and a Thermal Conductivity Detector. When H_2S was present in the liquid phase, the analysis technique of Jou et al. [2] quantified the amount of H_2S in the solution. Density measurements were performed with the gravimetric method. The viscosity of the liquid phase was measured with a Cambridge electromagnetic viscometer. Measurement of the IFT between the phases was carried out with the pendant drop method. A total of 17 experimental measurements for 98 weight percent (wt%) TEG with water system and 16 experimental measurements for 70 wt% MEG with water system were performed. Because the vapour pressure of the glycols was very low at the experimental conditions, no analysis of MEG or TEG in the vapour phase was performed. One example of the experimental results, out of the 33 experiments, is shown in Table 7.1.

Due to working limitations of the interfacial tension apparatus, measurement of the IFT was not performed when temperatures were below the ambient temperature or when systems contained H_2S. A total of 15 experimental IFT measurements for the 98 wt% TEG with water system and 4 experimental measurements for the 70 wt% MEG with water system were performed.

Table 7.1: Experimental VLE and Phase Property Data of the Methane + 98 wt% TEG with Water System I at 299.8 K and 6895 kPa

Mass % TEG	98			
Temperature (K)	299.8			
Pressure (kPa)	6895			
Component	Feed	Liquid	Vapour	K
	mole%	mole%	mole%	
Carbon Dioxide	--	--	--	--
Hydrogen Sulfide	--	--	--	--
Methane	52.254	1.791	99.982	55.839
Water	6.940	14.260	0.018	0.001
TEG	40.806	83.949	--	--
Volume %			75.2	
Density (kg/m³)			49	
Viscosity (mPas)			--	
IFT (mN/m)				

7.2.2 Vapour Liquid Equilibrium and Phase Density Modeling

The measured VLE and phase density data were compared to the results calculated with the simplified PC-SAFT (Perturbed Chain Statistical Association Fluid Theory) equation of state [3]. The PC-SAFT is an Equation of State (EoS) established upon modern molecular theory which has been shown to elegantly model the phase behavior of systems of association fluids. A simplified version of PC-SAFT (i.e. sPC-SAFT) was proposed by von Solms et al. [4] to reduce the computational effort without a significant sacrifice in performance. Grenner et al. [5] showed that the sPC-SAFT could satisfactorily model the phase equilibrium of glycol-containing systems; however, the use of binary interaction parameters was important to obtain a good representation of the phase behavior of these systems.

The essence of PC-SAFT is to sum various contributions from the hard chain, dispersion and association in terms of the residual Helmholtz energy,

$$\tilde{a}^{res} = \frac{A^{res}}{Nk_B T} = \tilde{a}^{hc} + \tilde{a}^{disp} + \tilde{a}^{assoc} \qquad (7.1)$$

where A^{res} is the residual Helmholtz energy; N is the total number of molecules; K_B is the Boltzman constant; T is temperature; $\tilde{a}^{hc}$ is the part of the Helmholtz energy from the hard-chain contribution; $\tilde{a}^{disp}$ is the part from the dispersive contribution; and $\tilde{a}^{assoc}$ is the part from the association contribution. A total of five pure-component parameters are required for associating fluids; the parameters are usually obtained by regression of the EoSto vapour pressure and liquid density data. Among the parameters, the segment diameter σ_i, segment number m_i and segment energy parameter ε_i are responsible for the non-associating effect; at the same time, the associating energy $\varepsilon^{A_i B_i}$ and effective associating volume $k^{A_i B_i}$ will characterize the association, i.e. hydrogen bonding sites.

In this study, sPC-SAFT parameters of the pure components, methane, carbon dioxide, hydrogen sulfide, water, MEG, and TEG, were obtained from Gross and Sadowski [3], Grenner et al. [6], and Grenner et al. [5] and are listed in Table 7.2. The acid gases, hydrogen sulfide and carbon dioxide, are assumed to be inertial components to keep the model simple. This assumption is justified by the fact that our interested pressure range in this work is relatively low (less than 13789.5 kPa). Meanwhile, a 4C associating scheme is assigned to all the solvents, including glycols and water.

When possible, binary interaction parameters were taken from Grenner et al. [5]; however, most of the binary interaction parameters in this study were not available in the open literature. The missing binary interaction parameters were obtained from the correlation of the literature VLE data of the binary systems. In this work, the binary interaction parameters between gases (methane, hydrogen sulfide, and carbon dioxide) were assumed to

Table 7.2: Pure-Component Parameters of sPC-SAFT Equation of State

Component	Associating Scheme	Ref.	m_i	σ_i (Å)	ε_i/k_B (K)	$k^{A_i B_i}$	$\varepsilon^{A_i B_i}/k_B$ (K)
CO_2	--	1	2.0729	2.7852	169.21	0.00	0.00
H_2S	--	1	1.6686	3.0349	229.00	0.00	0.00
CH_4	--	1	1.0000	3.7039	150.03	0.00	0.00
H_2O	4C	2	1.50000	2.6273	180.30	0.0942	1804.22
MEG	4C	3	1.90878	3.5914	325.23	0.0235	2080.03
TEG	4C	3	3.18092	4.0186	333.17	0.0235	2080.03

(1) Gross and Sadowski [3]; (2) Grenner et al. [5]; (3) Grenner et al. [6]

be zero due to the lack of Vapor Liquid Equilibrium data of these binary mixtures. Table 7.3 presents the non-zero binary interaction parameters. As shown in Table 7.3, most of these parameters were regressed in this work to match VLE data available in the open literature.

With the parameters in Tables 7.2 and 7.3, the sPC-SAFT EoS was used to predict the VLE and phase density data presented in Ng et al. [1]. The prediction of equilibrium constants from the sPC-SAFT has been plotted against the experimental data in Figure 7.1. For brevity, only the

Table 7.3: sPC-SAFT Equation of State Binary Interaction Parameters (k_{ij})

System	Temperature Range (K)	k_{ij}	Source	VLE Ref.
$CO_2 + H_2O$	285–383	−0.005	This work	5, 6
$H_2S + H_2O$	311–444	0.015	This work	4
$CH_4 + H_2O$	272–511	0.091	This work	2, 3
CO_2 + MEG	323–398	−0.0099	1	--
H_2S + MEG	298–398	0.0	This work	9
CH_4 + MEG	283–323	0.068	This work	7, 8
CO_2 + TEG	298–398	0.1	This work	10
H_2S + TEG	298–398	−0.013	This work	10
CH_4 + TEG	298–398	0.086	This work	10
MEG + H_2O	343–395	−0.046	1	--
TEG + H_2O	297–332	−0.147	1	--

(1) Grenner et al. [5]; (2) Olds et al. [7]; (3) Chapoy et al. [8]; (4) Selleck et al. [9]; (5) Coan and King Jr. [10]; (6) Spycher et al. [11]; (7) Folas et al. [12]; (8) Wang et al. [13]; (9) Jou et al. [14]; (10) Jou et al. [2]

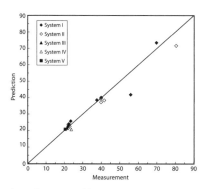

Figure 7.1: Experimental Methane Equilibrium Constants in 98 wt% TEG with Water Solutions Compared with Predicted Values Using the sPC-SAFT EoS

equilibrium constants for methane are plotted in this work. In Figure 7.1, the results are differentiated into five different systems. A system refers to the VLE measurement reported in Ng et al. [1] with the same gas mixture (See Appendix A for more details). As shown in Tables 7.4 and 7.5, the sPC-SAFT has generally provided a good prediction of the equilibrium constants of methane, carbon dioxide, and hydrogen sulfide in both glycol systems.

Relatively poor prediction of the experimental data is observed for some of the 70 wt% MEG with water solution data points. As can be seen in Appendix A, some of the poorly predicted data points are from those experiments performed at the lowest temperature in this study (272 K). These deviations may imply that the existing set of parameters, particularly the binary interaction parameters, were insufficient to model the VLE at these low-temperature conditions. To improve the predictions made by the sPC-SAFTEoS, new binary and multiple-components VLE datasets at low temperatures are required. Based on the results, the sPC-SAFT EoS appears to systematically under-predict the equilibrium constants of carbon dioxide in the 70 wt% MEG with water system. This under-prediction may be due to a poor CO_2-MEG binary interaction parameter. As shown in Table 7.3, this binary interaction parameter was obtained from VLE data

Table 7.4: AAPD Between the Experimental and Predicted Equilibrium Constants (sPC-SAFT) in the 70 wt% MEG with Water Solutions

Component	AAPD (%)
CO_2	33.7
H_2S	21.2
CH_4	38.1
H_2O	49.1

Table 7.5: AAPD Between Experimental and Predicted Equilibrium Constants (sPC-SAFT) in 98 wt% TEG with Water Solutions

Component	AAPD (%)
CO_2	34.2
H_2S	6.1
CH_4	11.2
H_2O	82.4

at temperatures in the range of 323 to 398 K, which is much higher than the temperature range (272 to 300 K) in this work. Because only constant binary interaction parameters were used, it is possible that a deviation in prediction could be due to the gap between the temperature ranges used to regress the parameters and the conditions of the experimental data.

Tables 7.4 and 7.5 indicate that only a fair agreement was achieved between the predicted and the experimental water equilibrium constants in both of the glycol systems. These deviations may be attributed to the gap between the temperature range under which the binary interaction parameters were obtained and the temperature range used in study. In addition, the experimental errors in the measurement of trace amounts of water in vapor phase may have also contributed to these deviations. Overall average absolute percentage deviations (AAPD) of 3.3% and 4.0% were determined for the liquid and vapor phase density calculation in the 70 wt% MEG systems, and 1.0% and 5.8% deviations, respectively, were obtained for the liquid and vapor phase density calculation in the 98 wt% TEG systems. The individual AAPD of the predictions are detailed in Appendix A.

7.2.3 Liquid-Phase Viscosity Modeling

Version 1 of the Expanded Fluid viscosity model [15] was used to predict the equilibrium liquid phase viscosity (μ). In this version of the model, experimental density values are used instead of those modeled from an EoS. The viscosity of a fluid phase is determined with (7.2),

$$\mu = \mu_o + c_1(\exp(c_2 \beta) - 1) \tag{7.2}$$

and is based on the fluid's dilute gas viscosity, μ_o, two parameters c_1 and c_2, and β which relates the viscosity to the expansion of the fluid.

$$\beta = \frac{1}{\exp[(\rho_s^*/\rho)^n - 1] - 1} \tag{7.3}$$

In (7.3), ρ is the fluid density, ρ_s^* is the density of the fluid in the compressed state and n is an empirical exponent. The density of the fluid in the compressed state was determined from (7.4),

$$\rho_s^* = \frac{\rho_s^o}{\exp(-c_3 P)} \tag{7.4}$$

which relates the compressed-state density in vacuum, ρ_s^o with the system pressure and a pressure dependency parameter (c_3).

In this investigation, values of n and c_1 were taken from Motahhari et al. [15]. Values of ρ_s^o for each component were taken from Motahhari et al. [15], except for methane which was taken from Yarranton and Satyro [16]. In the case of c_3 for carbon dioxide and water, the value for each component was taken from Motahhari et al. [15]. For the remaining components, c_3 values were determined with the correlation presented in Motahhari et al. [17]. Since the correlation for c_3 was based on hydrocarbons, an alternative approach where c_3 was set to zero for each component (including CO_2 and H_2O) was used.

The dilute viscosity term, μ_o, and the c_2 parameter (for each component other than methane) were also calculated from the correlation and parameters recommended in Motahhari et al. [15]. These properties for methane were determined with the procedure described in Yarranton and Satyro [16]. Because the fluid systems investigated in this work are multi-component mixtures, the mixing rules for μ_o, ρ_s^o, c_2 and c_3 recommended by Motahhari et al. [15] were used. In this approach (Version 1) no binary interaction parameters were used and the calculated results were based on pure predictions.

The overall AAPDs are 34.3% for the 70 wt% MEG with water system and 24.2% for the 98 wt% TEG with water system when $c_3 = 0$ and were, respectively, 35.5 % and 31.1 % when the tabulated values and the correlation for c_3 was used. As can be seen in Appendix A, some of the predicted values are significantly different from the experimental values. Version 1 of this model, based on experimental densities, is very sensitive to the input density, and any error in the density can have a significant effect on the calculated viscosity. Overall, the results are fairly reasonable, considering that the model was not fully set up for these types of systems. For example, the glycol parameters were determined at atmospheric conditions whereas this data-set is at elevated pressures. The individual AAPD of the predicted values when $c_3 = 0$ are detailed in Appendix A.

7.2.4 Interfacial Tension Modeling

The IFT, γ, was predicted with the Parachor method of Weinaug and Katz [18]. This approach relates the interfacial tension of a multi-component system, consisting of i components, with the bulk liquid and vapor phase densities (ρ_L and ρ_V) and molar compositions of the liquid phase, x_i and vapor phase, y_i:

$$\gamma^{1/P} = \sum_i P_i \left(\frac{x_i \rho_L}{M_L} - \frac{y_i \rho_L}{M_v} \right) \tag{7.5}$$

In (7.5), M_L and M_v are the respective molar masses of the liquid and vapor phases, p is the Parachor exponent, which was taken to be 4, and p_i is the Parachor of component i. In this investigation, the Parachor of each component was obtained from three different methodologies. The first method, Hugill and van Welsenes [19], was based on correlations of the Parachor with critical temperature, critical pressure, and acentric factor from Schlumberger [20].The second method determined the component Parachors with the Quantitative Structure – Activity Relationship (QSAR) method presented in Birdi [21]. This group-contribution method calculates the Parachor as an additive function of atomic and structure constants. The atomic and constitutional Parachor values used in the calculation were those presented in Birdi [21]. In the third method, the Parachor of the components were taken from Quayle [22]. The critical properties used in this work and Parachor values from all the three different methodologies are presented in Table 7.6. These Parachor values were also compared with those from other literature [23–28]. The comparison indicated that the Parachors from Hugill and van Welsenes [19], Birdi [21], and Quayle [22] were not only consistent with each other but also fairly close to those in the literature.

Table 7.7 summarizes the difference between the predicted and the experimental results. Unlike the other measurements, the averages are based on fifteen interfacial tension measurements for the 98 wt% TEG with water systems and on four interfacial tension measurements for the 70 wt% MEG with water system. In each case, the experimental molar composition, molecular mass and, density of each phase were used in (7.5). The results in Table 7.7 indicate that the Parachor values from Quayle [22] yield predicted IFTs values that compare closest with the experimental data. The individual AAPD of the predictions based on the Parachor values from Quayle [22] are detailed in Appendix A.

7.2.5 Commercial Software Comparison

In addition to the predicted values from the sPC-SAFTEoS, Parachor model, and the Expanded Fluid viscosity model, predicted values from Multiflash™ (version 4.0) software [29] were compared to the experimental results. In this investigation, two different model sets were used as a basis for the calculations. The first is based on Infochem's version of the CPA (CPA-Info) EoS which is based on a volume-translated Soave-Redlich-Kwong(SRK) EoS. The second is based on Infochem's advanced implementation of the SRK EoS (RKSA-Info). In both cases, Infochem's version of the SuperTRAPP and Linear Gradient Theory were used for the viscosity and the IFT calculations, respectively. In each case, the default parameters in each model set were used and no attempts were made to improve the results with parameter regression. The comparison between the predicted values and the experimental VLEdata (K_i values) are presented in Tables 7.8 and 7.9.

The CPA-Info model set was able to predict the equilibrium constant of each component in the MEG system (other than H_2S) better than the RKSA-Info model set. The performance for H_2S is comparable in both model sets. The performance comparison is the same for the TEG system except for water, in which significant deviations between the modeled results and the experimental data occurred. In the RKSA-Info model set, an AAPD between calculated and experimental data was 1.8% and 3.9%, respectively, for the liquid and vapor phase density (70 wt% MEG with water system) and 0.9% and 5.8%, respectively, (98 wt% TEG with water system). The CPA-Info model set showed an AAPD of 2.1% and 3.5%, respectively, for the liquid and vapor phase density in the 70 wt% MEG with water system and 0.6% and 5.8% in the 98 wt% TEG with water system. In this case, both model sets gave consentient results when compared with each other.

At almost all the conditions studied, the sPC-SAFT EoS appeared to yield the best match between predicted and experimental values of VLE

Table 7.6: Component Critical Properties and Parachors

				Parachor $(N/m)^{1/4} \cdot (m^3 mol^{-1})$		
Component	P_c (kPa)	T_c (K)	ω	(1)	(2)	(3)
CO_2	7382	304.2	0.228	13.64×10^{-6}	16.22×10^{-6}	13.78×10^{-6}
CH_4	4599	190.6	0.012	13.06×10^{-6}	13.02×10^{-6}	12.91×10^{-6}
H_2O	22055	647.1	0.345	11.94×10^{-6}	9.07×10^{-6}	9.07×10^{-6}
MEG	7530	645.0	1.137	21.28×10^{-6}	24.90×10^{-6}	26.48×10^{-6}
TEG	3320	769.5	0.759	59.96×10^{-6}	60.46×10^{-6}	62.40×10^{-6}

(1) Hugill and van Welsenes [19], (2) Birdi [21], (3) Qualye [22]

Table 7.7: AAPD Between Experimental and Predicted Interfacial Tensions in the 70 wt% MEG with Water Solutions and the 98 wt% TEG with Water Solutions

System	Hugill and van Welsenes [19]	Birdi [21]	Quayle [22]
MEG	16.1	22.6	10.9
TEG	15.4	15.9	13.3

and density of these multi-component aqueous glycol systems. Good agreement between the experimental data and predicted data from the sPC-SAFTEoS was achieved in medium- to high-temperature range, whereas only fair agreement was observed for results at low temperatures. The results suggest that temperature-dependent binary interaction parameters might improve the phase-behavior predictions of the studied systems. In addition, more experimental data (multi-component and binary systems) of glycol systems in the low-temperature range are needed to develop better phase-behavior models of these systems.

Liquid phase viscosities of the 70 wt% MEG with water and the 98 wt% TEG with water system, respectively, were predicted to within 8.3% and 18.2% with the RKSA-Info model set and to within 10.0 % and 17.2 % with the CPA-Info model set. Again, both model sets predicted this physical property to almost the same accuracy. In each case the results from Multiflash™ 4.0 software were significantly better than those obtained from the Expanded Fluid viscosity model. The results from the Parachor method (based on the Hugill and van Welsenes [19] method) were better than those obtained from the Linear Gradient Theory in both model sets. The IFTs of the 70 wt% MEG with water and 98 wt% TEG with water

Table 7.8: AAPD Between Experimental and Predicted Equilibrium Constants in 70 wt% MEG with Water Solutions

Component	RKS-INFO	CPA-INFO
CO_2	46.0	34.0
H_2S	41.5	42.7
CH_4	196.1	30.1
H_2O	40.8	27.6

Table 7.9: AAPD Between Experimental and Predicted Equilibrium Constants in 98 wt% TEG with Water Solutions

Component	RKS-INFO	CPA-INFO
CO_2	33.4	32.0
H_2S	37.2	37.7
CH_4	8.8	6.3
H_2O	55.0	100.0

system, respectively, were predicted to within 51.6% and 64.5%, respectively, with the RKSA-Info model set and to within 44.1% and 65.7% with the CPA-Info model set.

7.3 Conclusions

Multicomponent VLE, density, viscosity and IFT data for systems of methane and its mixtures with carbon dioxide and hydrogen sulfide in aqueous solutions of MEG and TEG at relevant operating conditions are rare. The sPC-SAFT EoS, Expanded Fluid viscosity model, and the Parachor method have been used successfully to predict the experimental data of these systems to within a reasonable agreement. The success of SAFT theory in association systems has been further strengthened through this comparison but further work with more multi-component and binary data at relevant conditions can improve the performance of this model. This is also true for the viscosity and IFT models used in this work.

The VLE predictions from the sPC-SAFT EoS are consistent with that obtained from Multiflash™4.0, however, the viscosity results from Multiflash are significantly better than those obtained from the Expanded Fluid viscosity model. The IFT results from the Parachor method are also better than those obtained from Multiflash 4.0.

Calculation of properties required in gas processing plant design requires data of multicomponent systems to validate models applicable to relevant operating conditions. The data set of Ng, et al. [1], helps to expand this data set, but additional data for binary systems of these components are required to further develop and extend the predictive capabilities of multicomponent VLE, equilibrium phase densities, liquid phase viscosity and interfacial tension modeling techniques.

7.4 Nomenclature

$\tilde{a}^{hc}$ Hard-chain contribution to the Helmholtz energy
$\tilde{a}^{disp}$ Dispersive contribution to the Helmholtz energy
$\tilde{a}^{assoc}$ Association contribution of the Helmholtz energy
A^{res} Residual Helmholtz energy, J
$AAPD$ Average Absolute Percentage Deviation
c_1 Parameter in EF correlation, mPa·s

c_2	Parameter in EF correlation
c_3	Pressure dependency parameter in EF correlation, kPa^{-1}
k_{ij}	Binary interaction parameter
k_B	Boltzmann constant, JK^{-1}
K	Equilibrium Constant
m_i	Segment number
M_L	Molar mass of liquid phase, kg/mole
M_V	Molar mass of vapor phase, kg/mole
n	Empirical exponent in EF correlation
N	Total number of molecules
T	Temperature, K
T_c	Critical temperature, K
p	Parachor exponent
p_i	Parachor of pure component I, (N/m)$^{1/4}$ m^3/mol
P_c	Critical pressure, kPa
x_i	Mole fraction of component i in the liquid phase
y_i	Mole fraction of component i in the vapor phase

Greek letter

β	Parameter between viscosity and fluid expansion in EF correlation
γ	Interfacial tension, N/m
ε_i	Segment energy parameter, J
$\varepsilon^{A_i B_i}$	Associating energy, J
$k^{A_i B_i}$	Effective associating volume
μ	Viscosity, mPa·s
μ_o	Dilute gas viscosity, mPa·s
ρ	Fluid phase density, kg/m^3
ρ_L	Equilibrium liquid phase density, kg/m^3
ρ_V	Equilibrium vapor phase density, kg/m^3
ρ_s	Compressed state density, kg/m^3
ρ_s^o	Compressed state density in vacuum, kg/m^3
σ_i	Segment diameter, Å
ω	Acentric factor

7.5 Acknowledgement

The financial support received from the Gas Processors Association is sincerely appreciated.

References

1. H.-J. Ng, J. Na, Y. Cheng, K.A.G. Schmidt, H. Schroeder, "Vapour-Liquid Equilibrium, Interfacial Tension and Transport Properties of Aqueous Solutions of Ethylene Glycol or Triethylene Glycol in the Presence of Methane, Carbon Dioxide and Hydrogen Sulfide", GPA Research Report 202; Gas Processors Association, Tulsa, OK, USA, September 2009.
2. F.-Y. Jou, R.D. Deshmukh, F.D. Otto, and A.E. Mather, *Fluid Phase Equilibria*, Vol. 36, p. 121, 1987.
3. J. Gross, and G. Sadowski, *Industrial & Engineering Chemistry Research*, Vol. 41, p. 1084, 2002.
4. N. von Solms, M.L. Michelsen, and G.M. Kontogeorgis, *Industrial & Engineering Chemistry Research*, Vol. 42, p. 1098, 2003.
5. Grenner, G.M. Kontogeorgis, N. von Solms, and M.L. Michelsen, *Fluid Phase Equilibria*, Vol. 261, p. 248, 2007.
6. Grenner, J. Schmelzer, N. von Solms, and G.M. Kontogeorgis, *Industrial & Engineering Chemistry Research*, Vol. 45, p. 8170, 2006.
7. R.H. Olds, B.H. Sage, and W.N. Lacey, *Industrial & Engineering Chemistry*, Vol. 34, p. 1223, 1942.
8. Chapoy, A.H. Mohammadi, D. Richon, and B. Tohidi, *Fluid Phase Equilibria*, Vol. 220, p. 113, 2004.
9. F.T. Selleck, L.T. Carmichael, and B.H. Sage, *Industrial & Engineering Chemistry*, Vol. 44, p. 2219, 1952.
10. C.R. Coan, and A.D. King Jr., *Journal of the American Chemical Society*, Vol. 93, p. 1857, 1971.
11. N. Spycher, K. Pruess, and J. Ennis-King, *GeochimicaetCosmochimicaActa*, Vol. 67, p. 3015, 2003.
12. G.K. Folas, O.J. Berg, E. Solbraa, A.O. Fredheim, G.M. Kontogeorgis, M.L. Michelsen, and E.H. Stenby, *Fluid Phase Equilibria*, Vol. 251, p. 52, 2007.
13. L.K. Wang, G.J. Chen, G.H. Han, X.Q. Guo, and T.M. Guo, *Fluid Phase Equilibria*, Vol. 207, p. 143, 2003.
14. F.-Y. Jou, R.D. Deshmukh, F.D. Otto, and A. E. Mather, *Chemical Engineering Communications*, Vol. 87, p. 223, 1990.
15. H. Motahhari, M.A. Satyro, and H.W. Yarranton, *Fluid Phase Equilibria*, Vol. 322, pp. 56-65, 2012.
16. H.W. Yarraton, and M.A. Satryo, *Industrial & Engineering Chemistry Research*, Vol. 48, p. 3640, 2009.
17. H. Motahhari, M.A. Satyro, S.D. Taylor, and H. W. Yarranton, *Energy & Fuels*, Vol. 27, p. 1881, 2013.
18. C.F. Weinaug, and D.L. Katz, *Industrial & Engineering Chemistry*, Vol. 35, p. 239, 1943.
19. J.A. Hugill, and A.J. van Welsenes, *Fluid Phase Equilibria*.Vol. 29, p. 383, 1986.

20. Schlumberger DBR Technology Center, *PVTPro Database CompBANK Version 3.0*, Edmonton, Canada, 2013.
21. K.S. Birdi, *Handbook of Surface & Colloid Chemistry*, Boca Raton, FL, USA, CRC Press, p. 9-10, 1997.
22. O.R. Quayle, *Chemical Reviews*, Vol. 53, p. 439, 1953.
23. D. Broseta, and K. Ragil, SPE Paper 30784, SPE Annual Technical Conference and Exhibition, Dallas, TX, USA, October 22-25, 1995.
24. J.R. Fanchi, *SPE Reservoir Engineering*, Vol. 5(3), p. 433, 1990.
25. F. Gharagheizi, A.A. Mohammadi, and D. Richon, *Chemical Engineering Science*, Vol. 66, p. 2959, 2011.
26. M. Liška, *Chemické Zvesti*, Vol. 16, p. 784, 1962.
27. S.M. Mousavi, G.R. Pazuki, M. Pakizehseresht, A. Dashtizadeh, and M. Pakizehseresht, *Fluid Phase Equilibria*, Vol. 255, pp. 24-30, 2007.
28. D.S. Schechter, and B. Guo, *SPE Reservoir Evaluation and Engineering*, Vol. 1(3), p. 207, 1998.
29. Infochem Computer Services Ltd., *User Guide for Multiflash Models and Physical Properties Version 4.0*, London, UK, 2011.

Appendix 7.A

Table 7.A1: Gas Mixtures Composition

	System ID	Mole %		
		CH_4	CO_2	H_2S
98 wt% TEG	I	100	-	-
	II	80	20	-
	III	80	10	10
	IV	80	-	20
	V	60	40	-
70 wt% MEG	I	100	-	-
	II	80	20	-
	III	80	10	10
	IV	60	40	-

Table 7.A2: AAPD Between the Experimental Data and the Predictions from the sPC-SAFT Model, Expanded Fluid Viscosity Model, Parachor Model for the 98 wt% TEG with Water System

Gas	Temperature (K)	Pressure (KPa)	KCH_4 (%)	KH_2S (%)	KCO_2 (%)	KH_2O (%)	ρ_L (%)	ρ_V (%)	μ_L (%)	γ (%)
I	299.8	6895	25.0	-	-	87.8	0.7	2.7	5.5	2.3
I	299.8	13790	9.5	-	-	90.6	0.3	2.0	3.4	22.9
I	316.5	3447	5.2	-	-	95.1	4.0	3.7	41.4	22.1
I	316.5	6895	0.6	-	-	90.6	0.7	1.0	0.6	49.1
I	316.5	13790	6.7	-	-	91.4	0.7	0.9	14.9	14.1
I	333.2	6895	2.8	-	-	88.2	1.2	3.0	15.1	1.9
I	333.2	13790	2.1	-	-	79.5	1.5	1.4	93.6	5.8
II	299.8	6895	0.7	-	14.9	81.5	0.8	2.8	9.3	4.7
II	299.8	13790	6.7	-	27.5	84.8	0.0	5.4	65.2	18.6
II	316.5	3447	11.1	-	3.6	74.8	0.5	3.5	13.6	7.6
II	316.5	6895	7.5	-	11.2	62.2	1.1	2.4	11.6	5.8
II	316.5	13790	2.1	-	19.3	69.2	1.2	22.5	20.9	2.3
II	333.2	6895	6.4	-	3.2	63.8	1.5	9.5	30.2	18.5
II	333.2	13790	2.9	-	13.0	70.7	0.5	5.3	3.1	
III	316.5	13790	0.0	39.5	6.9	91.3	1.0	5.8	47.9	
IV	316.5	13790	13.4	28.9	-	89.4	0.9	6.9	22.6	14.3
V	316.5	13790	1.2	-	1.5	89.5	0.5	20.4	12.3	10.2

Table 7.A3: AAPD Between the Experimental Data and the Predictions from the sPC-SAFT Model, Expanded Fluid Viscosity Model, Parachor Model for the 70 wt% MEG with Water System

Gas	Temp. (K)	Press. (kPa)	KCH_4 (%)	KH_2S (%)	KCO_2 (%)	KH_2O (%)	ρ_L (%)	ρ_V (%)	μ_L (%)	γ (%)
I	272.0	6895	216.2	-	-	5.6	4.0	0.3	49.8	
I	272.0	13790	113.8	-	-	60.9	4.1	0.7	56.1	
I	283.2	3447	34.2	-	-	36.1	1.7	5.8	117.4	
I	283.2	6895	76.2	-	-	27.0	4.3	0.3	46.6	
I	283.2	13790	41.1	-	-	44.2	3.1	0.5	0.3	
I	299.8	6895	9.8	-	-	56.6	3.3	1.3	0.7	3.6
I	299.8	13790	16.4	-	-	38.6	3.8	1.1	26.5	18.9
II	272.0	6895	9.9	-	35.0	52.0	2.5	3.4	57.0	
II	272.0	13790	20.5	-	29.6	58.5	3.2	4.5	1.9	
II	283.2	3447	3.0	-	39.1	51.3	4.0	2.6	28.5	
II	283.2	6895	1.4	-	37.9	49.8	1.9	4.6	102.9	
II	283.2	13790	7.3	-	31.0	58.3	3.4	12.7	6.3	
II	299.8	6895	0.8	-	37.8	53.5	3.8	2.0	16.2	7.1
II	299.8	13790	2.7	-	30.2	45.7	2.8	4.7	19.1	14.1
III	283.2	13790	26.3	21.2	28.1	80.1	3.3	8.1	1.8	
IV	283.2	13790	29.2	-	34.6	67.9	3.5	12.2	17.7	

8
Enhanced Gas Dehydration using Methanol Injection in an Acid Gas Compression System

M. Rafay Anwar, N. Wayne McKay and Jim R. Maddocks

Gas Liquids Engineering Ltd., Calgary, AB, Canada

Abstract

This chapter examines the effect of methanol injection on integrated gas dehydration using a typical five-stage acid gas compression train. Traditional glycol dehydration presents a number of issues including emissions, continuous maintenance due to leaks, safety, space requirements for installation, and power costs for glycol regeneration. The issue of emissions becomes especially significant when acid gas is being dehydrated. Maintenance issues can be circumvented using a thermodynamic gas dehydration method. In such systems, methanol is often used as a hydrate suppressant. This chapter examines the feasibility of using methanol injection to dehydrate acid gas when used in conjunction with thermodynamic methods to reduce the water content of acid gas. Three inlet compositions are examined to investigate the role played by methanol injection: (a) 100% CO_2, (b) 50% (mole basis) CO_2/50% H_2S, and (c) a typical Enhanced Oil Recovery (EOR) solution gas composition (with CO_2, H_2S, hydrocarbons, and BTEXs present). Methanol behaves differently depending on which thermodynamic method of dehydration is used. A number of cases will be reviewed and examined to illustrate the expected system performance.

8.1 Introduction

Methanol is used extensively to suppress or inhibit hydrate formation in both sweet and sour gas facilities and pipelines. In most pipeline applications, injected volumes are sacrificed and are not recoverable but the use of methanol as a hydrate suppressant within a scheme where it is recovered and reused has also been demonstrated through proven technologies such as IFPexol®[1]. The science behind methanol's hydrate suppression capabilities is well understood and will not be reviewed in detail here.

Apart from the issue of hydrates, water content needs to be carefully controlled in pipelines used for acid gas transport and injection to prevent corrosion. Safety against pipeline corrosion is achieved through gas dehydration. This paper concentrates on the possible advantage methanol could provide when introduced into a system using a thermodynamic method to dehydrate acid gas. The later sections of the paper address the role methanol can play in improving hydrocarbon recovery as part of an Enhanced Oil Recovery (EOR) scheme.

8.2 Methodology

8.2.1 Modeling Software

The major portion of simulation modeling for this study was conducted using VMGSim®, a commercial process simulator. Two of the key parameters being studied were water content in acid gas and prediction of acid gas hydrate temperatures. It was important to model accurately the polar behaviour of methanol. In light of these requirements careful consideration had to be given when choosing the thermodynamic property package to conduct the study. In VMGSim, simulations were modeled using the Advanced Peng-Robinson for Natural Gas (APRNG) Equation of State (EOS). This is a modified version of the basic Peng-Robinson EOS. It is particularly specialized for providing better predictions for solubility of hydrocarbons in water, hydrate temperature predictions and handling of polar components.

Water content in acid gas predictions were corroborated using AQUAlibrium®. This program is widely accepted as providing very accurate predictions for water content in acid gas especially at high pressures. Any deviations between water content as predicted by AQUAlibrium versus VMGSim were accounted for by minor adjustments of the VMGSim model to minimize the discrepancy.

Hydrate Formation Temperatures (HFT) for acid gas were corroborated using ProMax®. As per the recommendation found in the ProMax help file the modified polar version of the Peng-Robinson property package was utilized to provide better results when modeling acid gas with methanol injection. Where a discrepancy existed between ProMax and VMGSim in the predicted HFT, the VMGSim model was adjusted as needed.

8.2.2 Simulation Setup

Standard reference conditions for pressure and temperature were assumed to be 15.6°C (60°F) and 101.3 kPa (14.7 psi). In total three different inlet compositions were studied. The basic simulation setup for each inlet composition included a compression train with water, hydrocarbons or a combination of both being removed via a thermodynamic process implemented at the inlet to the suction scrubber of one of the stages. To allow for ease of comparison across various cases the initial inlet rate of saturated gas to the first stage in each model was specified as 283.2 e^{3}m^3/day (10 MMSCFD). Since this study does not go into the details of equipment selection compressor adiabatic efficiencies for all stages were set at 80% and pressure drops across heat exchangers and air cooled sections were neglected. Air cooler discharge temperature was fixed at 43.0°C (109.4°F) for summer and 25.0°C (77.0°F) for winter conditions. Pipeline ground temperature was assumed to be 4.0°C (39.2°F). The injection pressure at the final stage was fixed at 14,000 kPag (2,030.5 psig)

The thermodynamic method used to provide interstage cooling was based on the DexPro™ [2] process. This process utilizes a partial stream of fluid from the final discharge of a compressor train. This stream is then cooled through a Joule-Thomson (JT) expansion. The cold gas is mixed directly with warmer wet gas from the discharge of the previous stage causing water and hydrocarbons to condense out of the mixed stream. The advantage of using DexPro™ for this application is in the ability of this process to function at low temperatures where other traditional thermodynamic processes such as propane chilling, would run into issues with hydrate formation at various physical locations within the process, particularly the tubes in the chiller. The operating cost of using DexPro™ to provide cooling is the increased power load on the stages around which gas is recycled. However, as the suction temperature to these stages is lowered due to cooling the increase in power consumption is fairly low. Further detail on this process can be found in the paper [3].

In this study the DexPro™ recycle was designed around the final stage of compression for the first two compositions and around the final two stages for the EOR composition. In each case a target water content was determined. DexPro™ was activated to a recycle rate necessary to achieve this target The resulting power output of the final stage was matched by each of the previous stages in a given simulation using a series of process controllers to modify the pressure ratio of each stage. Once the simulation was balanced in this manner, the controllers were deactivated and the pressure ratios were fixed. This provided a base case against which comparisons could be made when the amount of recycle was modified or methanol was introduced into the system. An alternative method of balancing the compression train by matching cylinder discharge temperatures across the train was also considered. Temperature balancing caused the final stage discharge temperature to climb significantly to well above 150°C (302°F) when the DexPro™ recycle was reduced to zero. Hence, balancing the train using power was selected as the preferable method. The methodology for setting up the base simulation was the same for each of the three compositions. This information is summarized in Table 8.1.

8.3 CASE I: 100 % CO_2

8.3.1 How Much to Dehydrate

The inlet stream in the 100% CO_2 case was modeled after a waste CO_2 stream that could potentially be generated by a gas sweetening plant. Inlet temperature and pressure was specified as 41.4 kPag (6.0 psig) and 48.9°C

Table 8.1 Summary of Parameters for Simulation Setup.

Inlet Composition	100% CO_2	50%CO_2, 50%H_2S	EOR Composition
No. of stages in main Compression Train	5	5	3
First Stage Inlet Temperature	48.9°C (120.0°F)	48.9°C (120.0°F)	27.6°C (81.6°F)
First stage Inlet Pressure	41.4kPag (6.0psig)	41.4kPag (6.0psig)	1000kPag (145.0psig)
Reference Temperature and pressure	15.6 °C (60.0°F) and 101.3 kPa (14.7 psia)		
Pipeline Ground Temperature	4.0°C (39.2°F)		
Air Cooler Discharge Temperature	Summer: 43.0 °C (39.2 °F), Winter: 25.0°C (77.0 °F)		
Inlet Gas Rate	283.2 e³m²/day (10MMSCFD)		
Final Injection Pressure	14,000kPag (2,030.5 psig)		

(120.0°F) respectively to match typical conditions off the top of a amine regeneration tower reflux drum. To simplify and maintain clarity on how CO_2 dehydration using a thermodynamic method could be impacted by the introduction of methanol, the composition of the stream was assumed to be purely CO_2 for this case. This stream (indicated by "Dry_Gas" in Figure 8.1) was then saturated with water at compressor suction conditions. The saturated stream combined with the liquid dump from the second stage suction scrubber formed the inlet to the first stage of compression. The liquid dumps from each stage were recycled to the suction of the previous stage.

A dry stream of 100% CO_2 at 14,000 kPag and 4.0°C (39.2°F) was specified to simulate ground conditions for a CO_2 transportation pipeline. Using AQUAlibrium this stream was then saturated to determine the maximum amount of water that the dense phase injection stream could hold at these conditions yielding 2,060 ppm(mol) (97.8 lb/MMSCF). A small decrease in temperature, and a relatively much larger decrease in pressure, would cause water to drop out from this stream. Det Norske Veritas (DNV) is an independent classification society founded in Norway in 1864. In 2010 a report detailing practices for the safe design and operation of CO_2 pipelines was produced by DNV (DNV-RP-J202, April 2010)[1]. This report included a recommendation on the maximum water content of the fluid contained within a pipeline:

> "... ensure that no free water may occur at any location in the pipeline within the operational and potential upset envelopes and modes, unless corrosion damage is avoided through material selection."

For a normal operating pressure and temperature envelope DNV recommends a safety factor of 2. Hence, half the above determined water content of this dense fluid at 14,0000 kPag and 4°C was set as the upper limit for allowable water content of the fluid at injection, thus providing a minimum 2:1 safety margin from water dropout. Using the above recommendation the maximum water content for the 100% CO_2 case was set at 1,030 ppm(mol) (48.9 lb/MMSCF). VMGSim reports slightly higher values for water content in acid gas than does AQUAlibrium. To account for this discrepancy in the VMGSim models, 994 ppm (mol) (47.2 lb/MMSCF) was determined to be the water content that would provide a similar 2:1 protection from water drop out at the assumed pipeline conditions.

---

[1] Det Norske Veritas, "Design and Operation of CO_2 Pipelines, Recommended Practice DNV-RP-J202",Høvik, Norway (2010)

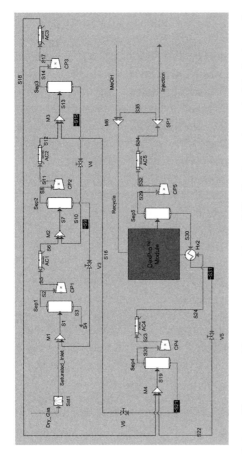

Figure 8.1 VMGSim simulation of five stage acid gas compressor with capability for thermodynamic dehydration and methanol injection at suction to fifth stage.

Note that in the case that such a pipeline is depressurized, reducing the water content to the above values would not prevent water drop out from occurring. However, controlled depressurization could ensure that water formation is transitory and the pipeline is exposed to water for a relatively short period of time. This paper does not address issues of possible corrosion during pipeline depressurization or upset scenarios and limits itself to normal operating scenarios.

Initially a few cases were studied without incremental cooling provided by a powered thermodynamic method to establish baseline conditions as discussed in the below sections.

8.3.2 Dehydration using Air Coolers

The initial compression train simulation was set up with cooling provided only by the air cooled after-coolers at each stage to a setpoint of 43.0°C (109.4°F) to simulate summer conditions. This cooling led to a small amount of water dropping out in the suction scrubber of each stage resulting in a final injection stream with a predicted water content in VMGSim of 2,776 ppm (mol) (131.8 lb/MMSCF). A quick investigation of the equilibrium results of this stream at lower temperatures indicated that at roughly 18.0°C (64.4°F) a separate aqueous phase consisting of approximately 97% water was predicted to start forming. As soon as this aqueous phase forms, the potential for significant corrosion exists. If this stream was to be injected into the pipeline and reservoir without any extra dehydration beyond that provided by air cooling, a corrosion mechanism could be initiated well before the contents of the pipeline reached the expected ground temperature of 4.0°C. Using ProMax the predicted HFT of this stream was determined to be 11.4°C (52.6°F).

The above simulation was modified slightly to set the cooler discharge temperature at 25.0°C (77.0°F) to simulate winter conditions. This resulted in an injection stream with a predicted water content in VMGSim of 1,108 ppm (mol) (52.6 lb/MMSCF) which does not quite meet the 2:1 water content requirement. Using ProMax, the predicted HFT of this stream was determined to be -5.6°(21.9°F) which indicates that using coolers alone in the winter could provide sufficient dehydration to protect against hydrate formation. Depending on gas composition, too much cooling can lead to the possible formation of hydrates in the cooler tubes. As such, using only air cooling for hydrate protection of the injection stream could only be achieved during winter and would be difficult to control.

8.3.3 Methanol injection for hydrate suppression

The summer case was further studied to investigate how much methanol would need to be injected to achieve sufficient hydrate protection at pipeline conditions. Methanol was injected at the suction to the final stage of the simulation at 43.0°C. To provide a safety margin enough methanol was injected to suppress the HFT to or below 0.0°C (32.0°F). To achieve this level of HFT suppression, the VMGSim model required the injection of 0.089 Sm^3/h (0.393 USGPM, 0.214 Sm^3/MMSCF process gas) of methanol. In this case the temperature at which a separate aqueous phase was predicted to form decreased to 10.7°C (51.3°F). While this stream would be protected against hydrate formation, the high water content and high dew point temperature would still lead to the possibility of corrosion well before ground temperature was reached.

8.3.4 Methanol Injection for Achieving 2:1 Water Content

Maintaining the cooler discharge temperature at 43.0°C the rate of methanol injection was increased using a controller until 994 ppm (mol) (47.2 lb/MMSCF) of water in the final injection stream was achieved. This required an injection rate of 0.209Sm^3/h (0.922 USGPM, 0.502 Sm^3/MMSCF process gas) of methanol. An interesting result of injecting methanol in large quantities was found when the temperature of the injection stream was lowered. At approximately -2.2°C (28.0°F) VMGSim predicted the formation of a separate aqueous phase thus potentially allowing for the initiation of a corrosion mechanism. The composition of this initial aqueous stream can be found in Table 8.2.

With such a high amount of methanol being injected to meet the 2:1 water content requirement, hydrate suppression is no longer an issue. However, the high methanol volumes combined with the remaining water

Table 8.2 Composition of initial aqueous phase with high methanol injection (14,000 kPag and –2.2°C).

Component	Mol percent
CARBON DIOXIDE	7.05
WATER	40.89
METHANOL	52.06
TOTAL	100.00

content could form a separate aqueous phase at a relatively higher temperature compared to the case where methanol was used for hydrate suppression alone. Apart from the economic barrier to injecting methanol in such high volumes, this phenomenon would disallow the use of methanol alone as the dehydrating agent for a pure CO_2 stream.

8.3.5 DexPro™ for Achieving 2:1 Water Content

The amount of cooling that can be provided by DexPro™ depends on two main factors. The first is the differential pressure between the discharge and suction of the stages around which gas is being recycled. The second factor is the amount of gas being recycled (stream labeled "Recycle" in Figure 8.1) around this stage. These two factors will determine maximum differential enthalpy achievable with this system. With the coolers held at summer conditions the simulation was modified using a controller to determine the amount of recycle gas needed to sufficiently dehydrate the final injection stream to achieve the required 2:1 water content ratio. The relevant information for the above cases is summarized in Table 8.3.

In the case where methanol was used to suppress the HFT to 0°C, the mass flow of stream S34 entering splitter SP1 was 20,224 kg/h (44,585 lb/h) with a total compressor power consumption of 2,002 kW (2,684 HP). With the DexPro™ recycle activated the mass flow rate of S34 increased to 24,400 kg/h (53,792 lb/h). Of this about 17.5% or 4,257 kg/h (9,385 lb/h) was recycled. The total power consumption increased to 2,015 kW (2,702 HP). Despite the significant increase in mass flow though the final stage of compression, the relative increase in the overall horsepower of the compression train was not as great due to the simultaneous reduction in suction temperature to the fifth stage and the improvement in compression efficiency resulting from the increase in density. A case study was set up to investigate the impact of injecting methanol on the recycle rate. The recycle controller was kept active so at each instance it would seek the same set point of 2:1 water content ratio of 994 ppm (mol) (47.2 lb/MMSCF) in VMGSim. The starting point of this study was with no methanol injection and only DexPro™ being used for dehydration. In the previous case where methanol alone was used to achieve this water content the required injection rate was 0.209Sm^3/h (0.922 USGPM). Hence this volume was used as an indicator of the upper limit of methanol injection. At this point it was expected that the required recycle rate would be close to or at zero. Table 8.4 summarizes the results of this study and uses a cost of US$0.08 per kWh for power and US$ 1.70/US gallon form ethanol to determine the costs of using either method or a combination of the two methods to achieve the desired water content.

Table 8.3 CASE I: Efficacy of dehydration methods for achieving target water content.

Case No.	Case Description	Cooler discharge temperature	Recycle	MeOH injection volume (std.)		Predicted hydrate formation temperature	Predicted aqueous phase formation temperature	Water content
		°C	kg/h	m³/h	m³/MMSCF	°C	°C	ppm(mol)
1	Air cooling: summer	43.0	0	0	0	11.4	18.0	2,776
2	Air cooling: Winter	25.0	0	0	0	-5.6	-11.2	1,108
3	MeOH: hydrate inhibition	43.0	0	0.089	0.214	0.0	10.7	2,168
4	MeOH: 2:1 water content	43.0	0	0.209	0.502	-17.2	-2.2	994
5	DexPro™	43.0	4,257	0	0	-7.9	-14.3	994

Table 8.4 Case I: Imact of methanol injection on DexPro™ recycle rates.

Point No.	MeOH injection Volume (std.)		Recycle	Pwr.	Pwr cost	MeOH cost	Total cost	Water content	Predicted aqueous phase formation temperature
	m³/h	USGPM	kg/h	kW	USD/yr	USD/yr	USD/yr	ppm(mol)	°C
1	0.00	0.00	4256.95	2015.0	$1,412,121	$0	$1,412,121	994	-14.3
2	0.02	0.10	3866.96	2012.9	$1,410,631	$89,354	$1,499,985	994	-13.9
3	0.05	0.20	3403.41	2011.0	$1,409,305	$178,708	$1,588,013	994	-13.3
4	0.07	0.30	2870.87	2009.3	$1,408,133	$268,062	$1,676,195	994	-12.4
5	0.09	0.40	2321.92	2007.7	$1,407,013	$357,417	$1,764,429	994	-11.3
6	0.11	0.50	1798.87	2006.0	$1,405,832	$446,771	$1,852,603	994	-9.9
7	0.14	0.60	1313.65	2004.5	$1,404,770	$536,125	$1,940,895	994	-8.3
8	0.16	0.70	867.66	2003.0	$1,403,675	$625,479	$2,029,154	994	-6.5
9	0.18	0.80	458.08	2001.4	$1,402,598	$714,833	$2,117,431	994	-4.6
10	0.20	0.90	78.87	1999.9	$1,401,554	$804,187	$2,205,741	994	-2.7
11	0.21	0.92	0.02	1999.7	$1,401,390	$822,038	$2,223,428	994	-2.2
12	0.23	1.00	0.00	–	–	–	–	812	-3.8
13	0.25	1.10	0.00	–	–	–	–	580	-4.2
14	1.14	5.00	0.00	–	–	–	–	0	-10.1
15	2.27	10.00	0.00	–	–	–	–	0	-9.8
16	3.41	15.00	0.00	–	–	–	–	0	-9.5
17	4.54	20.00	0.00	–	–	–	–	0	-9.2
18	5.68	25.00	0.00	–	–	–	–	0	-9.0
19	6.81	30.00	0.00	–	–	–	–	0	-8.8

This case study indicated that introducing methanol led to a sharp reduction in the amount of thermodynamic dehydration required to achieve a target water content in the injection stream as can be seen in from the decrease in kg/h of recycle. The presence of methanol in the system reduces the mass of water reaching the stage where thermodynamic dehydration is implemented as well as having an impact on the capacity of the stream to hold water. As a result, the load on the dehydration mechanism is greatly reduced. However, the costs associated with the volume of methanol required do not justify the savings from reduced power consumption or, in the case of glycol dehydration, lowered glycol circulation.

As noted earlier, and as can be seen from point no. 11 in Table 8.3, trading methanol dehydration for thermodynamic dehydration causes the predicted temperature at which an aqueous phase could form to rise. This phase consists of methanol and water. This temperature is as high as -2.2°C at the methanol injection rate where the need to recycle is just eliminated. At this point if the contents of the injection pipeline reach subzero temperature a corrosion mechanism could be started.

A few more points of interest were investigated. By point no. 11 the target water content of 994 ppm (mol) was achieved by methanol injection alone. Injecting methanol beyond point no. 11 led to a drop in water content below the target. With less water present the predicted temperature for the formation of an aqueous phase continued to fall. At point no. 15, the water content of the injection stream was negligible. Injecting further methanol beyond this point resulted in the predicted temperature at which an aqueous phase would form to begin rising again. A possible reason for this could be that the dense phase is completely saturated with methanol and a second aqueous phase consisting of methanol begins to form at higher temperatures as more methanol is injected. This behavior was not further investigated as part of this paper and could form the basis for further study.

8.4 CASE II: 50 Percent CO_2, 50 Percent H_2S

8.4.1- How Much to Dehydrate?

The simulation for this case was set up in a manner very similar to the simulation for the previous composition of pure CO_2. The mixed stream was assumed to be a byproduct of a gas sweetening plant. Again a dry stream with this composition was saturated and used as the inlet to the first stage of compression. Using AQUAlibrium the maximum water content that could be held at pipeline conditions of 14,000 kPag and 4°C yielded 4,654 ppm(mol) (221.0lb/MMSCF). The 2:1 water content ratio for this composition was 2,327 ppm(mol) (110.5 lb/MMSCF). In VMGSim the water content was fixed at 2,268 ppm(mol) (107.7 lb/MMSCF) to align the model reasonably closely with AQUAlibrium.

8.4.2 Dehydration using Air Coolers

In the summer condition of 43°C discharge for air coolers, the final predicted water content for this composition was 3,040 ppm(mol) (144.3 lb/MMSCF). At 5.6°C (42.1°F) the predicted HFT for this stream was found to be significantly lower than that for pure CO_2 which was at 11.4°C (52.6°F). The temperature at which a predicted separate aqueous phase formed was -3.9°C (25°F) which was again lower than for pure CO_2 at 18.0°C (64.4°F). Modifying the simulation for winter conditions by setting cooler discharge to 25.0°C (77.0°F) led to the water content dropping sharply to 139 ppm(mol) (6.61 lb./MMSCF) which is well below the required 2:1 water content target. As discussed in an earlier section, using air cooling to dehydrate would only be feasible in winter conditions and difficult to control. At the pipeline operating conditions the presence of H_2S enhances the ability of the stream to hold water. This provides a substantial benefit to the water handling and hydrate forming conditions within this scenario.

8.4.3 Methanol Injection for Hydrate Suppression

The summer case was modified to inject methanol to reduce the hydrate temperature of the injection stream to at or below 0°C (32.0°F). The temperature at which a separate aqueous phase formed was lowered to -9.3°C (15.3°F). This stream would be well protected against hydrate formation as well as corrosion but still contained 2,572 ppm(mol) (122.1 lb./MMSCF) of water.

8.4.4 Methanol Injection for Achieving 2:1 Water Content

The methanol injection rate was increased to achieve the 2:1 water content target for this stream. This required 0.068 Sm^3/h (0.454 USGPM, 0.247 Sm^3/MMSCF process gas) of methanol. The predicted HFT was -4.0°C (24.9°F) and the predicted temperature for formation of an aqueous phase was lowered to -13.0°C (8.6°F).

8.4.5 DexPro™ for Achieving 2:1 Water Content

DexPro™ was activated with the cooler discharge temperature held at summer conditions and no methanol being injected. The relevant information for the above cases is summarized in Table 8.5.

A case study was setup to investigate the impact of combining methanol injection with the recycle gas. The results are summarized in Table 8.6.

The predicted aqueous phase formation temperature follows a similar pattern as for the previous composition but the temperatures are all significantly lower. This could indicate that a mixture of H_2S and CO_2 in the dense phase may have a greater affinity for methanol than pure CO_2 in a manner similar to the dense phase mixture having a greater affinity for water.

8.5 CASE III: Enhanced Oil Recovery Composition

8.5.1 How Much to Dehydrate?

The composition studied for this case represents a typical stream from a field where CO_2 has been injected to enhance oil recovery (CO_2-EOR). The injected CO_2 accomplishes this through the twofold effect of increasing the pressure of the reservoir and producing larger volumes of oil due to the high miscibility of hydrocarbons in CO_2 at reservoir conditions. At surface temperatures and pressures, the bulk of the heavier hydrocarbons and water drop out from the production stream through a number of stages of separation. The vapour stream exiting the top of these separators is then recompressed for injection but is typically dehydrated to prevent hydrate formation and pipeline corrosion, as discussed in earlier sections. Using a thermodynamic method rather than an absorption method to achieve this dehydration, the liquid streams can be cascaded back to the upstream separation stages with a potential increase in the amount of recovered oil. Since the bulk of this stream consists of CO_2 the results of running this system with air cooling at the summer and winter conditions were not studied since this was covered in Section 8.3. Results from adding methanol were also expected to be similar to those established for Case I. The setup for this case was slightly different than for previous cases and can be seen in Figure 8.2.

In this case the main compression train consisted of three stages but the stream arriving at the suction to this train was increased in pressure by the Vapour Recovery Unit (VRU) and Booster compression. To make this case comparable to the previous two the flow rate of S23, the inlet stream to the first stage of compression, was maintained at 283.2 e^3m^3/day (10

Table 8.5 CASE II: Efficacy of dehydration methods for achieving target water content.

Case No.	Case Description	Cooler discharge temperature	Recycle	MeOH injection volume (std.)		Predicted hydrate formation temperature	Predicted aqueous phase formation temperature	Water content
		°C	kg/h	m³/h	m³/MMSCF	°C	°C	ppm(mol)
1	Air cooling: summer	43.0	0	0	0	5.6	−3.9	3,040
2	Air cooling: Winter	25.0	0	0	0	−52.9	−72.0	1139
3	MeOH: hydrate inhibition	43.0	0	0.068	0.163	−0.1	−9.3	2,572
4	MeOH: 2:1 water content	43.0	0	0.103	0.247	−4.0	−13.0	2,268
5	DexProTM	43.0	789.4	0	0	−1.4	−13.3	2,268

144 Gas Injection for Disposal and Enhanced Recovery

Table 8.6 Case II: Impact of methanol injection on DexPro™ recycle rates.

Point No.	MeOH injection Volume (std.)		Recycle	Pwr.	Pwr cost	MeOH cost	Total cost	Water content	Predicted aqueous phase formation temperature
	m³/h	USGPM	kg/h	kW	USD/yr	USD/yr	USD/yr	ppm(mol)	°C
1	0.00	0.00	789.40	1976.5	$1,385,108	$0	$1,385,108	2,268	−13.3
2	0.02	0.10	639.94	1975.5	$1,384,415	$89,354	$1,473,769	2,268	−13.3
3	0.05	0.20	478.55	1974.7	$1,383,861	$178,708	$1,562,569	2,268	−13.3
4	0.07	0.30	301.96	1974.1	$1,383,422	$268,062	$1,651,484	2,268	−13.2
5	0.09	0.40	109.37	1973.7	$1,383,165	$357,417	$1,740,582	2,268	−13.1
6	0.10	0.45	0.00	1973.6	$1,383,099	$405,207	$1,788,306	2,268	−13.0
7	0.11	0.50	0.00	–	–	–	–	2,165	−14.3
8	0.14	0.60	0.00	–	–	–	–	1,925	−17.5
9	0.16	0.70	0.00	–	–	–	–	1,670	−21.0
10	0.18	0.80	0.00	–	–	–	–	1,407	−24.8
11	0.20	0.90	0.00	–	–	–	–	1,145	−28.5
12	0.23	1.00	0.00	–	–	–	–	884	−32.2
13	1.14	5.00	0.00	–	–	–	–	5	−43.7
14	2.27	10.00	0.00	–	–	–	–	1	−42.1
15	3.41	15.00	0.00	–	–	–	–	0	−41.0
16	4.54	20.00	0.00	–	–	–	–	0	−39.6
19	6.81	30.00	0.00	–	–	–	–	0	−37.8

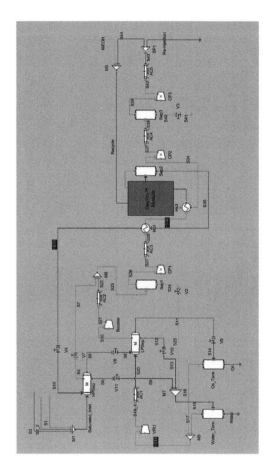

Figure 8.2 VMGSim simulation for compression and re-injection of an enhanced oil recovery well stream.

Table 8.7 Composition of S23.

Component	Mol percent
NITROGEN	0.01
CARBON DIOXIDE	94.21
HYDROGEN SULFIDE	0.04
METHANE	0.16
ETHANE	0.25
PROPANE	1.16
ISOBUTANE	0.44
n-BUTANE	1.51
ISOPENTANE	0.75
n-PENTANE	0.80
n-HEXANE	0.32
WATER	0.35
METHANOL	0.00
n-TRIACONTANE	0.00
TOTAL	100.00

MMSCFD). The composition of this stream can be found in Table 8.7. It should be noted that most EOR schemes have a composition that varies with production time, and miscible flood maturity.

In this case the DexPro™ assembly includes a Gas/Gas heat exchanger and recycles around two stages. The initial recovery of oil from the bottom of the Oil Tank was controlled to 1000STB/day to help illustrate how incremental barrels of oil may be obtained as a result of injecting methanol, adding thermodynamic dehydration or a combination of the two. The 2:1 water content for this case was determined in AQUAlibrium to be 756 ppm(mol) (35.9 lb/MMSCF) and was set to 731 ppm(mol) (34.7 lb/MMSCF)in VMGSim.

8.5.2 Enhanced Oil Recovery using Methanol

This portion of the paper concentrates on the suitability of using methanol to improve oil recovery from a CO_2-EOR stream in the presence of thermodynamic dehydration. The results for the initial scenarios studied are summarized in Table 8.8 with an additional column for barrels of oil recovered in each case.

Table 8.8 CASE III: Efficacy of dehydration methods for achieving target water content.

Case No.	Case Description	Cooler discharge temperature	Recycle	MeOH injection volume (std.)		Predicted hydrate formation temperature	Predicted aqueous phase formation temperature	Water content	Oil Recovery
		°C	kg/h	m³/h	m³/MMSCF	°C	°C	ppm(mol)	bbl/day
3	MeOH: hydrate inhibition	43.0	0	0.083	0.199	0.0	20.2	2,306	1,000
4	MeOH: 2:1 water content	43.0	0	0.539	1.294	-21.6	2.0	731	1,000
5	DexPro™	43.0	1,470	0	0	-11.0	-14.2	731	1,018

The amount of methanol required to suppress the hydrate formation temperature to 0.0°C in this case was comparable to the amount needed in CASE I. However, the amount of methanol needed to dehydrate the gas to 2:1 water content was over 2.5 times the amount needed in CASE I. This could be attributed to the presence of C6+ hydrocarbons in the stream which could reduce the dehydration ability of methanol. The addition of methanol for dehydration did not increase the amount of oil recovered.

The power consumption for this system without any recycle present was 1058.6kW (1419.0 HP). It was noticed that using thermodynamic dehydration to achieve the target water content led to the production of an extra 17.6 STB/day of oil from the oil tank with the system power consumption going up to 1091.1 kW (1463.0 HP). At this point 6.2% or 1,470 kg/h of the flow arriving at SP1 was being recycled. The simulation was modified to allow the recycle rate to increase. One of the consequences of increased recycle is a drop in temperature in the DexPro™ discharge stream feeding Sep2. Subsequently the amount of acid gas that condenses in this vessel increases as does the hydrate formation temperature in the dump line. As the system temperature drops, the hydrate formation temperature needs to be carefully monitored at various locations in the process. At the point where 2900kg/h or about 11.7% of the flow to SP1 was being recycled, the Sep2 liquid dump line was close to or at hydrate formation temperature. The number of incremental barrels of oil increased to 52.7 STB/day. The simulation was modified to add methanol so that hydrate formation in the liquid dump could be avoided and the recycle rate could be increased to33% of the mass flow arriving at SP1.The cost for power and methanol was maintained as for the earlier cases and a value of US$ 80/STB of oil was used in generating the results summarized in Table 8.9.

In applications such as the above where the system is operating very close to or near hydrate formation temperature DexPro™ provides an advantage over traditional shell and tube chillers. Since DexPro™ uses the process gas as its own refrigerant the hot and cold streams can be mixed directly and the issue of an approach temperature is eliminated. In a traditional chiller the fluid near the centre of the tubes is a few degrees hotter than the fluid on the shell side. However, the immediate film on the inside of the tubes is very close to the shell side temperature. When operating near the hydrate formation temperature of the tube side fluid this phenomenon can result in a hydrate layer forming on the inside of the tubes. This layer acts as an insulator and to maintain the tube side fluid temperature the process decreases the shell side temperature thus causing the hydrate layer to increase further in thickness. If sufficient methanol is not being added to combat hydrate formation this process can lead to tubes being blocked.

Table 8.9 Incremental Oil Recovery with increasing Thermodynamic Dehydration and Methanol Injection.

Point No.	Case description	MeOH injection Volume (std.)				SP1 Recycle %	Recycle	System hydrate approach temperature	Incremental oil	Incremental power	Incremental power cost	MeOH cost	Incremental oil revenue	Net revenue
		m³/h	m²/MMSCF (PROCESS GAS)	USGPM	USG/MMSCF (process gas)	%	kg/h	°C	bbl/day	kW	USD/yr	USD/yr	USD/yr	USD/yr
1	2:1 water content. No methanol added	0.00	0.00	0.00	0.00	6.2	1,470	6.5	17.6	32.5	$22,776	$0	$513,920	$491,144
2	Maximum recycle without hydrate formation. No Methanol added	0.00	0.00	0.00	0.00	11.7	2,900	0.4	52.7	73.6	$51,579	$0	$1,538,840	$1,487,261
3	33% recycle with Methanol for hydrate suppression	0.05	0.10	0.22	25.91	33.0	10,650	2.5	134.7	344.5	$241,405	$200,148	$3,3933,240	$3,491,687

8.6 Conclusion

In all three cases studied, using methanol alone to reduce water content to provide a margin of safety from corrosion is expected to be prohibitively expensive. In the absence of a suitable thermodynamic dehydration mechanism (e.g.: unscheduled maintenance) a gas plant producing an acid gas stream that cannot be flared without serious environmental impact or monetary penalty could conceivably use methanol to achieve the required target water content for injection. The observed increase in temperature at which a predicted aqueous phase could form when large volumes of methanol are injected, needs to be kept in mind. Even when using methanol for dehydration in an emergency/temporary situation, plant operators should be aware of the possibility of corrosion due to a mixture of water and methanol forming a separate aqueous phase.

Methanol can assist with dehydration when introduced into a compression system where temperatures are reduced using a thermodynamic method. The presence of methanol causes water to be removed from the system via inlet separator liquid dumps at earlier stages in the compression train thereby reducing the load on the thermodynamic dehydration system. The effect of injecting methanol in a compression train with cascaded liquid dumps is compounded since methanol from high pressure stages is recycled to the inlet side of previous stages instead of being removed to a water storage tank. Using methanol in conjunction with a thermodynamic dehydration method such as DexPro™ or a traditional refrigeration unit could provide an advantage in the scenario where there is a limit on available cooling. The possible savings would vary depending on the specifics of each case including, but not limited to the capital cost of replacing or improving existing equipment, and the cost of methanol and power. It should be noted that extremely rich gas streams, when combined with a cascaded dump system, may experience recycle trapping. This need to "break" the trap may also require "breaking" the methanol loop thus negating some of the advantages of a methanol based system.

Using methanol alone to dehydrate a hydrocarbon rich acid gas stream does not appear to increase the volume of oil recovered. Using a thermodynamic dehydration method such as DexPro™ could increase the oil recovery from such a stream in a significant way. In such an environment, the problem of hydrates and water freezing could be addressed with methanol injection with the increase in oil recovery justifying many times over, the increased cost of methanol and the extra power required.

For consistency throughout this study, predicted hydrate formation temperatures were determined using ProMax with the PR-Polar equation of state. Further work is warranted to increase confidence in predicted hydrate temperatures in methanol rich acid gas mixtures and under-saturated acid gas streams. A more thorough investigation of the predicted behavior of methanol in acid gas at high pressures and low temperatures could form the basis of a future study. Users of this work are encouraged to be cautious in interpretation of process simulation results as they can vary considerably.

8.7 Additional Notes

The following software packages were used in this study: Virtual Materials Group, Inc. VMGSim version 7.0.67; Calgary, Alberta, Canada (2013), FlowPhase , Inc. AQUAlibrium version 3.1, 2006, and Bryan Research & Engineering, Inc. ProMax version 3.2; Bryan, Texas, USA (2012).

References

1. Esteban, A., Hernandez V., Lunsford K., "Exploit the Benefits of Methanol"; 79th GPA Annual Convention, Atlanta, Georgia, USA (2000)
2. Maddocks, J., McKay, W., Hansen, V., "Acid Gas Dehydration - A DexPro™ Technology Update"; AGIS III, Banff, Alberta, Canada (2012)
3. McKay, W., Maddocks, J., "CO2 Dehydration: Why? How Much? How"; NACE Corrosion 2012, Salt Lake City, Utah, USA (2012)

9

Comparison of the Design of CO_2-capture Processes using Equilibrium and Rate Based Models

A.R.J. Arendsen[1], G.F. Versteeg[1], J. van der Lee[2], R. Cota[2] and M.A. Satyro[2]

[1]*Procede Gas Treating BV, Enschede, Netherlands*
[2]*Virtual Materials Group, Inc., Calgary AB Canada*

Abstract

The design of absorption processes with complex aqueous chemical reactions such as CO_2-capture, selective H_2S-removal as well as rate-limited physical separations like LNG pre-treatment is not simple or straightforward. Reaction kinetics, mass transfer, and thermodynamic-driven processes are coupled and must be taken into account simultaneously.

A new simulation tool developed by the Virtual Materials Group and Procede Process Simulations, VMG RateBase, is introduced. It is designed to provide consistent and accurate simulations for mass transfer limited, chemically reactive systems. The absorption of CO_2 from flue gas produced by a coal-fired power plant into an aqueous MEA solution is used as an example to show how a more rigorous model affects the process design and simulation.

VMG RateBase has a carefully developed thermodynamic package to support calculations for acid gas absorption together with databases for chemical kinetics, tray parameters, and random as well as structured packing. Several models for hydrodynamics and mass transfer, for example, Higbie penetration model, are available.

The case studies shown in this work demonstrate that the use of equilibrium-based models can cause significant undersizing of packing heights for carbon dioxide removal. Important mass transfer limitations can be observed only through the use of rate-based models, such as carbon dioxide removal limiting efficiencies as a function of solvent rate and effective mass transfer profiles as a function of packing height.

Other important phenomena for efficient design of separation systems, such as optimal solvent rates, to minimize reboiler duties and differences in vapor and liquid temperature profiles are only captured by the rate-based model.

The rate-based model also provides rich information related to the physics involved in the material and energy balances as expressed by the tower temperature profile, and this information will provide valuable information for process and controls engineers when developing new and more efficient processes or when doing intelligent retrofitting of existing facilities.

9.1 Introduction

The design of absorption processes with complex aqueous chemical reactions such as CO_2-capture, selective H_2S-removal as well as rate limited physical separations like LNG pre-treatment are not simple or straightforward. Reaction kinetics, mass transfer and thermodynamic driven processes are coupled and must be taken into account simultaneously. Traditionally these processes have been designed similarly to distillation processes where equilibrium models are constructed, solved and then corrected based on empirically determined tray efficiency. For non-reactive distillation processes this approach usually lead to satisfactory results for tray based contactors and to a certain extent to packed contactors discretized using the Height Equivalent to a Theoretical Plate (HETP) concept.

However, in the case of reactive absorption the concept of equilibrium based modeling cannot provide consistently reliable results. Due to the presence of simultaneous chemical reactions happening at different rates coupled with different mass transfer rates for key reactants and products to and from the gas and liquid phases, it is not possible to use the simple concept of overall tray efficiencies as commonly used for distillation. Individual key component efficiencies are commonly used for amine plant simulation and reliable results can be obtained for tray towers. Unfortunately the efficiency concept glosses over important phenomena such as different vapor and liquid temperatures of streams leaving a tray due to lack of thermodynamic equilibrium and make the handling of multicomponent systems always somewhat subjective.

The dissolution of gases in solvent solutions and their subsequent conversion into soluble salts is highly dependent on process conditions, temperatures, pressures and type of equipment. For example, observing the absorption of carbon dioxide in an aqueous solvent we have to contend with the absorption rate of carbon dioxide into the liquid governed by the activity gradient between the gas and liquid phases (thermodynamics), the speed at which carbon dioxide (and other species) diffuse to and from the vapor and liquid phases and the available area for mass transfer (mass transfer) and the speed at which carbon dioxide dissociates into ions in the liquid phase (chemical kinetics).

Therefore in reactive mass transfer limited separation processes the use of rate-base process models is preferred and leads to reliable process simulations and designs since it provides fundamental models for all the key simultaneous processes happening while the separation equipment operates

In this paper a thorough and systematic comparison will be presented between the equilibrium based and rate based modeling approaches using the absorption of CO_2 from flue gas produced by a coal-fired power plant into an aqueous MEA solution as a benchmark.

9.2 VMG Rate Base

The Virtual Materials Group developed in partnership with Procede Process Simulations a new flowsheeting tool, VMG RateBase, specifically designed for steady-state simulations of acid gas treating processes [1]. The process models include all features relevant for the design, optimization and analysis of acid gas treating processes, including post-combustion and pre-combustion carbon dioxide capture. The simulator consists of a user-friendly graphical user interface and a powerful numerical solver that handles the rigorous simultaneous modeling of thermodynamics, kinetics and mass transfer (the combination usually called a "rate-based" model). VMG RateBase also supports the main unit operations relevant for gas treating plants such as absorbers, strippers, flash drums, heaters, pumps, compressors, mixers and splitters as well as novel unit operations designed to make the process engineer's work more productive such as automatic ways to calculate water and solvent makeups. VMG RateBase was already validated and used for several carbon capture projects [2,3,4].

The program includes an extensive, carefully evaluated database of thermodynamic model parameters, binary interaction parameters, kinetics constants, chemical equilibrium constants, diffusivities. The models were optimized to accurately predict the vapor-liquid equilibria (VLE), thermodynamic and physical properties and the kinetically enhanced mass transfer behavior of acid gases in amine based capturing processes. Several models for hydrodynamics and mass transfer, e.g. Higbie penetration model [5], are available.

The thermodynamic model combines a Gibbs excess used to model the liquid phase with a cubic equation of state to model the gas phase. For the convenient prediction of column performance, the program includes an extensive database of various tray types, as well as a large collection of both random and structured packing data. Several mass transfer and hydrodynamic models were implemented that benefit from accurate physical property models for density, viscosity, surface tension, diffusivity and thermal conductivity specifically selected and validated for acid gas treating applications.

This attention to detail allowed for the construction of a simulator able to describe complete acid gas treating processes, including complex processes with multiple (mixed or hybrid) solvent loops. This provides significant understanding of the performance of potential new solvents current operations.

Figures 9.1 and 9.2 show the capability of VMG RateBase to model the CO_2 partialpressure as a function of the CO_2 liquid loading at fixed temperature and fresh solvent composition for MDEA and MEA aqueous solutions. Similar quality is observed at different solvent compositions and temperatures required to cover the industrial operating ranges of interest, with loadings between 0 and 1.5, temperatures between 0 and 150°C and partial pressures between 10 Pa to 100 bar. The differences between the experimental and calculated results are mainly caused by inconsistencies between the different data sources and experimental inaccuracies.

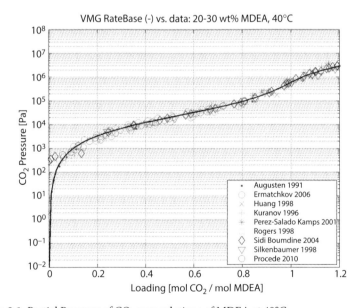

Figure 9.1 Partial Pressure of CO_2 over solutions of MDEA at 40°C.

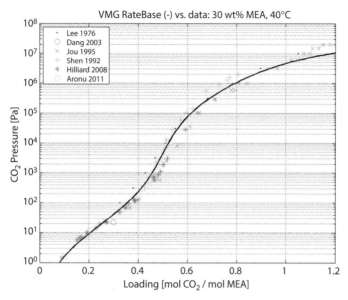

Figure 9.2 Partial pressure of CO_2 over solutions of MEA at 40°C.

9.3 Rate Based Versus Equilibrium Based Models

For rate based and equilibrium based modeling, of an absorber or regenerator, the contactor is discretized into different mass transfer units as shown in Figure 9.3. In counter-current operation the input of each transfer unit is the liquid from above and the vapor from below the unit. The output is the liquid to the unit below and the vapor to the unit above. The resulting number of transfer units (NTU) and the origin of these units are completely different depending on the way the model is constructed.

In equilibrium based modeling the liquid and gas phase are completely mixed and thermodynamic equilibrium is assumed through the use of a flash calculation. Therefore the temperature and pressure of the liquid and gas phase must be equal and the calculated vapor phase and liquid phase component mole flows are then sent to the next mass transfer unit. The NTU is then determined by the desired quality of the separation. For a binary separation this is vividly illustrated through the use of a McCabe-Thiele diagram where the number of ideal trays in the diagram corresponds to the number of transfer units. Similar calculations can be done for distributed (packed) systems through the calculation of differential mass balances and integration over a mass transfer height.

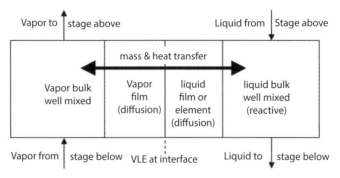

Figure 9.3 General mass transfer model for vapors and liquids.

In rate based modeling the gas and liquid phases are separated by an interface, the gas and liquid phases have different temperatures and the mass and heat transfer rates between the two phases is determined by the driving force between the two phases, the contact area, and the mass and heat transfer coefficients. The key difference between the models is that thermodynamic equilibrium is assumed to exist *only* at the interface. The bulk gas and liquid phases are not at equilibrium and therefore the heat transfer and mass transfer equations have also to be solved. The amount of mass transfer area is determined by the desired quality of the separation.

We use the mass transfer unit terminology for generality. A unit can be thought as a height of packing, a sieve tray, a valve tray or any other type of device where mass transfer between a gas and liquid phase is attempted.

9.3.1 Physical Absorption

In equilibrium based modeling the liquid mole fraction for a certain component (x) of the next unit is calculated based on the gas mole fraction (y) from the actual mass transfer unit using (9.1):

$$x_{n+1} = \frac{y_n}{K} \qquad (9.1)$$

Note that for simplicity the component indexes are omitted from the equations. In this discussion read the equations as applied to *one specific component* in the multicomponent mixture.

The constant-factor is determined from thermodynamics. K is only constant if the gas and liquid phase are ideal mixtures, the process is operated at constant temperature and pressure and if no chemical reactions occur.

The gas mole fraction from the next mass transfer unit is calculated based through a simple material balance:

$$y_{n+1} = y_n + \frac{L}{G} \cdot (x_{n+1} - x_n) \tag{9.2}$$

Where L is the total mole flow of the liquid phase and G is the total mole flow of the gas phase. These two equations suffice to calculate the change in mole fractions as a function of the transfer units and are shown in Figure 9.4. The equilibrium line represents (9.1) and the operating line is determined by (9.2). The dotted line represents the steps between the transfer units. This dotted line can actually be calculated from top down or bottom up. The figures show that the NTU is a function of the factor K*G/L and the desired gas mole fraction in the top of the absorber. Note that for a fixed amount of gas being treated (G) a decrease in solvent flow rate (L) corresponds to an increase in number of transfer units (or stages) to accomplish the same separation.

Subsequently, the results of these calculations are used to calculate the necessary interface area per transfer unit (A) shown in (9.3).

$$A = \frac{L \cdot (x_{n+1} - x_n)}{k_{ov} \cdot LMDF} \tag{9.3}$$

Where k_{ov} is the overall mass transfer coefficient and LMDF is the logarithmic mean driving force, analogous to the LMTD driving force used to design heat exchangers.

In rate base modeling the mole fractions of the gas and liquid phase are calculated by integration of the differential mass balance (9.4) and (9.5) across the height of the column (h).

$$L \frac{dx}{dh} = J \cdot A \tag{9.4}$$

$$G \frac{dy}{dh} = J \cdot A \tag{9.5}$$

In this approach the interface area is not calculated, as in equilibrium based modeling, but is set in advance. This area depends on the packing type or other mass transfer area present in the contactor such as the specific area for mass transfer used to model tray columns or bubble interfacial area present in a bubble tower as used in actual practice. The mass flux (J) in moles / (area * time) is calculated based on the driving force. If the driving force is defined as the concentration difference between the gas and liquid phase the flux is expressed in (9.6).

$$J = k_{ov} \left(\rho_G \cdot y - \rho_L \cdot x \right) \tag{9.6}$$

If the integration of this set of equations is done numerically the height of one transfer unit depends on the numerical discretization used for integration. In the case of a packed column, with negligible axial dispersion, the NTU is set at a value that results in plug flow. In case of trays, with the assumption that at each tray the liquid and gas phase are ideally mixed (of course not at thermodynamic equilibrium), the NTU can be set equal to the number of trays. This results in less plug flow due to axial dispersion. It should be noticed that in this way the axial dispersion is described by ideally mixed contactors in series.

In case of constant K values, Figure 9.4 show that equilibrium based modeling and rate based modeling almost result in the same mole fraction profile as a function of the absorber height. These figures also show that the rate based results only follow the operating line, as experienced by the actual operating equipment. The equilibrium line is never reached as in the equilibrium stage calculation.

9.3.2 Isothermal Absorption with Chemical Reactions

In the case of chemical absorption K is not constant as a function of x even in the case of an ideal solution and therefore the equilibrium line is curved. The operating line does not change since it represents the desired mass balance for the equipment. The mass transfer coefficient for the component of interest must be calculated taking into account the gas and liquid resistances (k_g and k_l) as well as an enhancement factor that results from the component of interest reacting and being "detained" in the liquid phase as a non-volatile compound shown in (9.7). For non-reactive systems the enhancement factor is one and only mass transfer enters in the calculations.

$$\frac{1}{k_{ov}} = \frac{1}{k_G} + \frac{1}{m \cdot k_L \cdot E_a} \tag{9.7}$$

In case of chemical absorption and the driving force is concentration based, k_{ov} is a function of the mass transfer coefficient of the gas phase (k_G) and liquid phase (k_L), the distribution coefficient based on concentrations

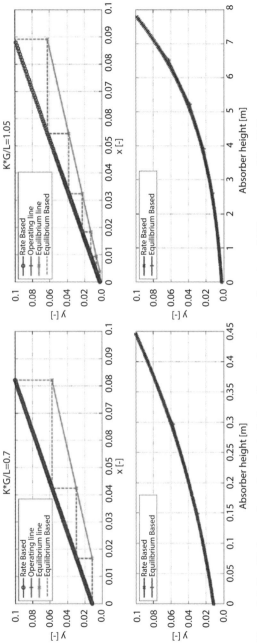

Figure 9.4 Equilibrium and rate based calculations for an absorber as a function of different gas and liquid flow rates.

(m) and the chemical enhancement. Subsequently, (9.3) can be used to calculate the necessary interface area per transfer unit.

In Figure 9.5 this approach is used using a 30 wt% MEA solution at 50°C. This figure shows that the packing depth, directly related to the interfacial area, per transfer unit is not constant, due to changes in the driving force. This figure also shows the rate based results.

Firstly, note that the equilibrium based CO_2 partial pressure is only calculated at 3 different depths, while the rate base profile is known at every integration length used for the solution of the differential material balances. It is pertinent to note that the conditions were set in such a way that the rate base CO_2 concentration is equal to the CO_2 concentration calculated with exactly 3 transfer units. The equilibrium based model is not able to calculate situations that present a fractional number of mass transfer units.

In general the energy balance must be taken into account when computing the material and energy balances along the tower. This will cause the temperatures of the gas and liquid phases to change and the operating line will not be straight but rather a curve (for the binary case) or a surface in general.

9.4 Process Simulations

A thorough and systematic comparison between the equilibrium and rate based approaches will now be presented using the absorption of CO_2 from flue gas produced by a coal-fired power plant into an aqueous MEA solution.

9.4.1 Configuration

The configuration as modeled in VMG RateBase is given in Figure 9.6. This is a standard CO_2 capture configuration.

The flue gas composition and flow is given in Table 9.1.

9.4.2 Absorber

The first simulations are performed with an absorber only. The lean loading of the solvent is set to be constant 30 wt% and the operation is assumed isothermal. Figure 9.7 shows the CO_2 capture using equilibrium and rate based models as a function of the solvent flow and NTU or height. Both figures show that the capture increases with the solvent flow and the NTU or height.

The results show that the equilibrium model capture is always greater than the rate based model capture. Even with two NTU (or equilibrium

COMPARISON OF THE DESIGN OF CO_2-CAPTURE PROCESSES 163

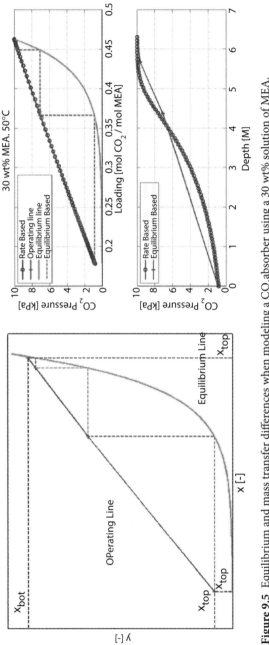

Figure 9.5 Equilibrium and mass transfer differences when modeling a CO_2 absorber using a 30 wt% solution of MEA.

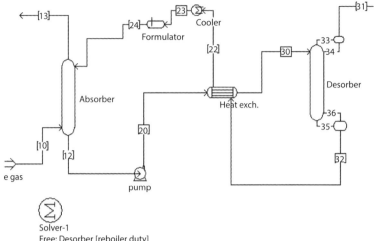

Figure 9.6 MEA plant used in this study. The flue gas contains 624 tons/h of CO_2 [stream 1] and it is desired to remove 90% of the carbon dioxide from the treated flue gas [stream 13].

Table 9.1 Main process variables for MEA CO_2 removal plant shown in Figure 9.6.

ITEM	Quantity	Units
Gross Output	827	MW
CO_2 Production	624	Tonnes/hr
CO_2 Recovered	90	%
Flue gas flow rate	855.2	m³/s
Temperature	40	°C
Pressure	1	bar
CO_2	12	mol%
H_2O	7	mol%
O_2	4	mol%
N_2	77	mol%
MEA percentage	30	wt%
Specific Surface Area	250	m²/m³

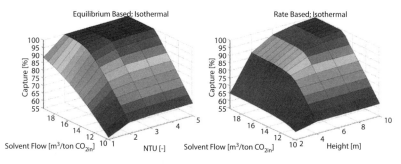

Figure 9.7 CO_2 capture as a function of equipment size (number of equilibrium stages or packing height) and MEA solution flow rate.

stages) it is possible to almost get 100% capture at high solvent flows. The results at 5 NTU or 10 m are comparable due to the fact that the length of the absorber is long enough to reach equilibrium conditions and at low solvent flows the amount of solvent is limiting the CO_2 capture.

Note that in this particular case if we were to design an absorption tower using the common concept of HETP the height of the tower could be grossly underestimated. For example if 5 equilibrium stages were selected and a common HETP height of about 0.6 m [6] a tower with a height of 3 m would be deemed to be capable of scrubbing essentially all CO_2 while in actually a much higher tower would be required.

In case of the equilibrium based calculations it is possible to calculate the height of the absorber with (9.3). Figure 9.8 shows that this height is a function of NTU, as expected. This figure shows also that the height is a function of the solvent flow. Especially at NTU above 3 there is a strong increase in height around 16 m³ / ton CO_2.

This explains the higher capture percentages shown in Figure 9.8. The main reason for this behavior is the fact that at high capture percentages the mean driving force for mass transfer decreases at the top of the absorber, resulting in higher required interface area for an equivalent CO_2 absorption. This effect is absent from the equilibrium based model and since the height of packing is fixed when analyzing actual plant performance the usefulness of an equilibrium model is limited when performing plant optimization studies.

Figure 9.9 shows the sensitivity of CO_2 percentage out or CO_2 captured as a function of solvent flow. According to the equilibrium base model the capture can increase from 90% to 99% with a solvent flow increase of only 10% for a number of equilibrium stages equal to 3. In case of the rate based

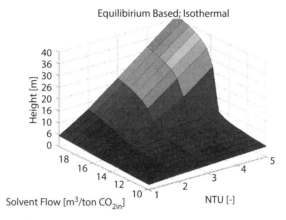

Figure 9.8 packing height versus NTU for various solvent rates.

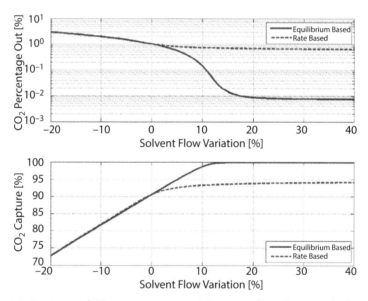

Figure 9.9 Sensitivity of CO_2 percentage out or CO_2 captured as a function of solvent flow.

model the capture increases only to 93% due to the limitation of the height assumed to be 10 m.

In practice, due to the heat of absorption and reaction of CO_2 in MEA the temperature in the absorber will increase. This effect is easily simulated by setting the tower operation to be adiabatic. The simulations show that the differences between equilibrium based and rate based modeling are even bigger. The results are shown in Figure 9.10.

Comparison of the design of CO_2-capture processes

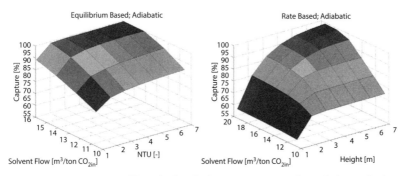

Figure 9.10 CO_2 capture profiles calculated taking into account the enthalpies of solution and reaction. Tower is assumed to be adiabatic.

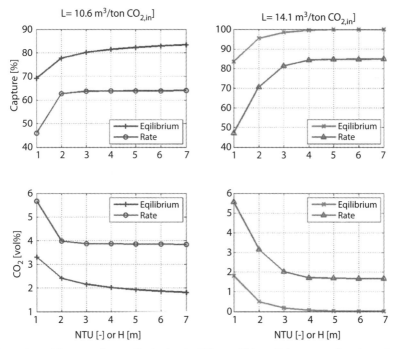

Figure 9.11 Mass transfer rate limitation in CO_2 scrubbing efficiency as a function of solvent flow.

Figure 9.11 shows the CO_2 capture out and CO_2 percentage removal as a function of the NTU or height at two different solvent flows. This figure shows that even at higher solvent flows and higher column lengths the CO_2 capture is not increasing. There is a mass transfer limitation happening that is not described by the equilibrium based model. Note the optimistic values obtained from equilibrium based model.

This limitation is show in Figure 9.12. The equilibrium lines are based on the actual temperature profiles in the absorber. For the equilibrium based approach three transfer units suffice to get high capture percentages. In the rate based case the decrease in CO_2 percentage is much less. Note that almost no mass transfer takes place at 40% of the absorber height.

Figure 9.13 provides a detailed view to what happens at the bottoms of the absorber and shows that the operating line and the equilibrium line are very close. This limits the mass transfer rate and is caused by the increase of the temperature in the absorber.

This phenomenon is also illustrated in Figure 9.14. The equilibrium based and rate based temperature profiles are quite different along the tower. They both increase as a function of the mass transfer units of depth, but the temperature in the equilibrium based case reaches a maximum and then drops as a function of the tower height thus making the separation easier. In the rate based case a temperature plateau is reached, which causes a temperature pinch.

9.4.3 Absorber and Regenerator

The next simulations were performed with the complete plant, now including a MEA regeneration tower. To demonstrate the difference between equilibrium based and rate based simulations the solvent flow rate is varied by a factor 6. The CO_2 capture is set to 90% by changing the reboiler duty. The rate based simulations are performed with an absorber and regenerator with a height of packing equal to 10.

The equilibrium based simulations were performed with 3 and 12 equilibrium stages. 3 is the minimum amount to reach 90% capture and 12 shows the effect of adding more stages. Figure 9.15 shows the results for different parameters of the process. First, the reboiler duty is shown as a function of the solvent flow. All simulations show a minimum value in the reboiler duty at lower solvent flows, but after that a steep increase of the duty for low solvent flow rates.

The fact that the duty is high at low solvents flow has to do with the fact that the number of transfer units or the height of the column becomes a limiting factor. To keep the capture at 90% the loading of the lean solvent has to be decreased in order to increase the driving force for the separation. At a certain flow this is not possible anymore and the reboiler duty shows an exponential increase that of course is limited in actual operations by the physical size of the reboiler and steam supply.

Increasing the solvent flow, increases the loss of energy in the lean/rich heat exchanger, therefore the reboiler duty slowly increases although the

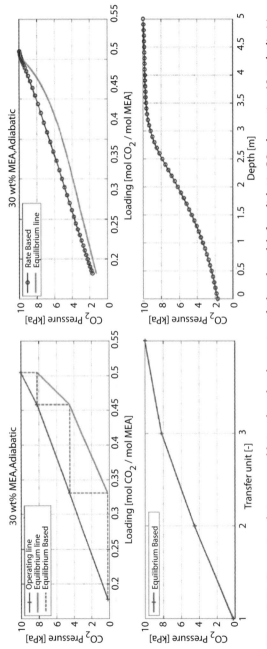

Figure 9.12 Comparison between equilibrium based and mass transfer based models for adiabatic CO_2 absorption. Note the limiting mass transfer that takes place at 40% of the absorber height.

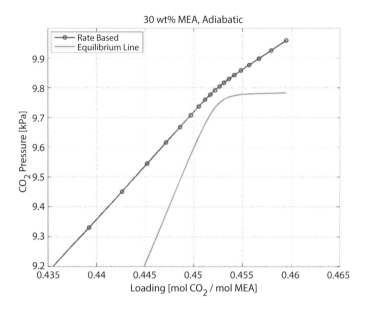

Figure 9.13 Mass transfer pinch point at bottoms of absorber. The closeness of operating line to equilibrium line limits the mass transfer rate.

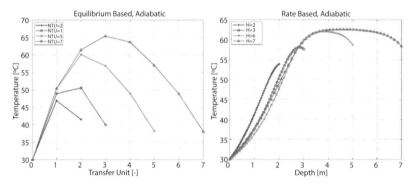

Figure 9.14 Different temperature profiles for absorption towers using equilibrium or mass transfer models as a function of equilibrium stages or packing height.

separation is easier. This is also limited in the actual plant due to the diameter of the tower and approach to flooding. The reboiler duty of the rate based simulations show a local maximum around 33 m^3/ ton CO_2. This phenomenon is not captured by the equilibrium based simulations and has to do with the temperature profiles in the absorber.

The main reason comes from the fact that the rich solvent temperature shows a maximum value at the same solvent flow. This higher temperature

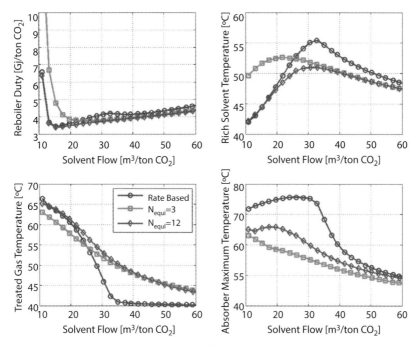

Figure 9.15 Simulations for complete MEA plant for a fixed CO_2 capture equal to 90% and MEA solvent flow rate changing from 10 to 60 m³/ton CO_2. Reboiler duty is allowed to float to meet CO_2 scrubbing specification.

increases the temperature of the solvent into the regenerator and causes an increase in the stripping ratio. This additional heat of vaporization must be compensated by an increase in the reboiler duty. The trends of the maximum temperature reached in the absorber for the three simulations are very different. The rate based simulations show a much higher temperature and this maximum temperature drops fast if the solvent flow is increased above 33 m³ / ton CO_2.

The treated gas temperature of the rate based simulations drop to the inlet temperature of the lean solvent until 33 m³ / ton CO_2. The drop of this temperature for the equilibrium cases is much slower. This is caused by the fact that the rate based model calculates the gas and liquid phase temperature separately.

The lean loading as a function of the solvent flow is increasing due to the fact that the amount of CO_2 captured is constant. Therefore the difference between the rich and the lean loading can be decreased at higher solvent flows. The trends between the difference simulations are comparable. This is not the case for the rich loading. The increase in rich solvent temperature at the bottom of the absorber cause the rich loading to drop in case of the rate based simulations.

Other important parameters are the MEA and water make-up needed to compensate for the losses due to evaporation in the absorber and regenerator. The evaporation of MEA from the absorber at low solvent flows is higher in case of the rate based simulations due to the higher temperatures in the absorber. For the evaporation of water the rate base simulations show a faster drop in the make-up flow as a function of solvent flow compared to the equilibrium simulations. This can be explained by the fact that water quickly condenses on the cooled lean solvent added in the top of the absorber.

For the equilibrium based simulations this effect is less due to the direct mixing of the gas and liquid phase at one temperature. These effects are shown in Figure 9.16.

Figure 9.17 shows different temperature and CO_2 profiles for the different simulations at different solvent flows. All figures show that only at very high solvent flows there is some overlap in both profiles for the rate based

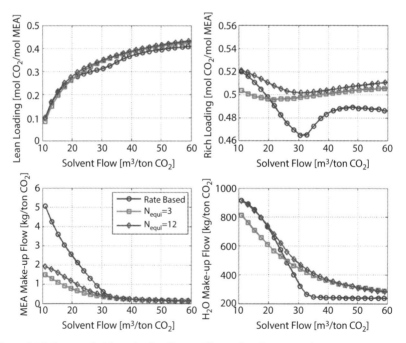

Figure 9.16 Lean and rich amine loading profiles and make up requirements as a function of solvent flow rates.

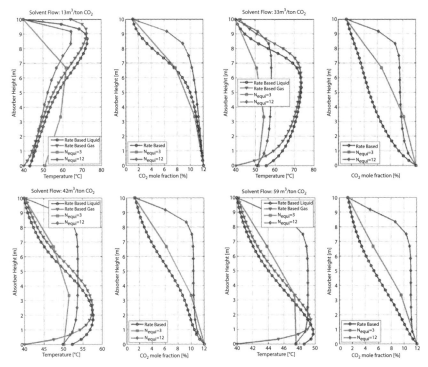

Figure 9.17 Different profiles for the simulations. Note the significantly different temperature profiles computed assuming equilibrium and rate based mass transfer.

and $N_{equi}=3$ simulations. The differences between the rate based liquid and gas temperature profiles is clearly visible.

The maximum temperature in the profiles for rate based is always higher than for the equilibrium simulations. The increase of the NTU from 3 to 12 mainly results in a plateau of constant temperature and CO_2 concentration in the middle of the absorber. This will not affect the overall material balance for the absorber but can play an important role on the development t of effective control algorithms for these systems.

9.4.4 Temperature Profile

Figure 9.18 shows the mass transfer flow rate from the gas phase to the liquid phase for CO_2 and H_2O for the rate based case corresponding to the maximum temperature profile at 33 m³ / ton CO_2. The CO_2 flux is almost constant along the height. The H_2O flux is at the maximum value, around ten times higher than the CO_2 flux. This water flux is the main cause for

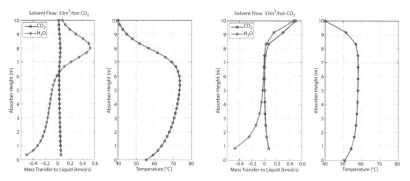

Figure 9.18 Mass transfer rates and temperature profiles for the CO_2 absorption tower showing the most extreme temperature profile.

the big temperature bulge observed in the tower. In the low part of the absorber a large amount of water is vaporized and condensed back to the top of the absorber. At around 6 meters the water flux is almost zero and the temperature is at its maximum. The constant amount of heat released by the absorption of CO_2 works like a heat pump that transfers heat through water vapourization and condensation to the top of the absorber.

The effect is much less in the equilibrium based simulations. Most transfer takes place in the top of the column. In the middle of the absorber not much happens, the amount of water that is transferred to the top is much less than in the rate based simulations. This is mainly due to the fact that the gas and liquid phase are supposed to coexist at the equilibrium temperature.

Figure 9.19 shows that the temperature profile is mainly caused by the heat of absorption of CO_2 and the heat of evaporation/condensation of water. Based on the fluxes and the heat of absorption and evaporation the amount of heat released in the liquid phase can be calculated. With the known liquid flow and heat capacity it is possible to calculate the temperature change of the liquid phase separately due to CO_2 or water. The sum of these two effects and the actual temperatures are also given. These two effects almost add up to the actual temperature profile.

The effect of the MEA evaporation and heat transfer due to the different temperatures explain the differences. At solvent flows equal to 13 m³/ton CO_2 more water evaporates than condensates, therefore the contribution of water to the temperature profile is negative. The CO_2 absorption compensates for this effect and overall results in a very small temperature increase for the rich solvent. At 33 m³/ ton CO_2 it clearly shows that the temperature bulge is due to the water flux. The net effect is zero and the

Comparison of the design of CO_2-capture processes

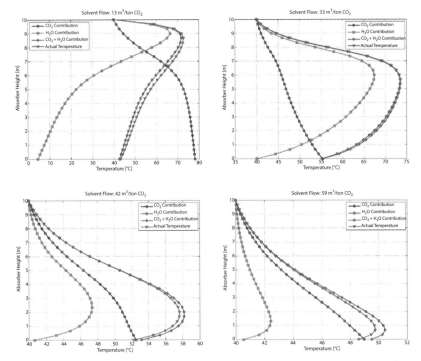

Figure 9.19 Different contributions from different fluxes for the calculation of absorber temperature profiles.

rich solvent temperature is due to the heat of absorption and reaction of CO_2 into the MEA solution. The other flows show that the net temperature increase for the rich solvent is decreasing as a function of the solvent flow due to the fact that the heat is released into more solvent.

9.5 Conclusions

A new rate-based process simulation tool for the steady-state simulation and optimization of acid gas treating processes, VMG RateBase was developed by the Virtual Materials Group and Procede Process Simulations Partnership.

The correct design of chemical absorption must be done using rate-based models supported by rigorous thermodynamics and accurate physical properties. The design of a column based on an equilibrium approach when chemical absorption and mass transfer limitations are important will result in incorrect results, including under sizing of the equipment.

The optimization of an absorber by changing the gas to liquid ratio with an equilibrium model is not possible. The design of a column based on an isothermal assumption will result in incorrect dimensions. The heat of reaction needs to be included. Adiabatic models based on an equilibrium approach will give different concentrations and temperature profiles compared to a rate-based model. The temperature profile is determined by the heat of absorption of CO_2 and for a major part by the heat of evaporation/condensation of water.

References

1. E.P. van Elk, A.R.J. Arendsen, G.F. Versteeg, A new flowsheeting tool for flue gas treating, Energy Procedia 1 (2009), 1481–1488
2. E.S. Hamborg, P.W.J. Derks, E.P. van Elk, G.F. Versteeg, Carbon dioxide removal by alkanolamines in aqueous organic solvents. A method for enhancing the desorption process, Energy Procedia 4 (2011), 187-194
3. J.C. Meerman, E.S. Hamborg, T. van Keulen, A. Ramírez, W.C. Turkenburg and A.P.C. Faaij, Techno-economic assessment of CO_2 capture at steam methane reforming units using commercially available technology, *To be published*, 2012.
4. A.R.J. Arendsen, E. van Elk, P. Huttenhuis, G. Versteeg, F. Vitse, Validation of a post combustion CO_2 capture pilot using aqueous amines with a rate base simulator, SOGAT, 6th International CO_2 Forum Proceedings, Abu Dhabi, UAE, 2012.
5. G.F. Versteeg, J.A.M. Kuipers, F.P.H. van Beckum and W.P.M. van Swaaij, Mass transfer with complex chemical reactions. I. Single reversible reaction, Chemical Engineering Science, 44, 2295-2310, 1989
6. Mendes, M.F.; HETP Evaluation of Structured and Randomic Packing Distillation Column; http://www.intechopen.com/books/mass-transfer-in-chemical-engineering-processes/hetp-evaluation-of-structured-and-randomic-packing-distillation-column

10
Post-Combustion Carbon Capture Using Aqueous Amines: A Mass-Transfer Study

Ray A. Tomcej

Tomcej Engineering Inc., Edmonton, AB, Canada

Gas absorption using aqueous amines is a technology that is often considered in the recovery of carbon dioxide (CO_2) from post-combustion gases in electric power plants. Certain amines have a high affinity for CO_2 at low partial pressures and the basic process has been proven through decades of use in the natural gas and hydrocarbon processing industries. Nevertheless, the use of the amine process in flue gas treating is met with considerable challenges.

The chemistry of the CO_2–amine reaction favors contacting conditions at high pressure and low temperature, opposite of those found in flue gas treating. Combustion gases, including sulfur dioxide and oxygen, and particulate matter require special handling to reduce reagent degradation and foaming. Low contacting pressure produces large volumetric rates of flue gas and results in extreme equipment sizes.

The basic amine process has been modified to adapt to these challenges. Specialty amines and solvent formulation aim to reach a compromise between solution reactivity and capacity. Novel mechanical design of contacting equipment and process configuration can be used to overcome limitations of pressure drop and heat effects.

In this work, fundamentals of gas absorption with chemical reaction are revisited to identify areas of the amine process that limit its effectiveness in flue gas-treating operations. A discussion of chemical reaction kinetics, contacting conditions, and absorber design is presented with examples to illustrate the impact on CO_2 recovery.

10.1 Introduction

Generation of electricity using coal-fired turbine units represents a significant portion of all electrical production in North America. There is a trend to replace coal with natural gas as abundant shale gas supplies emerge and stringent air quality regulations further restrict the amount of contaminants that may be releasedinto the atmosphere. Nevertheless both fuel types result in significant production of post-combustion carbon dioxide (CO_2), a known greenhouse gas. Its recovery from boiler flue gas for subsequent disposal by acid gas injection or commercial use in enhanced oil recovery is an industry objective.

In the United States, coal-fired power plants operate in a wide range of capacities ranging from a few megawatts (MW) to over 2000 MW. Older plants are systematically being retired or converted to natural gas but continue to emit large amounts of CO_2. Similar situations exist in other countries around the world.

Power plant facilities usually include downstream treatment of flue gas to remove nitrogen and sulfur compounds (NO_x and SO_x), fly ash and other particulates. An example process block diagram showing the flow of boiler flue gas is given in Figure 10.1.

The US Department of Energy's *Energy Information Administration*Form EIA-860 Database contains detailed information about all US power plants, however, actual flue gas emission data are scarce in the open literature [1]. reported actual capacity and emission data for several Wyoming coal-fired power plants ranging in size from 90 to 2120 MW. Ciferno [2] presented data for a case study of a 600 MW plant. Using the information from these sources, a representative average power plant generates 500 MW of electricity and releases 1.6 MMSm³/h of flue gas containing 14 mole % CO_2.

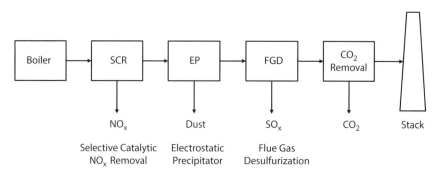

Figure 10.1 Process Diagram of Example Coal-Fired Power Plant

Table 10.1 Typical Flue Gas Composition and Conditions

Component	Composition, mole %
Nitrogen	65-70
Carbon Dioxide	10-20
Oxygen	5-15
Water	5-10
Sulfur Dioxide	200-2500 ppm
NO_x	100-1000 ppm
Other	50 ppm
Pressure	115-135 kPa(a)
Temperature	40-50°C
pCO_2 in feed	10-30 kPa(a)
pCO_2 in product @90% recovery	1-3 kPa(a)

Typical flue gas composition and conditions are given in Table 10.1. Target CO_2 recoveries for carbon capture are generally in the 85 to 95% range. At 90% recovery, the representative plant would capture 9100 tonne CO_2 per day. Chapel et al. [3] discuss the economics of various plant sizes and suggest that the largest feasible single-train plant from coal-fired flue gas is 4600 tonne/d.

Many innovativetechnologies have been proposed to recover this CO_2, including absorption, adsorption, membrane processes, cryogenic separation and others. One common technology is the use of aqueous alkanolamines to absorb CO_2 from the flue gas using a regenerative absorption cycle. There is considerable interest in deriving the ultimate amine-based absorption solvent to achieve this recovery.

In this work, mass transfer fundamentals are examined in an attempt to determine the important aspects of CO_2 absorption into aqueous alkanolamines at flue gas conditions.

10.2 Mass Transfer Basics

Absorption of CO_2 into aqueous alkanolamines is a mass transfer process accompanied by chemical reaction in the liquid phase. Alkanolamines are generally categorized according to the nature and extent of substitution of the amino group and are either primary, secondary or tertiary. The structural differences play a major role in governing their reactivity with CO_2. Detailed analysis of the reactions between CO_2 and primary-, secondary- and tertiary amines may be found in Astarita et al [4], Kohl and Reisenfeld [5] and Versteeg et al. [6].

In general, CO_2 reacts with alkanolamines in aqueous solutions according to the following scheme

Primary and Secondary Alkanolamines

$$R_1R_2NH + CO_2 \rightleftarrows R_1R_2NCOO^- + H^+ \qquad (10.1)$$

$$R_1R_2NH + CO_2 + H_2O \rightleftarrows R_1R_2NH_2^+ + HCO_3^- \qquad (10.2)$$

$$R_1R_2NH + HCO_3^- \rightleftarrows R_1R_2NH_2^+ + CO_3^= \qquad (10.3)$$

Tertiary Alkanolamines

$$R_1R_2R_3N + CO_2 \rightleftarrows \text{no direct reaction} \qquad (10.4)$$

$$R_1R_2R_3N + CO_2 + H_2O \rightleftarrows R_1R_2R_3NH^+ + HCO_3^- \qquad (10.5)$$

$$R_1R_2R_3NH + HCO_3^- \rightleftarrows R_1R_2R_3NH^+ + CO_3^= \qquad (10.6)$$

Where R_1, R_2 and R_3 represent substituted groups such as ethanol and R_2 is a proton in the case of primary amine.

CO_2 also undergoes two additional reactions in aqueous solvents

$$CO_2 + H_2O \rightleftarrows HCO_3^- + H^+ \qquad (10.7)$$

$$CO_2 + OH^- \rightleftarrows HCO_3^- \qquad (10.8)$$

A simplified expression for the rate of reaction of CO_2 in aqueous solutions of amine can be shown to be given by

$$r_{CO_2} = k_{H_2O}[CO_2] + k_{HO^-}[OH^-][CO_2] + k_{Amine}[Amine][CO_2] \qquad (10.9)$$

$$= k_{OV}[CO_2] \qquad (10.10)$$

Where the value of k_{OV} is an overall effective rate constant and $[CO_2]$ is the concentration of CO_2 in the liquid phase.

The two-resistance theory describing mass transfer across a gas-liquid phase boundary is represented in Figure 10.2. The interphase mass transfer is assumed to be confined to two stagnant layers on either side of the gas-liquid interface. All the resistance to mass transfer is contained in the two films. It is assumed that there is no resistance to mass transfer at the interface so that interfacial concentrations are in physical equilibrium. The concentration of the absorbing component in the gas phase, Y, decreases from the well-mixed bulk concentration to a value Y_i at the interface. Similarly, the concentration of the absorbing component in the liquid phase, X, decreases from a value X_i at the interface to the well-mixed bulk concentration.

The mass flux N across the interface can be represented by

$$N = I k_L^0 (X_i^{CO_2} - X_{bulk}^{CO_2}) \qquad (10.11)$$

Where the enhancement factor I is a measure of how much the rate of mass flux is increased by the presence of chemical reaction in the liquid phase and k_L^0 is the liquid-phase mass transfer coefficient in the absence of chemical reaction.

For low solution loadings (moles CO_2 per mole amine) and for conditions encountered in industrial applications, the CO_2 reaction with alkanolamine proceeds in the fast reaction regime and under pseudo-first order reaction conditions where

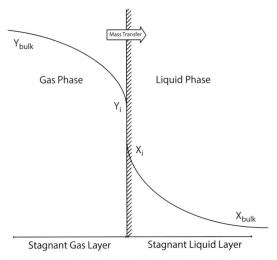

Figure 10.2 Two-Resistance Theory of Mass Transfer Across a Gas-Liquid Interface

$$Ik_L^0 = \sqrt{D_{CO_2} k_{OV}} \tag{10.12}$$

D_{CO_2} is the diffusion coefficient for CO_2 in the liquid amine solution. Actual mass transfer takes place across a finite interfacial area a so that

$$Mass\,transfer\ rate = aN = a\sqrt{D_{CO_2} k_{OV}}\,(X_i^{CO_2} - X_{bulk}^{CO_2}) \tag{10.13}$$

Therefore the overall rate that CO_2 is absorbed into alkanolamine solutions in flue gas applications is proportional to:

1. Overall concentration driving force;
2. The square root of the reaction rate constant;
3. Interfacial area available for mass transfer

10.3 Factors Influencing Mass Transfer

10.3.1 Concentration Driving Force

The overall concentration driving force is the difference between the interfacial concentration in the liquid and the concentration of CO_2 in the bulk liquid phase. Interfacial concentration of CO_2 in alkanolamine solutions cannot be measured directly because of the reaction and must be inferred using the nitrous oxide analogy [5, 6]. Since the interfacial concentration represents physical solubility of CO_2 gas in the alkanolamine solution, the actual value will generally be similar among all aqueous alkanolamine solutions of the same concentration at any given CO_2 partial pressure.

In the bulk liquid, under fast reaction regime conditions, the chemical reaction is fast enough to keep the CO_2 concentration at its equilibrium value. Aqueous solutions of alkanolamines can have considerably different equilibrium solubility properties. Figure 10.3 illustrates the equilibrium solubility of CO_2 in 4 kmol/m³ solutions of monoethanolamine (MEA), diethanolamine (DEA), methyldiethanolamine (MDEA), triethanolamine (TEA) and 4 kmol/m³ MDEA with 0.5 kmol/m³ MEA at 60°C in the partial pressure range present in flue gas operations. Assuming that the representative flue gas is 14 mole % CO_2 and at 135 kPa(a) the partial pressure of CO_2 is about 17 kPa. For 90% recovery of CO_2, the partial pressure in the treated gas would be about 2 kPa.

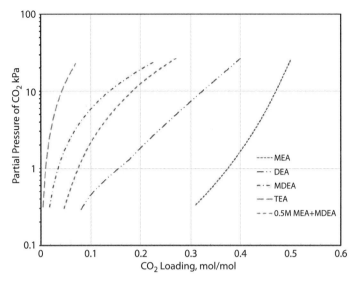

Figure 10.3 Equilibrium Solubility of CO_2 at 60°C in 4 kmol/m³ Solutions

Although Figure 10.3 does not give a direct indication of the concentration driving force, it does show qualitative differences between the different amine solutions. If the separation between the solubility locus and the left axis at 0 mol/mol loading is considered to be representative of the concentration driving force, it is apparent that the driving force is largest for primary and secondary amines with tertiary amines showing the smallest driving force. Some improvement is gained with the addition of a small amount of primary amine rate promoter to a tertiary amine solution.

Equation (10.13) shows that the mass transfer rate is directly proportional to the concentration driving force. Since the interfacial concentration X_i is obtained by a Henry's law relationship and will be similar in value among all aqueous alkanolamine solutions, mass transfer rates can only be increased by decreasing the concentration in the bulk phase. However the bulk phase concentration in the fast reaction regime is determined by chemical equilibrium and is, at best, near zero. From an engineering standpoint, mass transfer rates can be increased by using alkanolamine solutions with the highest concentration driving force.

As a secondary observation, consider the influence of equilibrium solubility on the required circulation rate of amine solution for the flue gas example. In operation, all of the recovered CO_2 leaves the contactor in the rich amine solution. A pinch analysis using the CO_2 partial pressure at rich amine conditions can be used to approximate the maximum obtainable loading. Using estimated lean amine residual loadings the net loading can be used to obtain the minimum solution circulation rate as shown in Table 10.2.

The weaker tertiary amines have unfavorable circulation rates compared to primary and secondary amines. It is interesting to note that under lean conditions at 60°C the equilibrium loading of MDEA at 2 kPa is calculated to be 0.052mol/mol and for TEA it is about 0.016. This would suggest that for these tertiary amines, there is a practical limit on the possible recovery of CO_2 and that this recovery is significantly affected by equilibrium solubility of CO_2 in lean amine solutions. This is unfortunate as tertiary amines have considerably lower heats of solution which translate to reduced energy requirements for regeneration.

The addition of CO_2 absorption rate promoters such as MEA or piperazine to tertiary amine solutions has been studied by Tomcej [7], Rochelle and Bishnoi [8], Rinprasertmeechai et al. [9] and others. Figure 10.3 shows the calculated effect of adding 0.5 kmol/m³ MEA to the existing 4 kmol/m³ MDEA solution. The addition of rate promoter to the solution increases the capacity, particularly at low loadings with moderate increase at higher loadings. The addition of rate promoter does improve the potential solution performance in flue gas treating.

10.3.2 Reaction Rate Constant

The overall mass transfer rate in (10.13) is affected by the square root term involving k_{OV}. The diffusion coefficient of CO_2 in alkanolamine solutions is a physicochemical property that must be inferred by the nitrous oxide analogy, in a manner similar to the interfacial CO_2 concentration. The remaining property, k_{OV}, is an effective overall rate constant for the reaction

Table 10.2 Approximate Minimum Solution Circulation Rate

	MEA	DEA	MDEA	TEA	MEA+MDEA
mol/mol, 17 kPa CO_2	0.49	0.36	0.19	0.06	0.23
mol/mol, lean	0.1	0.05	0.02	0.01	0.02
mol/mol, net	0.39	0.31	0.17	0.05	0.21
Relative Circulation Rate	1	1.3	2.3	7.8	1.9

of CO_2 with various species present in the liquid phase. Note in (10.9) that the rate constant for the amine contribution is a function of the amine type and the free amine concentration. Some manipulation of k_{Amine} is possible by using promoters or blends of amines. Values of k_{Amine} in 4 kmol/m³ lean solutions at 60°C were calculated using correlations presented in Hikita et al. [10] and Blanc and Demarais [11]. The values of Ik_L^0 relative to Ik_L^0 for MDEA are given in Table 10.3.

The chemistry of the CO_2 reactions with alkanolaminesare such that reaction rates are always fastest in lean solutions with low loadings. As CO_2 is absorbed into the liquid it reacts with the various ionic and molecular species and results in a consumption of free amine, a contributing factor in k_{OV}. Figure 10.4 shows the concentration of free amine in a 4 kmol/m³MEA solutionat 60°C as a function of CO_2 loading. The consumption of free MEA to form carbamate can be seen at loadings below 0.5 mol/mol. At an intermediate loading of 0.25 mol/mol the amount of free MEA available for reaction is only 2 kmol/m³ so the rate of absorption is reduced by a factor of $1/\sqrt{2}$.

The reaction rate constant is seen to influence the overall rate of absorption in a highly complex manner. Although it is desirable to have a solvent with a high rate of reaction, a high rate also increases the depletion of reactants in the liquid film. At higher rates, chemical reaction then occurs in the transition zone outside of the fast reaction regime in which the absorption rate is affected by gas diffusion to the reaction zone. In the extreme case, if the reaction rate is fast enough, the rate of absorption becomes totally controlled by the gas phase.

Thus for flue gas applications it is the combined effect of reaction rate promotion and increased solution capacity that is important in solution formulation. This combined effect must always be balanced by an analysis of the regeneration energy penalty that corresponds to the increased heats of solution in any formulation.

Table 10.3 Relative Rates of Reaction in 4 kmol/m³ Solutions at 60°C

Alkanolamine	Ik_L^0
MEA	32
DEA	11
MDEA	1
TEA	0.7

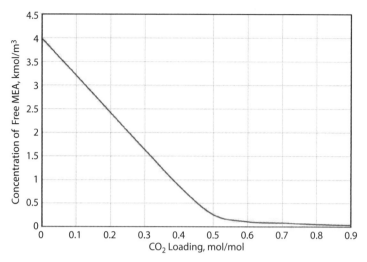

Figure 10.4 Concentration of Free Amine in 4 kmol/m³ MEA at 60°C as a Function of CO_2 Loading

10.3.3 Interfacial Area

The final factor in (10.13) that contributes to the overall mass transfer of CO_2 is the gas-liquid interfacial area. Mass transfer rate is directly proportional to a, which is completely independent of the amine chemistry. Interfacial area is a function of the mechanical equipment design, process configuration and the hydrodynamic conditions under which contact takes place.

The use of alkanolamines for gas conditioning largely began in the natural gas processing industry where the sour gas is at high pressure. Absorption is usually accomplished in countercurrent trayed or packed towers. In these systems, mass transfer occurs as a bubble of gas rises through liquid holdup on a tray deck or across the liquid film that flows through the contours of packing rings or structured elements. Contact is multistaged and pressure drop can be considerable, but not critical as the overall pressure is still high. Gas is the dispersed phase and liquid is the continuous phase.

Trayed distillation column performance has been studied for many decades in order to develop empirical correlations for a as a function of superficial phase velocities, active tray area, interfacial tension, phase viscosities and densities. Values of interfacial area per unit volume of gas-liquid dispersion typically lie between 100 and 600 m²/m³. Dispersion heights in the order of 0.2 to 0.4 m are representative of performance in trayed towers.

For packed towers, the interfacial area is usually expressed relative to the dry packing surface area and as a function of the same parameters as in the trayed column case. Standard packing materials such as Raschig rings, Intalox saddles and Pall rings have specific surface areas that range from 100 to 400 m^2/m^3 depending on the ring or saddle size.

Borrowing from the industrial experience of wet flue gas desulfurization units (FGD), consideration should be given to the use of alternate contacting devices for CO_2 recovery from flue gas. Of particular interest are spray tower and venturi or ejector contactors. Venturi/spray tower systems are already in use in FGD installations, do not include complex internal components which offer sites for deposition of amine degradation products and scale, provide low pressure drop and handle high liquid to gas rates. A typical venturi/spray tower design is shown in Figure 10.5. The device provides essentially two stages of contact. The converging section of the venturi accelerates the flue gas to a high superficial velocity. When the amine solution is injected at the throat of the venturi, the turbulence of the gas velocity atomizes the solution into small droplets which absorb CO_2.

The spray tower section provides additional contact with lean amine. Spray nozzles are capable of producing a wide range of spray patterns and

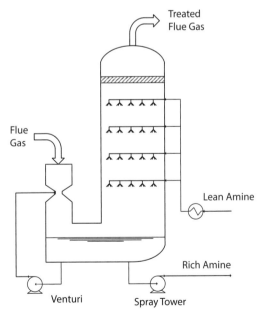

Figure 10.5 Venturi/Spray Tower System for Flue Gas Treating

drop size distributions in the range of 100 to 500 μm diameter. For an average drop size diameter of 200 μm, one cubic meter of liquid produces a surface area of 30000 m². Liquid droplets of diameter 200 μm are classified as drizzle in the sequence: drop, drizzle, mist, fog. The smaller the average drop size, the larger the surface area. For example if the average drop diameter is 150 μm, the surface area would be 40000 m². The specific interfacial surface area would be less because of the dispersion of the liquid droplets in the continuous gas phase. Mist elimination using demister pads and/or mellachevrons and downstream water wash would be required to recover vaporized or entrained amine and to reduce make-up requirements.

Regardless of the actual drop size diameter, suitable design of nozzle systemsin the spray tower should produce a significant increase in the interfacial surface area available for mass transfer over trayed or packed towers. In the limit, the mass transfer rates would be sufficiently high that equilibrium would be approached.

10.4 Examples

10.4.1 Venturi/Spray Tower System

Assume a 500 MW power plant emits 1.75 MMSm³/h (wet)(62 MMSCFH) of flue gas that contains 14 mole % CO_2 on a water-free basis. The water-saturated flue gas is available for CO_2 recovery at 135 kPa(abs) and 50°C. Use 24 weight % MEA (4 kmol/m³) in a venturi/spray tower system with a standard regeneration tower. Assume the atomized droplets in the venturi/spray tower are 200 μm in diameter and that superficial gas velocity is 1.5 m/s. Assume air coolers can cool the lean amine to 45°C. Split the full flow into eight trains.

Using in-house software, the results are summarized in Table 10.4. The calculated CO_2 recovery of 80% is below the target of 90% but illustrates that the process configuration warrants further study. A similar packed tower absorber configuration would be limited by the interfacial area available for mass transfer and would be unable to achieve this level of CO_2 recovery. Some improvement would be possible through the use of rate promoters or amine blends.

This example shows that increasing the interfacial area for mass transfer by using liquid spray dispersed in the gas phase directly increases the CO_2 absorption rate. It shows that modifying the characteristics of the gas-liquid contact through mechanical equipment design can affect CO_2 recovery to the same extent as solvent formulation.

Table 10.4 Summary of Example 1 Simulation Results

Treated Flue Gas		
	CO_2 Recovery	80 %
	tonne/day per train	1010
	°C	61
Lean Amine		
	L/min per train	12500
	mol CO_2/mol MEA	0.19
Rich Amine		
	mol CO_2/mol MEA	0.52
	°C	56
Reboiler Heat Duty		
	kJ/L lean amine	264

10.4.2 Amine Contactor with Pumparound

A flue gas that contains 10 mole % CO_2 on a water-free basis is available for CO_2 recovery at 105 kPa(abs) and 45°C at a rate of 3.28 MMSm³/d (wet) (116 MMSCFD). Use 30 weight % MEA (5 kmol/m³) in a 20-tray contactor. Assume 5800 L/min of lean amine at 45°C enters on the top tray and has a CO_2 residual loading of 0.2 mol/mol. Assume 12000 L/min of amine liquid is drawn off tray 16, cooled to 45°C and fed back to the contactor on tray 5. This configuration is shown in Figure 10.6.

Using in-house software, the results are summarized in Table 10.5. The calculated CO_2 recovery is 98% and illustrates the impact of process configuration on the recovery. The liquid pumparound CO_2 loading is 0.43 mol/mol. As shown in Figure 10.4, at these loadings, the concentration of free MEA has been substantially reduced. The corresponding rate of CO_2 absorption is also reduced. This effect pushes the absorption up the tower where the heat of absorption creates a large temperature bulge in the middle of the contactor [12]. At high loadings and elevated temperature, the equilibrium solubility of CO_2 in the MEA solution also limits the rate of absorption by reducing the concentration driving force.

By cooling the liquid pumparound, the heat of absorption is removed from the contactor. The high liquid rate in the middle of the tower also smoothes out the temperature bulge. Without the pumparound circuit, the contactor would be unable to recover more than a few percent of the CO_2 in the flue gas.

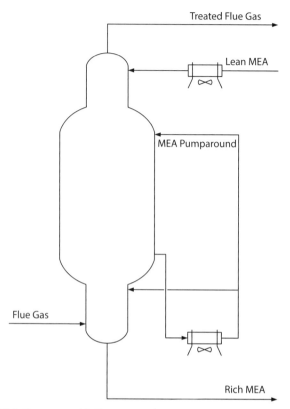

Figure 10.6 MEA Contactor with Pumparound

This example shows that an understanding of the heat- and mass-transfer characteristics of the amine process is imperative in the design of these types of absorption systems. It shows that this knowledge can be applied to explore alternate process configurations where standard design would fail.

10.5 Summary

Recovery of post-combustion CO_2 from power plant flue gas remains a major technological challenge in the attempt to reduce greenhouse gas emissions. One of the technologies in use is absorption of CO_2 into aqueous solutions of alkanolamines. The rate of absorption (or recovery) of CO_2 is hindered by low operating pressures. Using mass transfer fundamentals it can be shown that the mass transfer rate is proportional to: the available gas-liquid interfacial area; the square root of the CO_2 reaction

Table 10.5 Summary of Example 2 Simulation Results

Treated Flue Gas		
	CO_2 Recovery	98 %
	tonne/day	542
	°C	46
Lean Amine		
	L/min	5800
	mol CO_2/mol MEA	0.2
Pumparound Feed		
	L/min	12000
	mol CO_2/mol MEA	0.43
	°C	45
Pumparound Draw		
	°C	57
Rich Amine		
	mol CO_2/mol MEA	0.49
	°C	48
Pumparound Cooler		
	kJ/L lean amine	90

rate constant; and the overall concentration driving force. Low operating pressures, high temperatures and complex amine chemistry complicate the latter two factors. Novel contactor design and configuration to increase interfacial area and to reduce operating temperature are areas where further study is required.

References

1. Robertson, Eric P., "Analysis of CO_2 Separation from Flue Gas, Pipeline Transportation, and Sequestration in Coal", Idaho National Laboratory, September 2007.
2. Ciferno, Jared, "Carbon Capture Technology Options and Costs", Presented at the Indiana Carbon Capture & Sequestration Summit, Indianapolis, Indiana, September 3–4, 2008.

3. Chapel, Dan G., Carl L. Mariz and John Ernest, "Recovery of CO_2 from Flue Gases: Commercial Trends", Presented at the Canadian Society for Chemical Engineering Annual Conference, Saskatoon, Saskatchewan, October 4–6, 1999.
4. Astarita, Gianni, David W. Savage and AttilioBisio, "Gas Treating With Chemical Solvents", John Wiley & Sons, New York, 1983.
5. Kohl, Arthur L. and Fred C. Riesenfeld, "Gas Purification", 4th Edition, Gulf Publishing Co., Houston, 1985.
6. Versteeg, G.F., P.M.M. Blauhoff and W.P.M. van Swaaij, "The Kinetics Between CO_2 and Alkanolamines. A Study on the Reaction Mechanism", Paper 123d, AIChE Annual Meeting, Nov. 2–7, Miami Beach, Florida, 1986.
7. Tomcej, Ray A., "Mass Transfer With Chemical Reaction in Alkanolamine Treating Units", Ph.D. Thesis, University of Alberta, 1987.
8. Bishnoi, Sanjay and Gary T. Rochelle, "Thermodynamics of Piperazine / Methyldiethanolamine / Water / Carbon Dioxide", *Ind. Eng. Chem. Res.*, 41, 604–612(2002).
9. Rinprasertmeechai, Supitcha, SumaethChavadej, PramochRangsunvigit, and SantiKulprathipanja, "Carbon Dioxide Removal from Flue Gas Using Amine-Based Hybrid Solvent Absorption", Intl. J of Chem.AndBio.Eng., 6, 296–300(2012).
10. Hikita, H., S. Asai, H. Ishikawa and M. Honda, "The Kinetics of Reactions of Carbon Dioxide with Monoethanolamine, D Ciferno iethanolamine and Triethanolamine by a Rapid Mixing Method", *Chem. Eng. J.*, 13, 7–12(1977).
11. Blanc, C. and G. Demarais, "Vitesses de la Réaction du CO_2 Avec la Diéthanolamine", *Entropie*, 17(102), 53–61(1981).
12. Arnold, D.S., D.A. Barrett and R.H. Isom, "CO_2 Production From Coal-Fired Boiler Flue Gas By MEA Process", Paper A, Presented at the Laurance Reid Gas Conditioning Conference, Norman, Oklahoma, March 8–10, 1982.

11

BASF Technology for CO_2 Capture and Regeneration

Sean Rigby[1], Gerd Modes[1], Stevan Jovanovic[2], John Wei[2], Koji Tanaka[3], Peter Moser[4] and Torsten Katz[5]

[1]*BASF Corporation*
[2]*Linde LLC*
[3]*JGC*
[4]*RWE Power AG*
[5]*BASF SE*

Abstract

BASF's gas treating technology is a sophisticated, customizable technology portfolio complementary to the developing Carbon Capture, Utilization and Storage (CCUS) market. Of significance to high-pressure CO_2 applications are three BASF technologies: OASE™ purple, HiPACT™, and OASE™ blue.

OASE™ purple and HiPACT™ are marked as industry leaders for the efficient, stable, and cost-effective capture of CO_2 from applications such as natural gas treatment. Co-development of HiPACT™ by BASF and JGC yielded novel solvent chemistry and improvements to the process design resulting in faster kinetics of solvent-CO_2 absorption and increased loading capacity. Tests were conducted at INPEX's Koshijihara plant in Nagaoka, Japan. Results show reductions in solvent circulation rate and reboiler duty. Regeneration of CO_2 was demonstrated at 2.5 bara with conceptual regeneration pressures possible between 3 and 8 bara. In a case study of 1.5 Mtpa of gross CO_2 recovery, HiPACT™ conferred a 35% reduction in annual CO_2 compression costs by reducing the number of compression stages and requiring a lower power input for compression. Furthermore, the total annual expenditure, including CO_2 capture and compression, was reduced 28%. Consequently, CO_2 recovery benefits conferred by HiPACT™ include ultra-low CO_2 slip in the treated gas, relatively high CO_2 stripping pressure, and reduced recovery costs.

In addition to natural gas treating, BASF has developed, in cooperation with Linde, combustion flue gas CO_2 removal technology called OASE™ blue. As a step-change technology in the CCUS market, OASE™ blue offers significant improvements over state-of-the-art technologies and can be applied to flue gases generated by combustion of fossil fuels such as coal and natural gas. When compared against MEA, advantages of OASE™ blue are conferred from a novel, highly stable solvent and optimized process design. Laboratory analysis and testing at BASF's miniplant in Ludwigshafen, Germany, evaluated a total of 400 solvent candidates in various combinations based on multiple criteria including non-hazardous, low cost, thermally and chemically stable, fast reaction kinetics, high cyclic capacity, and low regeneration duty.

Selected solvent candidates and process design for OASE™ blue were tested in a slipstream pilot plant at RWE's Niederaussem 1000 MW_{el} lignite coal-fired power plant. In cooperation with Linde and RWE, BASF began the Niederaussem project in 2009 and it is still operational. The project consists of a flue gas slip stream of 1552 Nm^3/h with the potential to capture 7.2 tpd of CO_2. For approximately 26,000 hours of operation, the pilot plant availability was 97% with interruptions caused only by planned maintenance or power plant shutdowns. Since the commissioning, more than 6500 t of CO_2 was captured until August 2013. An MEA benchmark was measured to have a specific energy demand of 3.5 GJ/t of CO_2. OASE™ blue showed a specific energy demand of less than 2.8 GJ/t of CO_2, approximately 20% lower than MEA, at a much lower circulation rate than MEA. Contributing to the low specific energy demand were process improvements by Linde such as intermediate cooling in the absorber column and waste heat integration. Optimizations to the solvent, operating parameters, and process configuration reduced the specific energy demand from typically 3.5 GJ/t of CO_2 for MEA-based solvents to less than 2.8 GJ/t of CO_2. As part of the pilot plant demonstration at the Wilsonville, Alabama, Linde and BASF recently concluded a techno-economic analysis summarizing a rigorous simulation model of a 550 MWe pulverized coal power plant with subcritical Rankine steam cycle comparing the effects of OASE™ blue against MEA solvent (Case 10) [HYPERLINK \l "Cos07" 1]2]. Results show that implementation of OASE™ blue and process improvements developed by Linde increased power plant efficiency for 4.5%, reduced CAPEX 30%, and reduced the levelized cost of electricity 15%.

As a leader in gas treatment for over four decades, BASF has pioneered advances in CO_2 capture technology. This chapter reviews recent advances in CO_2 capture technologies with key features highlighted for HiPACT™ and OASE™ blue. In specific, key features include thermal solvent stability, CO_2 absorption performance, increased CO_2 stripping pressures, and significant reductions in the cost of electricity and of captured CO_2. Results from the studies examining these features are reviewed showing substantial advantages of BASF technologies above state of the art in the CCUS market.

11.1 Introduction

Utilization of carbon dioxide (CO_2) by means of enhanced oil recovery (EOR) confers the potential of producing an additional 5-15% of the original oil in place. The US Department of Energy (DOE) estimates that in the United States 137 billion barrels of oil can technically be produced by CO_2-EOR, however, only 67 billion barrels would be economically recoverable at $85 per barrel, 20% return on investment (ROI), and $40 per metric ton of CO_2 (MT_{CO2}).[3] Furthermore, approximately 18 billion metric tons of anthropogenic CO_2 would be needed in addition to the current CO_2 supply for the recovery of these 67 billion barrels of oil yielding an increase in the average CO_2 sales of $1.08 billion per year for 50 years.[3]

Anthropogenic CO_2 originates from various sources including natural gas processing, fossil-fueled power generation, and other industrial sources. Typical aqueous amine-based state-of-the-art (SOA) CO_2 capture technology, as shown in Figure 11.1, is used in such applications. Figure 11.1 shows a process diagram comprised of an absorber-stripper configuration. Feed gas (G1) in the form of natural gas or flue gas with previously removed SO_x components to the level of <5ppm enters the absorber and flows upward counter-current against the CO_2-lean solution(L4) entering at the top of the column. CO_2 depleted treated gas (G2) exits the absorber

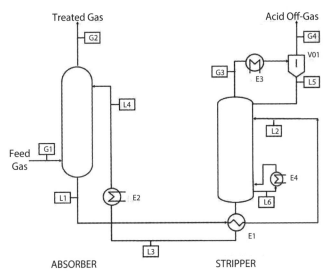

Figure 11.1 Process Flow Diagram of a Typical Absorber-Stripper Configuration Acid Gas Removal Unit

column. The CO_2 rich solution (L1) exits the absorber and is heated in the solvent-solvent heat exchanger (E1) with regenerated, CO_2 lean, solvent exiting the bottom of the reboiler. The heated solution (L2) is then fed to the top of the stripper column (also referred to as regenerator or desorber). The majority of CO_2 is released from the solution by cycling the solution (L6) through the reboiler (E4). This introduces the required heat for regeneration via low pressure steam condensation. The regenerated, lean solution is then cycled out of the stripper, cooled in the solvent-solvent heat exchanger (E1), further cooled in lean solution cooler (E2) and re-enters the absorber (L4). CO_2 released in the stripper exits at the top (G3), subsequently passes through a condenser (E3, V01) and is passed downstream for processing and compression (G4). Reflux condensate from the condenser (L5) is returned to the top of stripper column.

Cost-effective and robust purification of CO_2 from these sources requires advances over SOA technologies. For example, estimates for capture and purification of CO_2 from a coal-fired power plant using SOA CO_2-absorbtion technology and monoethanolamine (MEA) solvent result in the cost of recovered and compressed CO_2 to 153 bara at 49 2007$/ MT_{CO2} while adopting conditions and assumptions identical to those published in the DOE/NETL Report 2007/1281. [1]Release of absorbed CO_2 from the CO_2-rich solvent can considerably impact compression costs depending on solvent regeneration pressures. Typically, solvent regeneration is accomplished in a low-pressure environment yielding low-pressure CO_2 that must then pass through multiple costly compression stages. Next-generation technologies are needed to strip CO_2 at higher pressures to reduce the cost of compression while requiring less energy for solvent regeneration.

Reviewed in this paper are BASF CO_2-capture technologies that have been optimized for industry-specific applications and that yield CO_2 off-gas at pressures higher than SOA. Over 40 years of experience and greater than 300 reference plants have led to the development of BASF's gas treating technology portfolio called OASE™, which includes both the process design and chemical solvent. This vast experience confers the ability to adapt the CO_2 purification process to numerous industrial applications including natural gas treating (OASE™ purple) and fossil-fueled power generation (OASE™ blue). Moreover, BASF has partnered with both JGC in the development of High Pressure Acid gas Capture Technology (HiPACT™), a natural gas treating technology for relatively high pressure CO_2 generation (3-8 bara), and with Linde in the optimization of OASE™ blue for post-combustion CO_2 capture (PCC). Each industry has unique challenges to the CO_2 capture process, such as in coal-fired flue gas high amounts of oxygen cause rapid oxidative degradation of SOA solvents. This paper reviews the technological and chemical advancements of these technologies, the related pilot plants and reference cases, and associated techno-economic studies. It is shown herein that BASF technology is ideally suited for cost-effective capture of relatively high pressure CO_2 from multiple sources.

11.2 Materials and Methods

11.2.1 HiPACT™ Laboratory Screening [4]

Criteria for solvent evaluation and selection of novel CO_2 absorbing solvents included solvent stability, CO_2 absorption capacity, and CO_2 absorption kinetics. Solvent stability was evaluated at high pressure and temperature, 162°C and 5.3 barg, in an autoclave in the presence of CO_2. The solvent decomposition rate for solvent candidates (Results shown in Figure 11.2) was calculated and compared against MEA, DEA and MDEA. CO_2 absorption capacity of HiPACT™ and OASE™ purple was obtained using a thermostated high-pressure optical cell. CO_2 absorption kinetics of CO_2 rich and lean solutions were determined with a laminar jet absorber.

11.2.2 HiPACT™ Pilot Plant [4]

Pilot plant tests were conducted at both BASF's pilot plant in Ludwigshafen, Germany and at JGC's pilot plant at Oarai, Japan. The HiPACT™ pilot

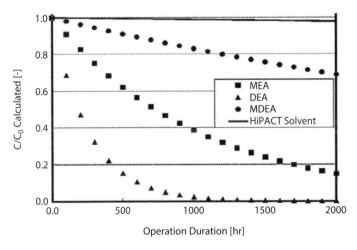

Figure 11.2 HiPACT™ Solvent Stability Compared Against Common Gas Treating Amine Solvents [4]

plant was a continuous absorption-regeneration column design. Feed gas entering the absorber bottom was a combination of treated gas exiting the absorber top, CO_2 regenerated from the stripper, and a synthetic gas blend. The pilot plant served a dual function of validating laboratory results and providing insight into commercial scale conditions. Operating conditions of the pilot plant were tailored to more accurately represent full-scale conditions. Specifically, in the regenerator the operating pressure was kept between 4 – 6 barg, and the solvent residence time in the regenerator bottom was maintained with control of the solvent buffer tank (V-8) in the reboiler loop (Figure 11.3). Moreover, holding volumes in each tank, operating temperatures and CO_2 loadings were kept within the operating parameters reflective of actual plant conditions. The CO_2 concentration in the feed gas and the hot-oil temperature were adjusted as needed so that the CO_2 loading in the solvent was kept at 0.75 mol_{CO2} / mol_{amine} for rich loading and 0.01 mol_{CO2} / mol_{amine} for lean loading.

In addition to testing at the laboratory scale, CO_2 loading capacity was measured in the pilot plant tests. Long-term stability testing was conducted for approximately 2000 hours with alternating absorption and regeneration. The regenerator operating pressure was maintained at 5.2 barg and

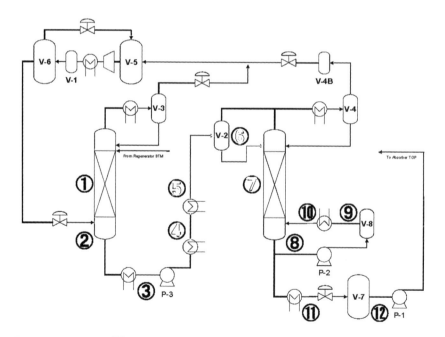

Figure 11.3 HiPACT™ Pilot Plant Simulation Modeling for Solvent Stability [4]

the regenerator bottom temperature was 161°C. During the long term testing, solvent samples were periodically taken for analysis.

Material corrosion tests using test coupons of various materials were sampled in the pilot plant for 47 days. Materials that were tested included carbon steel, Type 304L, and Type 316L. Locations of the samples in the pilot plant are shown in Figure 11.3.

Simulation of solvent stability was determined as a function of CO_2 loading, temperature and residence time. The basic pilot plant configuration was represented by 12 blocks (Figure 11.3) and modeling was performed with variations in process configurations and operating conditions. The simulation tool was designed to predict solvent degradation rate and was subsequently optimized for commercial-scale design of the HiPACT™ process.

11.2.3 HiPACT™ Demonstration Plant [5]

Train B of the INPEX Koshijihara Natural Gas Production Plant was temporarily converted from OASE™ purple to HiPACT™. The Koshijihara plant underwent modifications when the AGRU was upgraded from DGA® to OASE™ purple. Modifications included an increased maximum solvent circulation rate with increased lean solvent pump capacity. Also, the total gas pressure was increased and a new absorber with a larger diameter was installed. DGA® operating conditions, feed gas composition and flow rate, and consequently CO_2 partial pressure, were also changed in the transition to OASE™ purple. Temporary modifications to the OASE™ purple design were introduced for the HiPACT™ trial primarily for solvent handling and storage. There were no major changes to operating conditions. Furthermore, all exchangers were not replaced or modified, which limited the extent of testing for maximum regeneration pressures of 2.5 bara. Of significance, temporary amine was stored in two ISO containers; OASE™ purple was stored in the Train-A collecting pit with a temporary drain pit pump; and HiPACT™ was stored in the Train-B sump drum.

Parameter test runs were all conducted with a treated gas specification CO_2 slip of less than 1000 ppm, however operations were maintained at a CO_2 slip between 25–100 ppm. Parameters that were evaluated included amine strength, reboiler duty, circulation rate, lean amine temperature, and regeneration pressure (Table 11.1). Results were obtained based on process simulations and operational data.

Solution absorption capacities were examined with a commercially available rate-based design tool for DGA®, and a BASF in-house tool

Table 11.1 HiPACT™ Demonstration Plant Operating Parameters for Test Runs [5,12]

Run	Run Description	Operating Parameters					
		Amine Strength	Reboiler Duty	Solvent Circ. Rate	Lean Amine Temp.	Regenerator Pressure
#1	High Strength	Base +10%	Base +5%	Base +13%	Base	Base
#2 Base	HiPACT base	Base	Base	Base	53degC(Base)	1.4 bara (Base)
#3	Reduced reboiler duty	Base	Base- 13%	Base	Base	Base
#4	Lean temp. 50degC	Base	Base	Base	Base – 3degC	Base
#5	Lean temp. 45degC	Base	Base	Base	Base – 8degC	Base
#6	Lean temp. 55degC	Base	Base	Base	Base + 2degC	Base
#7	Increased reboiler duty	Base	Base+ 11%	Base	Base	Base
#8	High regenerator pressure	Base	Base	Base	Base	bara*

* 2.5 bara was the maximum allowable pressure of the existing equipment at the site

was used for OASE™ purple and HiPACT™. An operational baseline was established for HiPACT™ and is labeled as test run #2 (Table 11.1). The CO_2 equilibrium loading of all solutions was first examined with a temperature of 75°C. Absorber bottom temperatures were then varied to reflect actual optimized conditions for each solution.

Absorber temperature profiles for each of the solvents were obtained by temperature measurements in the absorber column during operation and were compared against process modeling. Simulations for temperature profiles and CO_2 concentration profiles were based on feed gas properties and data obtained from both laboratory-scale testing and pilot plant operation.[4]

Reboiler duty parametric testing was conducted by changing the reboiler duty -13% (test run #3) and +11% (test run #7) relative to baseline operating conditions (test run #2) (Table 11.1). Simulations for all three solutions were performed and compared against parametric testing data. In addition, select operating data points were also obtained; specifically, reboiler duty was obtained for both a high regenerator pressure of 2.5 bara (test run #8) and a CO_2 slip of 23 ppmv.

The effects of variations in regenerator pressure were evaluated using two regenerator overhead pressures, 1.4 bara (test run #2) and 2.5 bara (test run #8). The solution circulation rate and other process parameters were kept constant, and the reboiler duty was adjusted to meet CO_2 specification in the treated gas.

Solvent losses were modeled using both an in-house BASF thermodynamic tool and a rate-based simulation program. In addition, solvent losses during operation were monitored for all the HiPACT™ test runs #1 – 8 (Table 11.1). Four methods of analysis were used, including (1) on-line gas sampling of treated gas in the absorber overhead, (2) solvent accumulation in the di-ethylene glycol (DEG) dehydration unit and solvent emissions downstream of the DEG unit, (3) solvent inventory loss and make-up rate, and (4) changes in the ratio between base solvent and activator system. The third method, monitoring make-up rates, required monitoring solvent make up volumes over the operation time and dividing the quantities by the produced amount of gas. Additional simulations were performed based on a configuration optimized for reduced solvent losses.

11.2.4 HiPACTTM Case Study [4,5]

The HiPACT™ case study was conducted based on data from both plant operations and process modeling. Conditions and parameters are

Table 11.2 HiPACT™ Case Study Conditions [4]

Study cases	
Case -1 : OASE™ purple with low pressure regeneration	
Case -2 : HiPACT with high pressure regeneration	
CO_2 Recovery rate (Gross)	1.5 million tons CO_2 per year
CO_2 content in raw natural gas	6.8 vol%
Raw natural gas pressure/temperature	60 barA/ 20 degC
CO_2 Compressor discharge pressure	200 barA

described in Table 11.2. The study assumed that the natural gas producing plant recovered 1.5 million tons per year of CO_2 by an acid gas removal unit (AGRU) and CO_2 compression. In addition, it was assumed that major utilities were available including hot oil for the reboiler, electricity for air-fin coolers, pumps, and fuel gas for the gas turbine. In the case of revamp, major items include replacement of the regenerator system and installation of a new solvent-solvent heat exchanger. The number of AGRU trains was determined by the upper limit set on the weight of the CO_2 absorber and the availability of cranes for construction. The number of CO_2 compressors was dependent on the upper limit of the compressor BHP and the first stage inlet volume based on the installation records of centrifugal CO_2 compressors. Process parameters, such as regeneration pressure, rich solvent CO_2 loading and reboiler duty were optimized for each case.

CAPEX was determined by JGC's in-house cost data. Annual fixed costs of 13% were a sum of 10% for depreciation and 3% for operation and maintenance. OPEX was based on estimated utility requirements and annual solvent refill volumes. Annual expenditure was the sum of annual fixed costs and OPEX in which the following net CO_2 recovery cost equations were used:

$$\text{Net } CO_2 \text{ recovery cost [USD/ton}_{CO2}] = \text{(Fixed cost + OPEX) [USD/y] / Net } CO_2 \text{ recovery rate [ton/y]} \quad (11.1)$$

Where,

$$\text{Net } CO_2 \text{ recovery rate [ton/y]} = \text{Gross } CO_2 \text{ recovery rate} - CO_2 \text{ emission from AGRU and } CO_2 \text{ compression} \quad (11.2)$$

11.2.5 OASE™ blue Laboratory Screening [6, 7, 8, 9]

Multiple criteria for solvent selection were considered, such as lowest possible health and environmental toxicology, viscosity, vapor pressure, cyclic capacity and specific reboiler duty. CO_2 cyclic capacity and reboiler duty were estimated by vapor-liquid equilibrium (VLE) testing at conditions relevant to flue gas CO_2 capture in an AGRU. CO_2 absorption capacity and mass transfer rates were evaluated in a double-stirred contactor at pressure and temperatures representative of an absorber column. Thermal degradation of solvent was measured by heating solvents with a residual CO_2 loading at 160°C for 150 hours. Solvent content was measured before and after heating and the solvent degradation reported as percent of solvent remaining after the 150 hours.

11.2.6 OASETM blue Miniplant [7, 9]

Solvents identified in the laboratory were subsequently analyzed in BASF's 0.015 MWe at 0.24 t_{CO2}/d miniplant at Ludwigshafen, Germany. The miniplant was comprised of a synthetic flue gas, an absorber-stripper configuration, and all necessary equipment such as pumps, heat exchanges and automated process controls. The synthetic flue gas was composed of nitrogen, CO_2, H_2O, and optional O_2. Solvent performance data obtained in the laboratory was validated in the miniplant and compared against a 30%-weight MEA benchmark.

11.2.7 OASE™ blue Pilot Plant:Niederaussem [7,8,10]

Solvent formulations that met or exceeded the selection criteria in the miniplant were further tested in a 0.5 MWe, 7.2 t_{CO2}/d slipstream plant at RWE's lignite-fired power plant in Niederaussem, Germany. Data from the miniplant was validated at the pilot plant, and continued operation is ongoing for greater than 26,000 hours to enable commercial application of OASE™ blue.

The pilot plant is equipped with more than 275 process measurement points and 10 sampling points as shown in the process flow diagram in Figure 11.4.[6] Design parameters and feed gas conditions are shown in Table 11.3. A direct contact cooler (DCC) operated upstream of the pilot plant functioned to cool the flue gas slipstream to between 30 - 35°C, and to prescrub the flue gas to reduce SO_2 and other compounds. The combined pressure drop of the DCC and the absorber column was counteracted by a flue gas booster fan positioned between the DCC and the absorber column.

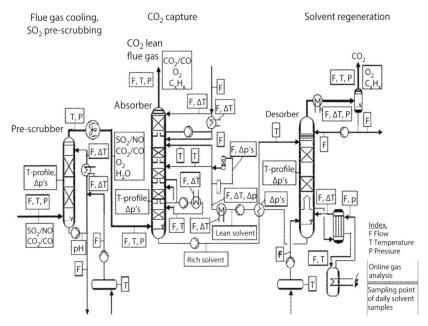

Figure 11.4 OASE™ blue Pilot Plant Process Flow Diagram with Key Measuring Points and Sampling Points [7]

Table 11.3: OASE™ blue Pilot Plant Parameters

Design and operational parameters		
Flue gas flow	[m_N^3/h]	1,550
Flue gas temperature (after FGD)	[°C]	65
CO_2 content at flue gas absorber inlet	[vol.-%,dry]	14.2
NO_x content in flue gas (after FGD)	[mg/m_N^3, 6% O_2corr, dry]	190
SO_2 content in flue gas (after FGD)	[mg/m_N^3, 6% O_2corr, dry]	93
O_2 content in flue gas (after FGD)	[vol.-%,dry]	5.0
CO_2 captured (maximum)	[tCO_2/day]	7.2
Pilot plant availability	[%]	97

PCC process improvements included a gravity flow interstage cooler in the absorber column, advanced emission control in the absorber overhead, advanced column material design, optimized blower concept, and optional interstage heater in the stripper column.

Samples of MEA and OASE™ blue solvent candidates were drawn during 4000 hour testing campaigns, and analyzed for O_2 degradation products.

Materials tested at different heights in the DCC, absorber and stripper columns included austenitic stainless steels, austenitic-ferritic stainless steel, and polypropylene. A concrete module with a polypropylene in-liner was positioned in the lean-solvent stream between the stripper and absorber columns. Located throughout the process configuration between major pieces of equipment were flange-tube-flange components FRTP/vinyl ester resin and FRP/epoxy resin, in addition to EPDM, PTFE and NBR rubber gaskets. Total materials testing time in the pilot plant with MEA as the solvent was between 3,875 – 4,920 hrs. Non-destructive x-ray and ultrasonic testing was conducted after three months and six months. Material analysis at the conclusion of the test run was done by weight loss measurements, visual inspection with the binocular microscope (25x magnification), bending the sample 45 degrees to show stress corrosion cracking, geometric measurements, elasticity, tensile strength at yield and break, and change in elongation at yield and break. Moreover, the concrete module was cut and visually inspected. The polypropylene in-liner was tested for solvent absorption, polymer degradation, stabilization agent extraction, and weldability.

11.2.8 OASE™ blue Case Study [1,2]

A techno-economic analysis of the OASE™ blue technology with Linde PCC process enhancements was compared against an MEA process, specifically reference Case 10 of DOE/NETL_2007 study (Case 10). [1,2] The analysis was based on a 550 MWe subcritical pulverized coal power plant as described in Case 10. All site characteristics, raw water usage, and environmental targets were described in Case 10. Simulation tools used for the study included Honeywell's UniSim Design platform, Bryan Research and Engineering's Promax® software, and BASF's proprietary in-house process modeling tool. Energy demand, net higher heating value (HHV) efficiency, incremental capital costs, and levelized cost of electricity (LCOE) were determined for both OASE™ blue and MEA based PCC plant consistent with DOE/NETL methodology. [1] Capital cost elements were estimated based on in-house proprietary tools in combination with commercial experience. The LCOE value and the related cost of captured CO_2 were expressed in 2007\$ for operation of the plant over a period of 20 years, and was calculated using the following equation used for MEA based Case 10 [1]:

$$LCOE = \{(CCF)(TPC) + \Sigma[(LF_{Fi})(OC_{Fi})] + (CF)\Sigma[(LF_{Vi} * OC_{Vi})]\} / [(CF)(aMWh)] \quad (11.3)$$

Where,

> aMWh annual net Mega Watt-hours of power generated at 100 percent capacity factor
>
> CCF Capital Charge Factor for a levelized period of 20 years
>
> CF plant Capacity Factor (0.85 in this study)
>
> LF_{Fi} Levelization Factor for category i Fixed operating cost
>
> LF_{Vi} Levelization Factor for category i Variable operating cost
>
> OC_{Fi} category i Fixed Operating Cost for the initial year of operation (but expressed in "first-year-of-construction" year dollar)
>
> OC_{Vi} category i Variable Operating Cost for the initial year of operation (but expressed in "first-year-of-construction" year dollar)
>
> TPC Total Plant Cost, $

Two cases for OASE™ blue integration with the power plant were considered [2]:

> LB-1 subcritical pulverized coal power plant integrated with OASE™ blue and Linde PCC process improvements
>
> LB-2 subcritical pulverized coal power plant integrated with OASE™ blue and Linde PCC process improvements using waste heat recovery and power cogeneration for internal heat and power requirements

11.3 Results

A typical carbon capture utilization and storage (CCUS) project is a mosaic of industries. Each CO_2 capture technology must be tailored for integration into other units upstream and downstream. To that end, BASF has partnered with both Linde and RWE to develop OASE™ blue for flue gas CO_2 capture, and with JGC and INPEX in the development of HiPACT™ for CO_2 capture in natural gas treating. Both OASE™ blue and HiPACT™ are technologies optimized for CCUS projects, but individually tailored for their respective industries. For both technologies there was a two-fold approach to development, (1) rigorous chemical screening and selection, and (2) optimization of process configuration and operation. Reviewed in this section are the results of developing these novel CO_2 capture technologies from laboratory screening, through pilot plants, to commercial readiness.

11.3.1 HiPACT™ CO_2 Capture Technology for Natural Gas Treating

The general concept of BASF's High Pressure Acid-gas Capture Technology (HiPACT™) is described in Figure 11.5.[4] Development of HiPACT™ began at the laboratory scale in which solvent candidates were screened for high pressure regeneration of CO_2.[4] Screening results of the novel HiPACT™ solvent were then validated along with further testing at pilot plants located at the JGC facility at Oarai, Japan and BASF's laboratory in Ludwigshafen, Germany. 4] Pilot testing was followed by a demonstration plant at the INPEX Koshijihara Natural Gas Production Plant, Japan.[5] The Koshijihara plant was originally operated with DGA® for 15 years. The plant was then swapped to OASE™ purple and operated for 10 years. The HiPACT™ demonstration was a temporary swap from OASE™ purple, and lasted for 2 months. Throughout testing at the pilot plant and demonstration plant, multiple process simulations were conducted to validate the accuracy of modeling and plant design.

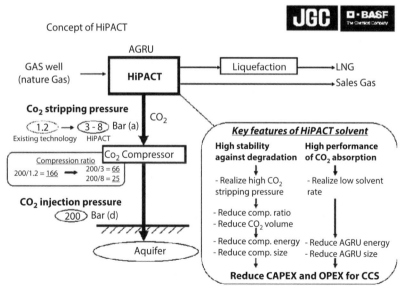

Figure 11.5 General Concept of HiPACT™ Deployment [4]

11.3.2 HiPACT™ Solvent Stability and Losses

Solvent stability and losses were key properties studied in all stages of HiPACT™ development. Results from batch tests show that relative to MEA, DEA and MDEA, HiPACT™ is substantially more stable with minimal degradation during 2000 hours of operation (Figure 11.2).[4] Moreover, simulations compared against solvent degradation indicate accurate modeling of solvent losses. Solvent losses were simulated based on equipment in the demonstration plant (Figure 11.6) and with optimizations to the backwash sections above the absorber and stripper columns (Figure 11.7).[5] Results show the majority of losses occurred in the treated gas at 3.9 kg/hr; also measured as 91 mg/Nm³ solvent concentration in the treated gas. Addition of an optimized backwash section substantially reduced solvent losses to 0.05 kg/hr. Validation of the simulations was performed by conducting 4 different experiments for measuring solvent losses in the absorber overhead (Table 11.4). Each experiment differed in expected levels of accuracy. For example, the least accurate method (1) by on-line sampling of the treated gas was difficult to reproduce due to minor condensation effects in the sampling system. Whereas, method (3) which measured the decrease of solvent inventory and solvent make-up rate over the entire operation time made the most accurate reflection of solvent losses. Results indicated that simulation of solvent losses was relatively accurate fitting well to the range of 60 - 140 mg/m³ as shown in experimental methods (2) – (4).[5]

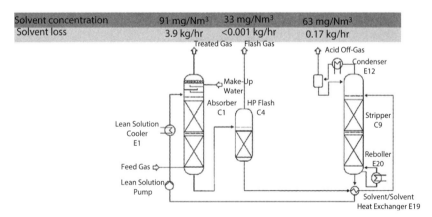

Figure 11.6 HiPACT™ Solvent Losses Simulated for the Existing Equipment at the Demonstration Plant [5]

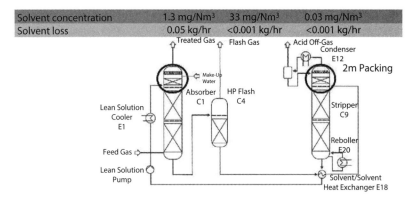

Figure 11.7 HiPACT™ Solvent Losses Simulated for the Demonstration Plant with Optimized Backwash Sections [5,12].

11.3.3 HiPACT™ Solvent CO_2 Absorption Capacity and Kinetics

Solvent evaluation also included analysis of both the capacity and kinetics of CO_2 absorption. Laboratory tests with HiPACT™ solvent indicated that the new solvent has a high capacity for CO_2 loading as compared to state-of-the-art natural gas treating solvents.[5] Further testing in the pilot plant shows similar results in which HiPACT™ absorption capacity was measured relative to OASE™ purple loading.[4] HiPACT™ CO_2 loadings ranged between 1.2 – 1.4 times greater than OASE™ purple. Moreover, results from the simulations and operational data from the demonstration plant further validate these findings.[5] At a fixed absorber bottom temperature of 75°C, HiPACT™ absorption capacity is greater than OASE™ purple, and is similar to DGA® CO_2 loading of solvent in mol/ton. However, under real plant conditions for the INPEX plant configuration and operating conditions, in which the absorber bottom temperatures are set at real operating conditions (Table 11.5), DGA® shows a lower absorption capacity than HiPACT™ (Figure 11.8). In addition, although DGA® and OASE™ purple show a similar CO_2 loading capacity, the INPEX plant was switched to OASE™ purple in 2002 due to the improved corrosion properties and lower specific reboiler duty for OASE™ purple. Similar improvements are also shown with HiPACT™, discussed in detail below.

A temperature profile of the absorber column was generated and shows a significant bulge for DGA® and indicates that the operating conditions were close to CO_2 breakthrough (Figure 11.9).[5] Breakthrough of CO_2 in the treated gas can lead to corrosion and entrainment downstream of the AGRU. Conversely, both HiPACT™ and OASE™ purple displayed a smaller temperature bulge with significantly more stable conditions at the top of the absorber. The CO_2 profile of the absorber column shown in Figure 11.10 further confirms the CO_2 absorption capacity even at high temperatures seen in the range of the temperature bulges.[5]

Table 11.4 Methods and Results for Evaluation of HiPACT™ Demonstration Plant Solvent Losses [5]

		Expected accuracy	Measured solvent loss [mg/m³]
1.	On-line sampling of treated gas	–	10–25
2.	Solvent accumulation in the DEG unit + emissions solvent from the DEG unit	0	60–100
3.	Decrease of solvent inventory/ solvent make-up rate	+	75
4.	Change in the ratio between base solvent and activator	0	70–140

Table 11.5 HiPACT™ Demonstration Plant Real Absorber Bottom Conditions [5,12]

		DGA	OASE purple	HiPacT
Temperature	[°C] [°F]	92 197	75.5 168	80.6 177
CO_2 partial pressure	[bara] [psia]	3.9 57	5.2 75	5.2 75
Feed gas flow rate	[Nm3/day] [MMSCFD]	800,000 29.8	1,033,000 38.5	1,033,000 38.5
Specific leam solvent circulation rate*	[%]	100	135.8	108.1

$$^*\text{specific lean solvent circulation rate} = \frac{\left(\frac{solvent\ flow\ [t/hr]}{captured\ co_2\ [t/hr]}\right)_{solvent}}{\left(\frac{solvent\ flow\ [t/hr]}{captured\ co_2\ [t/hr]}\right)_{DGA}}$$

11.3.4 HiPACT™ Materials Compatibility

Corrosion rates of various materials were tested throughout the pilot plant. Results indicate that HiPACT™, whether rich with CO_2 or lean, induced negligible corrosion of all the materials tested (Figure 11.11). Corrosion rates were measured at less than 0.06 mm/yr.[4]

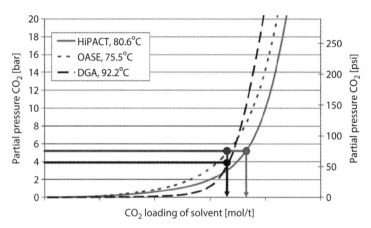

Figure 11.8 HiPACT™ Equilibrium Loading at Real Absorber Bottom Temperature Specific to INPEX plant Configuration and Operational Conditions [5,12]

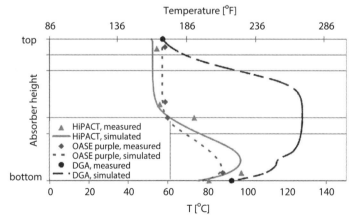

Figure 11.9 Temperature Profile of the HiPACT™ Demonstration Plant Absorber Column [5,12]

11.3.5 HiPACT™ Energy Requirements

Test runs with HiPACT™ were performed in which reboiler duty was raised (+11%) or lowered (-13%) relative to baseline (Table 11.1). Process simulations were a close fit with operational data. In particular, simulations showed slightly higher reboiler duty requirements than is necessary to meet treated gas CO_2 slip with HiPACT™ or OASE™ purple[5] In addition, as compared against OASE™ purple, HiPACT™ has a 10 - 20% lower reboiler duty and 20 – 27 % lower solvent circulation rate. Interestingly, DGA® displays the highest specific reboiler energy, but as a consequence the lowest solvent circulation rate.

11.3.6 HiPACT™ CO_2 Stripping Pressure

Regeneration of the solvent in the regenerator column of an AGRU releases CO_2 in a more purified acid-off gas. To facilitate release of the CO_2, however, regenerator operating conditions are typically a combination of rapid pressure drop and increased temperature. HiPACT™ development was specifically targeted at regeneration of the solvents at relatively higher pressures, approximately 3 - 8 bara. Due to equipment constraints at the demonstration plant, the maximum allowable pressure at the regenerator overhead was 2.5 bara. Consequently, test runs were performed at low regeneration pressure (1.4 bara) and high regeneration pressure (2.5 bara). Results shown comparing test runs #2 and #8 in Figure 11.12 and Table 11.6

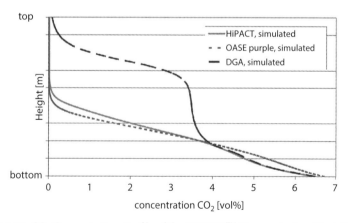

Figure 11.10 CO_2 Concentration Profile of the HiPACT™ Demonstration Plant Absorber Column [5,12]

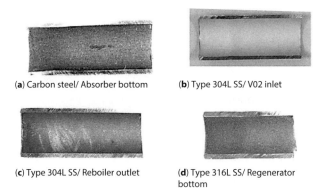

(a) Carbon steel/ Absorber bottom
(b) Type 304L SS/ V02 inlet
(c) Type 304L SS/ Reboiler outlet
(d) Type 316L SS/ Regenerator bottom

Figure 11.11 Corrosion Test Coupons from HiPACT™ Pilot Plant After 47 Days [4]

indicate that regeneration at higher pressure leads to reduced solvent circulation rate. [5] The circulation rate was reduced as a consequence of two factors. First, there was deeper solvent regeneration in the stripper leading to higher absorption capacity in the absorber. Second, there were different feed gas compositions between test runs #2 and #8, which caused a higher CO_2 partial pressure in the feed gas. In addition to a decreased circulation rate, the specific reboiler energy decreased, which was a consequence of higher pressures reducing the amount of water vapor in the stripper overhead. [5] Less water vapor in the stripper overhead drove more water to remain in the stripper to facilitate a more energy efficient regeneration of the solvent.

11.3.7 HiPACT™ Economics

A case study of a grassroots AGRU and compressor plant was conducted to contrast the economics of OASE™ purple (case-1) against HiPACT™ (case-2).[4] Results showed that due to the higher absorption capacity and lower circulation rate, a single AGRU train was required for HiPACT™, whereas two trains were required for OASE™ purple (Table 11.7). Furthermore, the higher pressure CO_2 from the HiPACT™ AGRU led to an increased capacity of the CO_2 compressor. Consequently, HiPACT™ required one compression train, whereas two trains were required for OASE™ purple. Results of the cost parameters indicate that, when compared against OASE™ purple, fixed costs for the HiPACT™ AGRU are reduced 26%, compression expenditures are reduced 35%, and there is a total cost saving of 28% (Table 11.8). Moreover, HiPACT™ displayed an

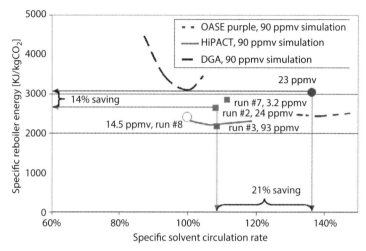

Figure 11.12 HiPACT™ Energy Duty and Solvent Circulation Rate Compared against DGA® and OASE™ purple [5,12]

Table 11.6 ReboilerDutIes of HiPACT™ Demonstration Plant Operated at Low and High Stripper Overhead Pressures [5]

		Test run #2	Test run #8
Stripper top pressure	[bara]/[psia]	1.4/20	2.5/36
T stripper bottom	[°C]/[°F]	112.8/233	127.1/261
Spec. lean solvent circulation rate	[%]	108.1	99.8
Specific reboiler energy	[KJ/kgCO$_2$]	2649	2419
Absorber CO$_2$ slip	Measured [ppmv]	24.3	14.5

11% improvement of net CO_2 recovery rate. Cumulative economic benefits of HiPACT™ yield a 35% reduction in net CO_2 recovery costs. Since CAPEX is dependent on the number of trains, and the CO_2 recovery rate determines the number of trains needed it is critical to account for the recovery rate. In the case study described, the gross CO_2 recovery rate is 1.5 million tons/year. When the CO_2 recovery rate is reduced to 1.0 million tons/year both technologies require only single trains for both the AGRU and CO_2 compressor. Despite both requiring only single trains, the CO_2 recovery cost for HiPACT™ is still 25% lower.

Another case study examined retrofitting an existing OASE™ purple plant with either CO_2 compressors newly installed (case-3) or revamping the AGRU to use HiPACT™ with CO_2 compressors newly installed (case-4). [5] Due to no additional investment to the AGRU and only maintenance costs, the net CO_2 recovery cost for case-3 as compared against case-1 is 34% lower (Table 11.9). For case-4, the HiPACT™ revamp costs yield an annual expenditure 2% higher relative to case-1. Despite the increased annual expenditures the net CO_2 recovery is reduced 42% for case-4 relative to case-1 due to the impact of higher CO_2 stripping pressure lowering the compressor costs.

11.3.8 OASE™ blue CO_2 Capture Technology for Flue Gas Treating

Unique to flue gas treating are the challenges of low CO_2 partial pressure, high O_2 content, high volumetric flowrates, and project economics. Overcoming these obstacles was the focus of OASE™ blue development. Presented in this section is a review of results from laboratory screening and miniplant operations in Ludwigshafen, Germany, ongoing operations at the pilot plant at Niederaussem, Germany, and a preliminary techno-economic analysis from a Department of Energy (DOE)-sponsored pilot project in Wilsonville, Alabama.

11.3.9 OASE™ blue Solvent Stability and Losses

High O_2 content in the flue gas and high temperatures in the AGRU contribute to solvent degradation. The impact of O_2 on degradation of OASE™ blue was tested during 4000 hours under real flue gas conditions at the

Table 11.7 HiPACT™ Case Study Process Parameters (Only Case-1 and Case-2 are Shown) [4]

	Case-1 (OASE™ purple)	Case-2 (HiPACT)
Train Number (AGR)	2	1
Train Number (CO_2 Comp.)	2	1
Regeneration Pressure	1.7 barA	5.7 barA
Rich Solvent CO_2 Loading	1.0 (base)	1.6
Solvent Circulation Rate	1.0 (base)	0.6
Reboiler Duty	1.0 (base)	0.9
Compressor BHP	1.0 (base)	0.7

Table 11.8 HiPACT™ Grassroots Case Study Results for Case-1 and Case-2 [5]

Grassroots AGR with CCS		Case-1 (OASE™ purple)			Case-2 (HiPACT™)		
		AGRU	Comp.	Total	AGRU	Comp.	Total
Annual fixed cost	[relative value] *1	0.44	0.19	0.64	0.29	0.13	0.42
OPEX	[relative value] *1	0.31	0.05	0.36	0.26	0.03	0.30
Owner's annual expenditure	[relative value] *1	0.75	0.25	—	0.56	0.16	0.72
Cost Saving from OASE purple	[%]	—	—	—	−28%	−35%	−28%
CO_2 emission	[mil ton/year]	0.34	0.15	0.49	0.28	0.10	0.38
Net CO_2 recovery	[mil ton/year]			1.01			1.12
Improvement of net CO_2 recovery	[%]			—			11%
Net CO_2 recovery cost	[relative value] *2			1.00			0.62(−35%)

"Annual fixed cost" represents an annual cost corresponding to investment cost (e.g. depreciation) plus maintenance cost, in [mil-USD/year]. Annually 13% of investment cost was counted for "Annual cost".

"OPEX" represents utility costs to operate AGRU and CO_2 compressor, in [mil-USD/year].

(Owner's annual expenditure)=(Annual fixed cost)+(OPEX)[mil-USD/year]

"CO_2 emission" corresponds to energy consumption within AGRU and CO_2 Compressor.

(Net CO_2 recovery cost)=(Owner's annual expenditure)/{(Gross CO_2 recovery)−(CO_2 emission)}

*1 Owner's annual expenditure for OASE purple = 1.0

*2 Net CO_2 recovery cost for OASE purple = 1.0

Table 11.9 HiPACT™ Revamp Case Study Results for Case-3 and Case-4 [5]

Revamp for CCS		Case-3 (OASE™ purple) AGRU: existing OASE purple plant is continuously used Comp. newly installed			Case-4 (HiPACT™) AGRU: exisiting OASE plant is revamped for HiPACT Comp. newly installed		
		AGRU	Comp.	Total	AGRU	Comp.	Total
Annual fixed cost	[relative value] *1	0.10	0.19	0.29	0.22	0.13	0.35
OPEX	[relative value] *1	0.31	0.05	0.36	0.26	0.03	0.30
Owner's annual expenditure	[relative value] *1	0.41	0.25	0.66	0.49	0.16	0.65
Cost Saving from OASE purple	[%]	–	–	–	17%	−35%	−2%
CO_2 emission	[mil ton/year]	0.34	0.15	0.49	0.28	0.10	0.38
Net CO_2 recovery	[mil ton/year]			1.01			1.12
Improvement of net CO_2 recovery	[%]			–			11%
Net CO_2 recovery cost	[relative value] *2			0.66			0.58

*1: Owner's annual expenditure for grassroots OASE purple case = 1.00
*2: Net CO2 recovery cost for grassroots OASE purple case = 1.00

pilot plant. Results showed that significantly less O_2-induced degradation products were formed with OASE™ blue as compared to MEA (Figure 11.13).[7] In addition, thermal degradation of OASE™ blue was examined in batch testing and results show that the nearly 90% of OASE™ blue solvent remained after 150 hours at high temperature whereas just over 75% of MEA remained.

11.3.10 OASE™ blue Process Materials Compatibility

Further testing at the pilot plant examined the corrosive impact of process contaminants such as O_2, flue gas trace elements, and MEA-based degradation compounds. Various materials were placed at key locations with a high potential for corrosion such as at the stripper inlet and outlet.[8]

Results showed that for the metal coupons there was no local corrosion such as pitting and cracking, and the corrosion rate was less than 0.001 mm/yr. Polypropylene samples were determined to be corrosion resistant and absorption of the solvent was negligible. All other materials were corrosion resistant with two exceptions, (1) EPDM gaskets which were mechanically squeezed and showed brittle disruptions, and (2) NBR gaskets increased from 1.8 mm to 1.9 mm, and were mechanically squeezed and showed brittle disruptions.[8] Results from testing with MEA will be used as a benchmark for data collected from OASE™ blue operation in the pilot plant. It is hypothesized that results from material testing with OASE™ blue will be

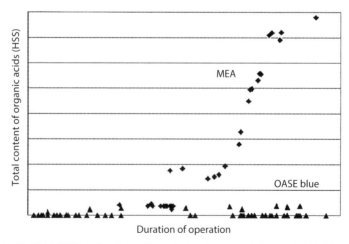

Figure 11.13 OASE™ blue Oxidative Degradation Compared Against MEA [8]

similar or improved as compared to MEA due to the minimal formation of degradation products with OASE™ blue (Figure 11.13).

11.3.11 OASE™ blue Solvent Capacity, Kinetics, Energy Requirements, and CO_2 Stripping Pressure

Fast reaction kinetics, high cyclic capacity and low specific energy demand for absorption-desorption of CO_2 with solvent candidates for OASE™ blue are shown in Figure 11.14.[9] Additional testing at the pilot plant showed similar benefits for solvent candidates as compared against MEA.[10] However, these criteria are inherently contradictory; for example, rapid mass transfer typically is associated with higher specific reboiler duty. Consequently, the solvent formulation finally selected for OASE™ blue culminated in a formulation optimized for efficient flue gas CO_2 removal. During testing at BASF's miniplant under synthetic flue gas conditions, the reboiler duty for OASE™ blue was initially shown to be 25% lower than MEA. After further testing under real flue gas conditions at the pilot plant in Niederaussem, OASE™ blue was shown to reduce reboiler duty 20 – 24% depending on CO_2 regeneration pressure (Table 11.10).[6] Interestingly, with OASE™ blue the regeneration energy reflects no significant dependence on pressure on the stripper overhead, whereas, MEA displays a 4% difference in energy demand between 1.5 – 1.75 bara. In

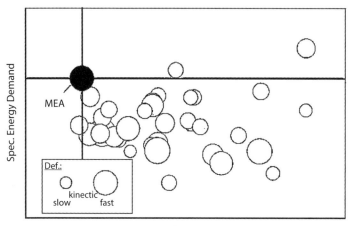

Figure 11.14 OASE™ blue Candidate Solvents Screened for Physical Properties of CO_2 Absorption Relative to MEA [6]

Table 11.10 Regeneration Demand for MEA and OASE™ blue at Varying Desorber Pressures [7]

Desorber pressure [bara]	Regeneration demand [MJ/tCO$_2$]	
	MEA	OASE™ blue
1.5	3,640	2,760
1.75	3,480	2,770
1.90/2.00	3,520	2,790

addition, although published data show that regeneration of solvent and release of CO$_2$ in the OASE™ blue process is possible up to 2.0 bara in the stripper overhead (Table 11.10), experience at the pilot plant indicates operation of the desorber column is possible at higher pressures.

11.3.12 OASE™ blue Economics

With enhancements such as higher CO$_2$ regeneration pressures and lower energy requirements OASE™ blue offers significant economic benefits as compared to MEA. An examination of the techno-economics of post-combustion capture (PCC) and compression of CO$_2$ was performed for OASE™ blue and MEA.[2] As illustrated in Figure 11.15, the net higher heating value (HHV) efficiency of the pulverized coal power plant with a PCC plant was most impacted by the novel OASE™ blue solvent formulation, which improved efficiency by 1.76%. The study examined two cases, LB-1 and LB-2. Case LB-1, as compared against DOE/NETL Case 10, consisted of more efficient heat management and lower energy requirements for the flue gas blower and solvent circulation pumps. In addition, solvent regeneration was increased to 3.4 bara and condensation lowered to 20°C. Consequently, LB-1 increased the HHV efficiency by 1.39%. Additional benefits from Case LB-2, which incorporated heat and power integration options with the power plant and PCC unit, led to significantly reduced energy requirements for PCC operation and CO$_2$ compression. Consequently, the HHV efficiency was further increased for 1.35 % by Case LB-2. In total, OASE™ blue with process improvements increased the HHV efficiency 4.5 % over MEA based Case 10.

Total PCC and compression costs for MEA and OASE™ blue were compared.[2] In general, results showed significant reductions for both OASE™ blue options, LB-1 and LB-2. Solvent and process improvements (LB-1) yielded a 34.6% Total Plant Cost (TPC) reduction, from 484.5

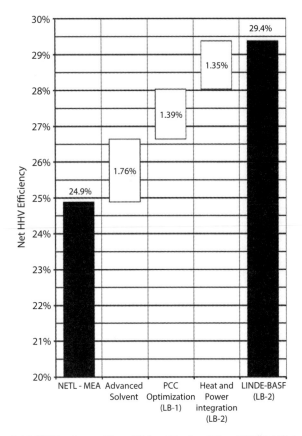

Figure 11.15 OASE™ blue Power Plant Higher Heating Value (HHV) Efficiency Improvements over MEA [2]

2007MM$ with MEA based Case 10 to 316.9 2007MM$ with LB-1 (Figure 11.16). While further enhancements in waste heat integration (LB-2) slightly increased TPC to 337.2 2007MM$, i.e. reduced TPC 30.4% below that of MEA based Case 10, they measurably increased net power plant efficiency and consequently led to further reduced cost of electricity as detailed below.

The levelized cost of electricity (LCOE) was calculated for MEA-based Case 10 and for OASE™ blue-based LB-1 and LB-2 cases over a period of 20 years.[2] To be consistent with the DOE/NETL-2007 study, LCOE values have been expressed as 2007$/MWhr. Althoughthe addition ofwaste heat integration to OASE™ blue process slightly increased the total plant cost (Figure 11.16), the LCOE decreased incrementally with each improvement from MEA to OASE™ blue with LB-2 (Figure 11.17).

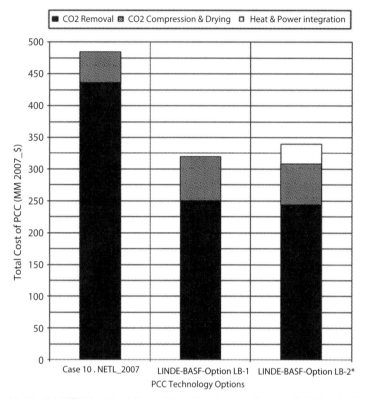

Figure 11.16: OASE™ blue Total Cost of Post Combustion Capture (PCC) and CO_2 Compression [2]

Compared to MEA-based Case 10, the OASE™ blue solvent-based BL-1, with related process enhancements and PCC optimization yield a combined $16.1/MWhr reduction in LCOE. Waste heat integration (LB-2) further reduced the LCOE for $2.3/MWhr to a final LCOE of $101.2/MWhr for the OASE™ blue, 15% lower than MEA. Furthermore, results show that the cost of CO_2 captured with OASE™ blue was estimated at $39.4/$MT_{CO2}$ and $39.0/$MT_{CO2}$ for options LB-1 and LB-2, respectively, are more than 20% less than $48.8 MEA (Figure 11.18), assuming all conditions as in DOE/NETL Report 2007/1281.[1]

11.3.13 OASE™ blue Emissions

Environmental impact of OASE™ blue was also considered in the techno-economic analysis.[2] Results showed that with OASE™ blue, emissions of CO_2, NOx, PM and Hg were reduced compared to MEA 11.1% and 15.2%

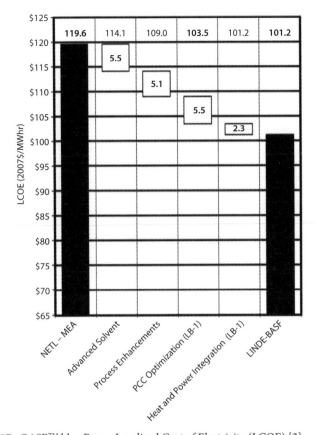

Figure 11.17: OASE™ blue Power Levelized Cost of Electricity (LCOE) [2]

for LB-1 and LB-2, respectively. In addition, compared against MEA, OASE™ blue fresh water makeup requirements were reduced 22.4% and 25.7% for LB-1 and LB-2, respectively.

11.4 Conclusions

Development of HiPACT™ and OASE™ blue has culminated in novel solvents and optimized processes, each tailored for their respective industries. Key solvent properties and process parameters, such as stability and losses, CO_2 absorption capacity and kinetics, materials compatibility, economics, and CO_2 stripping pressure were all optimal for high pressure AGRU operation.

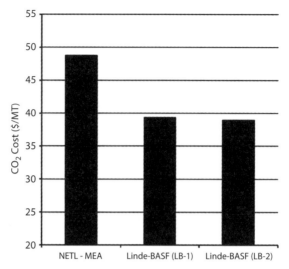

Figure 11.18: OASE™ blue Cost of CO_2 Capture [2]

Compared to existing technology, the technical advancements of HiPACT™ confer economic benefits to the plant owner. For example, the increased stripping pressure of CO_2 yields a lower compression ratio conferring reduced energy consumption costs for compression. Additionally, the advanced solvent formulation is highly stable against thermal degradation culminating in minimal solvent refill expenses. Process configuration and operating parameters for HiPACT™ yielded 20 – 27% lower solvent circulation rate and 10 – 20% lower specific reboiler energy. The net CO_2 recovery cost for HiPACT™ was 35% lower than a grassroots SOA natural gas treating plant and 42% lower for a revamp of an existing SOA natural gas treating plant. Although demonstrated at a 2.5 baraCO_2 stripping pressure due to existing equipment constraints, HiPACT™ has been developed for operation at 3 – 8 bara making it ideally suited for integrated CCUS projects in which natural gas treating is the source of CO_2.

OASE™ blue technology in combination with Linde PCC process enhancements significantly improves the techno-economics of a post-combustion CO_2 capture and compression. Benefits derive from solvent regeneration and CO_2 off-gas pressures as high as 3.4 bara, and condensation at 20°C. Consequently, OASE™ blue economic benefits are found in total plant costs reduced by 30–35% and the levelized cost of electricity is reduced 16%. In addition, the environmental impact is reduced with the

OASE™ blue process requiring 22% less makeup water. Through rigorous techno-economic analysis, BASF's OASE™ blue technology and Linde's PCC process enhancements are shown to be less than $40 per metric ton of CO_2 while utilizing the DOE/NETL cost estimation methodology detailed in reference [1]. As described previously, the target estimated by the US-DOE in which 67 billion barrels of oil would be economically recoverable at $85 per barrel with 20% ROI is estimated at $40/$MT_{CO_2}$.[3] Although OASE™ blue is still undergoing further optimization at the pilot plants located in Niederaussem, Germany and Wilsonville, Alabama, it is evident that this cost-effective and robust technology is far superior to MEA and is currently capable of meeting technical and commercial demands of PCC-CCUS projects.

11.5 Acknowledgements and Disclaimer

The experimental stage and pilot plant testing for HiPACT has been sponsored by the Japanese Ministry of Economy, Trade and Industry (METI). We wish to thank METI for their support to BASF/JGC joint development.

The Niederaussem pilot plant project mentioned in this report is supported by funding from the German Ministry of Economics and Technology (BMWi), whom we would like to thank for their commitment, under sponsorship codes 0327793A to F for RWE Power, BASF and Linde. The responsibility for the contents of this publication rests with the authors. With its crucial financial contribution within the scope of its CCORETEC initiative, the BMWi is pursuing the principles of safeguarding resources, assuring security of supply and supporting the competitiveness of German industry.

This document was prepared as a review of work sponsored by an agency of the United States Government. Neither the United States Government nor any agency thereof, nor any of their employees, makes any warranty, express or implied, or assumes any legal liability or responsibility for the accuracy, completeness, or usefulness of any information, apparatus, product, or process disclosed, or represents that its use would not infringe privately owned rights. Reference herein to any specific commercial product, process, or service by trade name, trademark, manufacturer, or otherwise does not necessarily constitute or imply its endorsement, recommendation, or favoring by the United States Government or any agency thereof. The views and opinions of authors expressed herein do not necessarily state or reflect those of the United States Government or any agency thereof.

References

1. Cost and Performance Baseline for Fossil Energy Plants – Volume 1: Bituminous Coal and Natural Gas to Electricity, DOE/NETL-2007/1281 Study, Final Report, Rev. 1. Cost and Performance Baseline for Fossil Energy Plants – Volume 1: Bituminous Coal and Natural Gas to Electricity, DOE/NETL-2007/1281 Study, Final Report, Rev. 1 (2007)
2. Jovanovic, S., Stoffregen, T., Clausen, I., Sieder, G.: Slipstream Pilot-Scale Demonstration of a Novel Amine-Based Post-Combustion Technology for Carbon Dioxide Capture From Coal-Fired Power Plant Flue Gas. (2012)
3. Kuuskraa, V. A., Van Leeuwen, T., Wallace, M.: Improving Domestic Energy Security and Lowering CO_2 Emissions with "Next Generation" CO_2-Enhanced Oil Recovery (CO_2-EOR). (2011)
4. Tanaka, K., Fujimura, Y., Katz, T., Spuhl, O.: HiPACT – Advanced CO_2 Capture Technology for Green Natural Gas Exploration. In : LRGCC 2010 Conference Proceedings and Fundamentals Manual, Norman, Oklahoma (2010)
5. Komi, T., Tanaka, K., Fujimura, Y., Contreras, E.: Demonstration Test Result of High Pressure Acid-gas Capture Technology (HiPACT). In : LRGCC 2012 Conference Proceedings and Fundamentals Manual, Norman, Oklahoma (2012)
6. Moser, P., Schmidt, S., Sieder, G., Garcia, H., Stoffregen, T., Stamatov, V.: The post-combustion capture pilot plant Niederaussem – Results of the first half of the testing programme. Energy Procedia 4, 1310–1316 (2011)
7. Stoffregen, T., Rosler, F., Moser, P., Schmidt, S., Sieder, G., Garcia, H.: Scrubbing Up. World Coal Power Technology 1(1), 41–46 (2011)
8. Moser, P., Schmidt, S., Uerlings, R., Sieder, G., Titz, J.-T., Hahn, A., Stoffregen, T.: Material Testing for Future Commercial Post-Combustion Capture Plants - Results of the Testing Programme Conducted at the Niederaussem Pilot Plant. Energy Procedia 4, 1317–1322 (2011)
9. Moser, P., Schmidt, S., Sieder, G., Garcia, H., Ciattaglia, I., Mihailowitsch, D., Stoffregen, T.: Optimized Post-Combustion Capture Technology for Power Plants. In : Clean Coal Technologies, Dresden (2009)
10. Mangalapally, H., Notz, R., Asprion, N., Sieder, G., Garcia, H., Hasse, H.: Pilot plant study of four new solvents for post combustion carbon dioxide capture by reactive absorption and comparison to MEA. International Journal of Greenhouse Gas Control 8, 205–216 (2012)
11. Katz, T., Spuhl, O., Contreras, E., Komi, T., Tanaka, K., Fujimura, Y.: HiPACT – An Improved Acid Gas Capture Technology.

12

Seven Deadly Sins of Filtration and Separation Systems in Gas Processing Operations

David Engel and Michael H. Sheilan

Sulphur Experts International Inc., Filtration Experts & Amine Experts Divisions, Kemah, TX, USA

Abstract

Separation systems play a fundamental role in gas processing, both for reliability and as an enabling device for enhanced throughput and process stability. In many cases, these systems are the only and/or last line of defense to protect the plant from unwanted and detrimental contaminants. As plants are required to process gas with more contaminants (such as certain Shale Gas feeds), these devices are increasingly being required to perform under progressively difficult conditions. As we evaluate, troubleshoot, and improve many of these systems at a number of worldwide plant operations, we have successfully identified many different failure modes in these systems, with various degrees of severity. These can range from incorrect design, to poor choice of physical location, to errors in instrumentation, and most critically, erroneous or non-performing internals. A large set of these cases were combined and compiled into a list of the seven most common errors in filtration and separation systems applied only to gas processing operations. The effects of not having correct contamination control in gas processing operations quite often cause profound impacts in the vast majority of gas processing plants and are manifested as solvent contamination and degradation, foaming, fouling, low reliably, low efficiency, increased maintenance, and undesirable environmental emissions. All this leads directly to high operational costs and/or the inability to meet sweetening specifications. Each deadly sin will be analyzed and presented with actual plant cases. Each case is also analyzed for its effects, consequences, or ramifications, and in some cases how the problem was successfully corrected.

12.1 Gas Processing and Contamination Control

For almost any gas processing plant, contamination is a reality that can cause a numerous associated problems, and if not solved properly, will become a chronic problem. Some of the problems caused by contamination include: foaming, fouling, corrosion, solvent degradation, deposition, undesired side-reactions and downstream impacts (SRU, Flare and treated products). These problems invariably cause capacity reductions, efficiency decay, inability to meet specifications, equipment failures, loss of solvent, high maintenance and operational costs, and ultimately leading to undesired environmental emissions and loss of profitability. Most, if not all of these detrimental effects can be mitigated by using proper contamination control measures. These situations can be addressed by the use of chemical additives or mechanical separators. In this paper, however, only mechanical separator solutions will be discussed.

Mechanical separation in gas processing, as related to contamination control, is undertaken generally in two of the following areas of the process. These are:

a) **Feed and Effluent Separation.** This is referred to as contamination removal in feed streams and in effluents streams, such as sour gas or wet gas (feed) and treated/sweet gas or dry gas (effluent).

b) **Unit Internal Separation.** This relates to contamination removal within the unit, such as amine solvents, glycol solvents and other gas processing solvents.

Both of the preceding categories can demand different separation approaches, and in many cases different mechanical separation systems. In essence, Feed and Effluent Separations are typically associated with gas streams with the exception of Liquefied Petroleum Gas and Natural Gas Liquids. While Unit Internal Separation is generally associated with liquid streams in the unit. In both cases, the predominant mechanisms for contaminant separation are as follows:

- Filtration: For suspended solids in a liquid or gas stream
- Coalescence: For liquids in a gas stream or emulsifiers in a liquid stream
- Adsorption: For dissolved, predominately organic components in a gas or liquid stream

12.1.1 Feed and Effluent Separation

Proper feed gas contamination control, prior to any gas processing, is essential for plant stability, performance and low operational costs. This is often achieved when there is thorough understanding of virtually all inlet feed contaminants in the gas stream and the information is taken into consideration when determining what separation process and system is used. Inlet contamination in gas processing can vary drastically and depends on a number of factors, including the following:

- Geographical location and geological formation
- Gas exploration and production operations and equipment types
- Chemical additive use (type and dosage)
- Contaminant types and concentrations

Aside from heavy hydrocarbons and acid gases (H_2S and CO_2), feed streams can contain other sulphur bearing species (COS, CS_2, RSH, etc.) and many other contaminants. The most common contaminants in gas feeds are: asphaltenes, waxes, water, oxygen, mercury, sulphur (elemental), ammonia (NH_3), methanol, salts, compressor lubrication oils, chemical additives, iron sulphides/oxides, silica and sand. This does not include any heat stable salt precursors typically associated with refineries.

Feed gas and effluent gas (or treated gas) contamination control is often conducted through the use of knock-out drums equipped with a demister section, using a mesh pad or a vane pack. Some plants use horizontal filter separators with a vane, or cyclonic elements or stages. None of these systems are entirely adequate for an effective inlet contaminant removal from sour gas feeds. These systems are typically designed for either bulk liquids removal or large aerosol droplet size removal as opposed to the fine sub-micron size liquids found in many gas streams. In addition, none of these devices is really designed for solids separation (usually done effectively by a wet scrubber or a particle filter). With the exception of cyclonic systems and some filter separators that could remove certain solid particles and some liquid aerosols, most contaminants often enter the gas processing units untouched. In essence, demisters are just as effective as slug catchers. Unfortunately, the most difficult and challenging contaminants in any gas stream are small aerosols. These aerosols are finely divided liquid particles with diameters ranging from a few hundred microns to less than 0.1 microns. Separation of these liquid contaminants is done using Microfiber Sub-Micron Gas-Liquid Coalescers (MSC). These MSCs should be capable of removing, on average>99.97% of all aerosols with diameters between 0.1 to 1.0 microns (and also larger). In practice, this represents the majority of the liquid aerosol contamination in a gas stream. These devices should be protected with a suitable particle filter separator (equipped with the correct separation media) in order to extend the on-line life of the coalescer and to minimize operational costs since the replacement filter elements for particle separation are much less expensive than coalescing elements. Correctly designed sub-micron coalescer vessels have two stages: the bottom section designed to remove bulk liquids and an upper 'high efficiency' stage for aerosol removal. In certain situations, the bottom section can be fitted with a mesh pad or vane pack, or may be designed in such way as to have cyclonic action. The gas then leaves the bottom chamber, flowing into the second stage immediately above via the interior of the coalescing elements. The gas is then directed across the microfiber coalescing media from the internal core to the exterior surface. The fine aerosols are intercepted, coalesced and finally drained from the elements by gravity. Like the lower stage, the upper stage has a liquid removal system comprised of a level control and drain valves.

12.1.2 Unit Internal Separation

This is related to contamination removal within the unit, recirculating solution such as amine solvents, glycol solvents and other gas processing solvents. These are generally particle filters for removing suspended matter, coalescers to promote liquids contamination separation (typically hydrocarbons, both free and emulsified) and adsorption beds such as activated carbon to remove dissolved components. Other adsorption systems are molecular sieve beds, alumina/silica beds and salt beds. All three common separation processes are in place in gas processing operations. These include: filtration, coalescence and adsorption. Other processes are also available. However, these are less common and their use has limited applications. One such technology is heat stable salts removal in amine units. This can be done with ion-exchange, dialysis or vacuum distillation. However, their use is limited to amine units containing these contaminants and where feed control is deficient or not feasible (typically refinery amines in connection with Fluidized Catalytic Cracking, Visbreaker and Coker units).

12.1.3 Seven Sins of Separation Devices in Gas Processing Facilities

All of the previously discussed concepts are fairly generic. Certain gas processing facilities may operate well with high levels of contamination, while others will be disrupted at minimal contamination ingress. Hence, each unit is almost a separate case in terms of contamination control and what measures are required for efficient contamination removal. Throughout our involvement in evaluating gas processing units and mitigating deficient or negative situations, it was apparent that a large number of these facilities do not have the appropriate contamination control devices.

They are actually either deficient or non-existent. During this process of troubleshooting gas processing facilities, a set of general failure categories covers the majority of the root-causes. This set of failure modes were later branded as the "seven deadly sins of separation devices in gas processing facilities" and is the center point of this paper.

12.2 The Seven Deadly Sins of Filtration and Separation Systems in Gas Processing Operations

12.2.1 Sin 1. Unsuitable Technology for the Application

This is related to the use of devices that are not capable of properly functioning for the application they operating. This is generally found in cases where poor understanding of the application leads to incorrect selection of the equipment. In some cases perceived capital cost savings also lead to incorrect equipment selection. Examples of this can be found in the use of automatic filters with metal-based filter media in amine units. In spite of being a good idea, these devices do not preform adequately in amine units mainly because of the contamination types found in the streams. The highly fouling amine streams (rich and lean) with strongly adherent and sometimes gel-like solids cause back-washable or any other self-cleaning mechanical systems to perform well below their expected efficiency. The contaminants just cannot be removed to any significant extent. The use of backwash liquid volumes can be large and the cleaning frequency can be high. The efficiency of the metal filter element seldom returns to values near its original state. In addition, these solids, in combination with hydrocarbon contamination, are perhaps one of the most challenging mixtures to separate.

A secondary effect is that now there is a new stream with high solids content that has to be treated somewhere. If this is not taken into consideration prior to equipment selection, often this becomes a problem rather than a solution. Similarly, this also applies to pre-coat systems with the use of diatomaceous earth as a filter aid. Although these devices are important and have their applications in the process industry, they can be very difficult to maintain, and waste generation and high amine consumption can be a problem. For amines units and other gas processing systems, the solid waste residues coupled with solvent losses is highly problematic. If H_2S is present, the waste is also toxic and has to be disposed as hazardous material. Some of these systems can be expensive and large in size with considerable capital cost and maintenance expenses. Finally, the use of cyclonic devices can, to some extent, assist in solids separation provided they have the right particle size and density. These are more for bulk separation. Application of cyclonic devices is quite specific and the operational windows are rather small. Hence, process variations are not supported well by these devices without changes in their configuration. However, the lack of moving parts and high temperature operation capabilities make them attractive in many potential applications. The industry is full of such technology misapplication examples. However, disposable filtration still is one of the best alternatives for filtration of suspended solids in gas processing operations.

To illustrate this, Figure 12.1 shows the drawing of an Amine Unit Flash Tank (UAE). This vessel was designed for only five minutes residence because it was equipped with internal mesh-pads for hydrocarbon coalescing (note that these pads were not removable). The concept of internal coalescing pads leads to the notion that there was no need for > 30 minutes residence time in the flash tank. Hence, the tank was built significantly

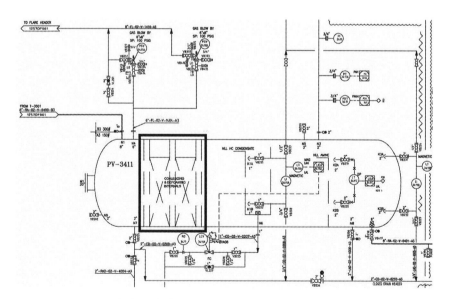

Figure 12.1: Amine Flash Tank drawing showing the internal metal-based coalescing pads (red box).

smaller in size than normally expected. The basic concept might have been somewhat correct if coalescing was to be done on large liquid droplets in a non-fouling stream. However, the lack of understanding with respect amine units caused these internal pads to foul rapidly, leading the plant to suspend operations (see Figure 12.2). The pads have since been removed, but the short residence time has become a problem since it can only remove flashable or free hydrocarbon. No emulsified hydrocarbons (to any extent) can be removed.

12.2.2 Sin 2. Incorrect Compatibility (thermal, chemical, mechanical)

There are several aspects to understand in materials compatibility. In general terms, the effects on media and associated materials such as metal parts, screens, epoxy adhesives, end-caps, are complex. These can include: chemical degradation of the media (media erosion and distortion), media disassembly (media fiber release), media solubility (loss in media material), media modification (incorporating contaminants in other fiber) and media leaching (residues being released form the media material). Thermal compatibility is related to melting point (or softening point) of the material. High temperatures generally lead to deformation of the material matrix and likely enhancement of chemical degradation. Mechanical compatibility is directly related to the tensile strength of the material. In

Figure 12.2: Plugged Coalescing Pad removed from the inside if an undersized Flash Drum.

other words, how strong is the material at the actual process conditions? Chemical and thermal incompatibilities can also lead to rapid mechanical degradation and vice-versa. An example of such a situation is the use of polyester filter media in any process that contains amines (either as a solvent or as contaminant). Polyester undergoes chemical reaction with amine solutions, essentially causing the fiber to fail and eventually rupture.

To illustrate this, Figure 12.3 shows foam formation and stabilization in an amine solvent caused by the materials in a filter element (epoxy seam). An amine unit (USA) experienced foaming in their absorber. After extensive investigation, the filter element was evaluated for chemical compatibly. A properly conducted soak test of all materials in the filter element indicated that the adhesive used to connect the media was causing foam stabilization. A change to compatible filter elements eliminated the foaming.

12.2.3 Sin 3. Deficient Vessel Design

This is associated with the vessel itself. As seen in real cases of poor contamination control, perhaps the lasting cause is a defective vessel design. This can occur in many forms such as the following: undersized vessels, unbalanced array of internals causing preferential flow, incorrect internal flow geometries, incorrect placement of inlet/outlet, erroneous vent or drain location, incorrect support thicknesses causing vessel internal

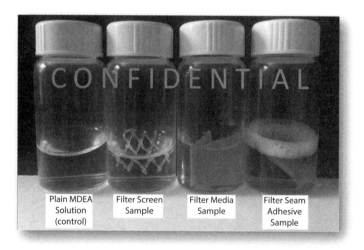

Figure 12.3: Soak/Foam test of all materials in a filter element using lean MDEA. Soak test: 77°F/48 hrs. Foam test: All vials agitated together for 1 minute (manually). Image taken after resting for 10 minutes.

failure, presence of inlet baffles that cause liquids shattering, lack of internal baffles causing lateral impacts into the internal elements (causing media rupture) and others. In some cases, certain vessels can be modified, upgraded or improved. However, for vessels that are undersized, there is no real practical solution. Filters will have exponentially high operational costs. Coalescer vessels will have considerable liquids carry-over. and other separation systems will simply lack the separation efficiency with excessively high operational costs.

To illustrate this, Figure 12.4 shows a vessel drawing for a particle filter in a rich amine stream (UAE). As can be seen, there is a vapour vent (N4) and a liquid drain (N5)on the left (dirty) side of the tubesheet. There is neither a vapour vent nor a liquid drain on the clean side of the tubesheet. There should at least be a vent in the clean side of the filter. The lack of this vent produces a pocket of gas (usually high in H_2S and CO_2 concentrations). This pocket cannot be eliminated as there is no place for the gas to escape. An internal plate separating the clean and dirty sides prevents the gas from venting. There is then the formation of agas-liquid interface leading to a high potential materials failure in the near future.

12.2.4 Sin 4. Inappropriate Sealing Surfaces

Often overlooked are the sealing surfaces. This small detail results in a fundamental difference in the performance of the elements. The sealing surfaces are present at the interface where the internal element, responsible for actual contamination separation, meets the vessel. These parts

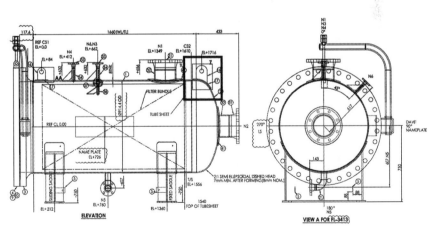

Figure 12.4: Rich Amine particle filter design showing the location where a vent is missing.

play the critical role of ensuring that the fluid is properly routed into the separation media, effectively producing a seal between the contaminant-laden stream and the clean stream. These are usually gaskets or O-Rings produced from so-called "elastomers" (polymers with flexible elasticity). Each elastomer has its own compatibility just as indicated previously. Some sealing surfaces actually consider the contact between the media itself with metal parts at the vessel (called seat cups). These are poor in performance and prone to by-pass. Flat gaskets are also deficient since they have to be attached to the element itself using an adhesive. These, often times, fail to offer proper seal and cause contaminant by-pass.

To illustrate this, Figure 12.5 shows filter elements with flat gaskets that have been degraded (UAE). The edge of the material is eroded and, in parts, sections of the elastomer material are missing. It can be also noted that on the elements behind, the gasket is actually missing. Modification of the vessel to use elements with improved O-Ring seals eliminated the by-pass.

12.2.5 Sin 5. Wrong Internals & Media

This general area relates to incorrect filter element design and encompasses incorrect media selection. Elements with poor design and less than optimum media surface area will have a reduced contamination capture capacity and low on-line life, thus requiring more frequent maintenance. This

Figure 12.5: Coalescer Elements with sealing surface materials that have been degraded along with failed attachment to the metal surface.

also generates higher waste volumes and also results in higher operational costs. However, excess media surface area in a filter element will also cause reduced contamination capture capacity as a phenomenon called media "blinding" takes place. This occurs when parts of the media experience ineffective exposure to the fluid stream. Media efficiency selection is also an area where a number of failures occur. This is because of a poor understanding of the trade-off sin any process separation, in terms of separation cost versus the downstream effects of contamination penetration. It is always critical to understand why a given contaminant is required to be removed and what are the operational expectations of the filter at the location it will be installed.

To illustrate this, Figure 12.6 shows a filter that was designed with the expectation to operate as a depth filter (Europe). The expectation was for the filer to accumulate contaminants not only at the media surface, but also at the inner media layers. It can be observed that only the outer media layer is actually being used leaving the inner sections contaminant free. New media array, with different materials and efficiencies, produced an element capable of accumulating contaminants in all layers.

12.2.6 Sin 6. Lack of or Incorrect Maintenance Procedures

Essentially, all separations systems, as is so with virtually all other equipment, have the necessity of human intervention for maintenance. It is

Figure 12.6: Filter Element with incorrect media. The internal layers of the media are free of contaminants. All contaminants are being accumulated at the surface of the filter element.

surprising how many filters and other separation systems go without proper maintenance. Many systems operate for years with no differential pressure increase, just to open the vessel and discover no internals. Several other vessels are found using non-original low cost internals to save money. These typically are the culprit of many processing upsets as they lack the robustness to handle variable contaminant concentrations. Proper maintenance of any separation system has to start with a thorough internal inspection of the vessel to detect any possible anomaly and damage. The sealing surfaces must to be inspected so they are clear of solid deposits, minimizing the possibility of deposit by-pass. Periodic monitoring of internals replacement procedures is critical to ensure the correct accommodation of the internals into the vessel, leading to correct operation.

To illustrate this, Figure 12.7 shows the end-cap of a coalescer element with missing O-Rings used for sealing (USA). Maintenance manually removed the O-Ring, indicating that this helped the element to fit. Also, it can be noted that there is a white residue at the base of the end-cap. This is hardened material originating from the grease used to facilitate element accommodation into the vessel. The grease material reacted with the hydrocarbons in the gas stream forming a solid material, cementing the element into the vessel. Element removal caused internal support damage. The use of inert mineral oil led to a much smoother element installation with no incompatibility issue.

Figure 12.7: Filter Element with one of its O-Rings removed. The oil used to lubricate the O-Rings reacted with the process stream forming a hard cement-like material.

12.2.7 Sin 7. Instrumentation Deficiencies

Another area many times disregarded is related to instrumentation. Some systems never even have instrumentation, while others have incorrect instrumentation. For example, level controls have to properly consider the density of the interface fluid that it is supposed to be measured (liquid-liquid or gas-liquid). The location of instrumentation is also important since some units in cold locations tend to have no protection. They often freeze and provide incorrect readings. Proper monitoring of differential pressures is important since it is the only way some vessels communicate. Filters have differential pressure devices that must be verified for accuracy. Coalescer vessels are much more complex. They use differential pressure gauges, level controls and dump vales. All of these components are required to operate properly for effective liquids removal. If any become damaged, are missing or defective, liquid carry-over will likely occur.

To illustrate this, Figure 12.8 shows a gas conditioning skid (pre-filter and liquids coalescer) with missing instrumentation (Latin America).Level controls are not connected to any automatic dump valve to remove liquids from the vessel interior. The level control is not wired to the Control Room. Liquids removal is done periodically and manually. Often, this will lead to the entire coalescer vessel being filled with liquids, causing slugs of liquids

Figure 12.8: A gas coalescing skid showing the large coalescer vessel equipped with improper liquids removal capabilities (red box). No dump valve available and no connection from the level control.

to enter the downstream unit (usually an amine absorber) generating uncontrollable upsets, foaming because of high hydrocarbon ingression.

In another location (Canada), excessive pressure differential across the coalescer elements caused the elements to rupture (Figure 12.9). Upon inspection of the vessel internals, damage to the support arm of one of the risers was also detected (yellow arrow). The high pressure differential was a result of excessive plugging of the outer coalescing layer by solids and salts, leading to excessive pressure drop in a matter of only days. Operations did not catch the high pressure differential because the remote pressure transmitter was frozen. The transmitter is now insulated, allowing for constant readings in the Control Room.

12.3 Concluding Remarks

Invariably, the most important observation made during our many years in the field is that the vast majority of process upsets, lack of profitability and impacts on efficiency are related to contamination control. Additionally, perhaps the most important factor in process control is proper contamination control. Most plants that do not take this seriously and rigorously struggle with high operational costs and low systems reliably, followed by a number of other undesired technical, economic and environmental aspects. In our experience, the lists of the seven areas discussed in this paper comprise the bulk of the reasons as to why these systems do not operate as expected. By focusing on these areas, cost of the problems will subside and proper contamination control leading to effective process control will be achieved. Also important to note is the fact that often any

Figure 12.9: Ruptured coalescing elements and damaged riser support arm caused by excessive pressure drop across the vessel: undetected by operations because of a frozen transmitter.

capital savings in low cost separation and filtration will ultimately lead to exponentially high processing costs, low reliability, frequent unit upsets and loss in profitability. Finally, each unit and process has their own equilibrium point where the cost of contamination control is acceptable with tolerable residual contamination level breakthrough. This is where users, engineering firms and suppliers, have the responsibility to be involved in finding such balance with the objective of supplying the right separation and filtration solution for each individual plant.

13
Development of Management Information System of Global Acid Gas Injection Projects

Qi Li, Guizhen Liu, and Xuehao Liu

*State Key Laboratory of Geomechanics and Geotechnical Engineering (SKLGME),
Institute of Rock and Soil Mechanics (IRSM), Chinese Academy of Sciences, Wuhan China*

Abstract

As the operation of first acid gas injection (AGI) was started in 1989 on the outskirts of Edmonton, Alberta, and Canada, there are nearly 80 AGI projects spread all over the world currently. An intelligent system was developed to manage the large number of information of global acid gas injection projects. Architecture of the system was designed based on prevalent standards and guidelines of knowledge and information management. The major contents of AGI database include basic information, reservoir geology, reservoir properties and fluids, licensed operations, engineering characteristics, and leakage events. The metadata were managed via a background ArcGIS engine. Finally, an interactive program was developed via Microsoft Visual Studio in Visual Basic environment to enhance the system with a strong flexibility and reliability for data analysis and information visualization.

13.1 Background

Acid gas is a mixture of H_2S and CO_2 as a waste stream from natural gas or oil production, which has to be removed before the final product is sent to markets. Because surface desulphurization is costly and the price of produced sulphur is fluctuant, more operators are turning to acid gas disposal by injection into deep geological formations[1, 2]. In addition to providing a cost-effective alternative to sulphur recovery, acid gas injection (AGI) reduces atmospheric emissions of noxious substance and alleviates the public concern resulting from sour gas production and flaring[3].

The first acid gas injection operation was started in 1989 on the outskirts of Edmonton, Alberta, Canada. To date, there are nearly eighty AGI projects all over the world[4, 5]. The acid gas was disposed into different reservoirs at the depth from hundred meters to thousand meters. Operations of existing AGI projects provide a lot of experiences and some confidence for the design of a new project. So, an information management system (MIS) was developed to mining the global acid gas injection projects information. The management information system may provide some insights into site selection and engineering operation for a new AGI project[6], thereby it is very useful to provide a reference to promote implementation of AGI in China.

13.2 Architecture of AGI-MIS

An AGI project can be characterized by items of basic information, reservoir, operation and leakage event, so the architecture of the AGI-MIS was designed according to the following characteristics:

(1) Basic information: such as the start-up and end time, the location of the project, acid gas source, fraction of CO_2 and H_2S, the injection volume, etc.
(2) Reservoir geology: including lithology, thickness and other information of the injection formation, cap rock and underlying formation.
(3) Reservoir properties and fluids: including the porosity, permeability, formation temperature, geothermal gradient, formation pressure, formation water salinity, oil gravity, gas specific gravity, etc.
(4) Licensed operations: including the maximum approved, fracturing, etc.
(5) Engineering characteristics: including the information of the wells, casing, tubing, pipelines, compressors, etc.
(6) Leakage event: leakage event and corresponding answer-up.

Characteristic (1) listed above is the basic information of an AGI project, characteristic (2) and (3) are the reservoir properties to describe a site characteristic, and much more knowledge could be obtained according to data mining of details. It can be considered as a reference for sites selection and so on. And the characteristic (4) and (5) are the operations of AGI projects, they provide the technology and equipment information. The last characteristic (6) is the experience of the leakage event, failures of wells used in AGI are included. The MIS frame with the six characteristic of AGI projects were listed in Table 13.1. Detail information were collected and organized to a standard GIS tool.

Table 13.1: MIS Frame of global AGI projects[4].

	Characteristic	Category	Attribute
AGI-MIS	Basic Information	General briefing	Number/Name
			Start-end time
			Location: Latitude/Longitude/Country/County
			Investment/Company/Sponsor/Principal Scientist
		Acid gas	Source of acid gas
			CO_2/H_2S fraction (%)
			Total injection Volume ($10^3 m^3$) (ton)
	Reservoir Geology	Injection formation	Reservoir type
			Injection depth: Average/Min/Max
			Lithology/Group name
			Thick/Net pay (m)
			Reservoir volume ($10^3 m^3$)
		Cap rock	Lithology/Group/Thick (m)
		Underlying formation	Lithology/Thick (m)
	Reservoir properties and fluids	Permeability	Porosity (%)
			Permeability (mD)
		Temperature	Temperature (°C)
			Geothermal gradient (°C/m)
		Pressure	Original formation pressure (kPa)
			Formation startup pressure (kPa)
		Fluids characteristic	Salinity (mg/l)
			Oil gravity (oAPI)
			Gas specific gravity
	Licensed operations	Maximum approved	BHIP/WHIP (kPa)
			Maximal approved ($10^3 m^3$/d)
			Total approved ($10^3 m^3$)
			Actual injection ($10^3 m^3$/d)

Table 13.1: (Continued)

AGI-MIS	Characteristic	Category	Attribute
	Engineering characteristics	Fracturing	Hydraulic fracturing
			Fracturing gradient
			Fracturing radius (m)
			Life time (years)
			EPZ (km)
		Well group	Injection wells number/distribution
			Pumping well
			Pressure relief well
		Casing	Depth of production casing (m)
			Size of production casing (mm)
			Density of production casing (kg/m)
		Tubing	Size of Production tubing (mm)
			Density of production tubing (kg/m)
			Packer Depth
		Pipelines	Pipe length (m)
			Pipe grade
			Diameter of pipe (mm)
			Pipe wall thickness (mm)
		Compressors	Max operating pressure (kPa)
			Max allowable stress level (%)
			HP rating
			Design rate (103m3/d)
			Discharge pressure (kPa)
			Dehydration method
	Leakage event	Leakage event	
		Answer-up for leakage	

13.3 Data management

Basic information of AGI projects includes the start-up and end time, the sponsor information, the location of the project, acid gas source, fraction of CO_2 and H_2S, the injection volume, etc. Exact location of an AGI project was described not only the country and county, but also the latitude and longitude, so that project could be easily depicted in the world map via a GIS tool. As shown in the upper-right insert of Figure 13.1, most of the AGI projects in the world are located in British Columbia and Alberta,

Development of management information system 247

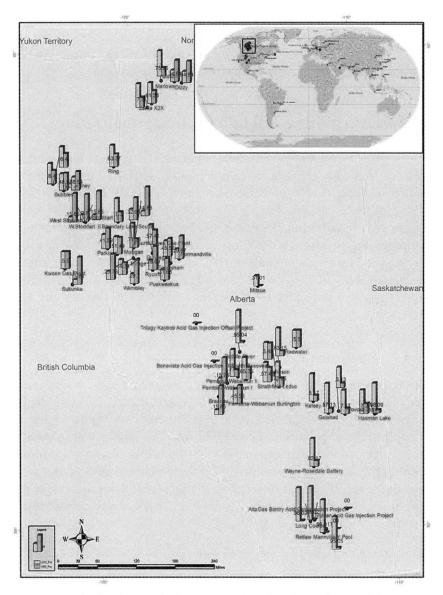

Figure 13.1: The distribution of AGI projects in Canada with visualization of their CO_2 and H_2S fraction.

Canada[7, 8], and other projects are distributed in different countries, such as USA[9], Qatar[10], and Poland[11, 12]. After that, all attributes will list in the attributes table in the ArcMap, and attributes of global projects can be labeled in the map and visualized in different symbols.

Figure 13.1 also shows the fractions of CO_2 and H_2S of the injection acid gas of projects in Alberta and British Columbia in Canada. The minimum of CO_2 fraction is 0.14, and the maximum is 0.98, while the fraction of H_2S of the injection projects is between 0.02 and 0.85 in Canada.

13.4 Data mining and information visualization

All the related data and information of global AGI projects are collected with yearly update. The data are categorized and organized in detail information of attributes in one database, named IRSM AGI DB. After data mining of related AGI projects, a GIS tool is used to do data analysis and information visualization. Because of the large numbers of AGI projects in Canada, more attentions are put forward on them, and elaborate analyses are conducted with many kinds of statistical methodologies according to the database[4, 7, 13–18]. In the following sections, some major information is visualized and addressed according to the data mining of some AGI projects in Canada.

13.4.1 Injection formation

Injection formation is one of the categories of reservoir geology characteristic. This category includes attributes of reservoir type, injection depth, group name, thickness, net pay, and reservoir volume. Figure 13.2 is an exmaple to show the injection depth and reservoir lithology of some AGI projects located in Alberta, Canada. 46 AGI projects data were visualized in the graph. The injection (net pay) depth of each AGI project, the average injection depth of all projects, and the lithology of disposal reservoirs were investigated. As shown in Figure 13.2, acid gas was disposed into

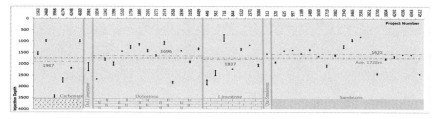

Figure 13.2: Reservoir lithology and injection depth of some AGI projects in Alberta, Canada. (Black dots or lines: injection depth; Red dash-dot-dot-line: average injection depth according to each type of reservoirs; Green dash-dot line: average injection depth according to all the reservoirs.)

carbonate, dolostone, limestone, and sandstone formations mostly, and little was storaged in Dol. Limestone and Qtz-Sandstone formations. The average injection depth of all the AGI projects in Alberta, Canada is about 1728 m, and the average injection depth of carbonate, dolostone, limestone and sandstone formations is about 1967 m, 1696 m, 1837 m, and 1622 m respectively.

13.4.2 Pipeline

Pipeline is the main option to transfer acid gas from the source to the disposal destination in the AGI and CCS projects, and the parameters of pipes such as diameter and wall thickness are key engineering characteristics. From the data mining, four types of diameters have been chose for AGI projects in Alberta, Canada (Figure. 13.3). Pipes with diameters of 60.3 mm and 88.9 mm were used extensively, while the diameter of pipes of Pembina-Wabamun I project is 48.3 mm, and the one of Bistcho and Brazeau projects is 114.3 mm. As shown in Figure. 13.3, the wall thickness of pipes are different accordingly for the 26 statistical projects. The minimum wall thickness of the pipe is 3.2 mm, and the maximum is 11.13 mm. Most of them were varied from 3 to 6 mm, except for three projects, i.e., Boundary Lake South, Brazeau and Strathfield Leduc projects. Thicker pipes were uesd in those three projects with wall thickness of 8.74, 8.56, and 11.13 mm respectively. Finaly, there is no significant correlation between a diameter and a wall thickness of pipes.

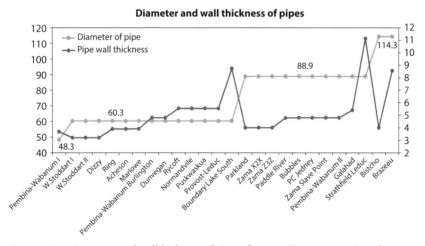

Figure 13.3: Diameter and wallthickness of pipesof someAGI projects in Canada.

250   GAS INJECTION FOR DISPOSAL AND ENHANCED RECOVERY

13.4.3   Injection rate

Figure 13.4 shows the approved maximal injection rate of the reservoirs and design rate of the compressors. It is an synthesis graph of licensed operations and engineering characteristics. As shown in the figure, injection rates in different AGI projects show significant correlations with design rates of the compressors without doubt. In the 21 statistical projects, the approved maximal injection rate of the most reservoirs is less than 1.0 million cubic meter per day of acid gas. The reservoirs of the Galahad project has the best injectivity with the approved maximal injection rate of 35 million cubic meter per day.

13.4.4   Leakage events

On April 26, 2011, a hydrogen sulphide leak was discovered at an acid gas injection well northwest of Tumbler Ridge, BC, Canada. The leak was detected at 3:45 p.m. at a Talisman Energy site at kilometre 25 of the Sukunka Forest Service Road. As the site is connected to a Spectra Energy pipeline, Spectra initiated an emergency shutdown and cut off flow to the injector. There was no gas in the pipeline at the time. Air quality was monitored along the Sukunka road overnight. Roadblocks were set up, and the road stayed open, since no gas was detected. The Oil and Gas Commission was notified at 4:10 p.m., but the National Energy Board has jurisdiction to review the incident. It's not yet clear how much hydrogen sulphide leaked from the site, or what concentration of hydrogen sulphide was in the leak[19]. No more information of the incident can be got till now.

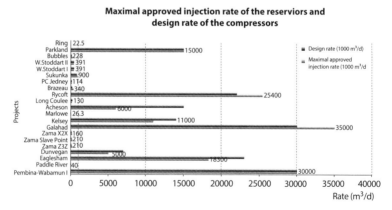

Figure 13.4: Some known approved maximal injection rates of the reservoirs and design rate of compressors ofsome AGI projects in Canada

Currently in Alberta, Canada there is no leakage incidents of AGI projects, however, some operational problems (Figure 13.5) were encountered by some AGI operations such as acid gas breakthrough at producing wells, pressure buildup beyond approved limits, and well tubing damage caused by water freezing as a result of sub-zero temperature of the injected gas[20, 21]. Except for operational problems, well failure is another key problems of AGI projects in Canada. Well failures include five modes: surface casing vent flow, casing failure, tubing failure, packer failure and zonal isolation failure[22].

13.5 Interactive program

After the database were combined with the GIS tool, an interactive executable program was developed by Microsoft Visual Studio in Visual Basic environment. In the programming development, LicenseControl MapControl, ToolbarControl, and ToCCControl of ArcGIS Windows Forms were added, and graph functions were tailored. Users without a GIS background can get some information of AGI projects simply. Customer demand will be further investigated to enhance the program in the next step (Figure 13.6).

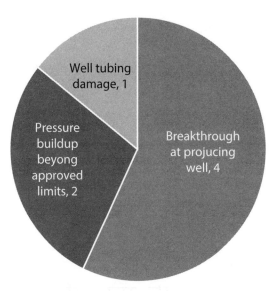

Figure 13.5: Some operational problems encountered by some acid gas injection operations in Canada (Until 2009).

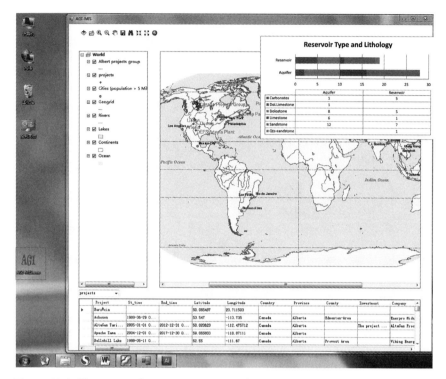

Figure 13.6: Illustration of interactive program of AGI-MIS

13.6 Conclusions

The management information system (MIS) of global acid gas injection projects contains a large number of data including basic information, reservoir, operation and leakage event, and etc. An interactive executable program was developed for the management information system of global AGI projects, so a user without a GIS background can get information of interested projects simply. Integrity of data mining and information visualization should be added in the MIS, and it may be provide a strong reference for a site selection and an engineering operation design of a new AGI project. Thereby it promotes the early implementation of the AGI project in China.

13.7 Acknowledgements

Q.L. thanks the hundred talent program of Chinese Academy of Sciences and NSFC (Grant No. 41274111). G.Z.L. thanks National Key

Technology R&D program (Grant No. 2012BAC24B05). Guiju Li (Wuhan Documentation and Information Center of Chinese Academy of Sciences) is acknowledged for data retrieval and initial arrangement. We also thank J.J. Carroll for very good suggestions and comments to raise great improvements of this manuscript.

References

1. X. Liu; Q. Li; L. Du; X. Li. Economic comparison of both acid-gas reinjection and sulfur recovery in high-sour gasfields. *Natural Gas Technology and Economy* 6, (4), 55–59, (2012).(in Chinese)
2. Q. Li; X. Liu; L. Du; B. Bai; M. Jing. Economics of acid gas reinjection with comparison to sulfur recovery in China. *Energy Procedia*, 10.1016/j.egypro.2013.06.132, (2013).
3. J.J. Carroll. Acid gas injection: Past, present, and future. In *Acid gas injection and related technologies*, Y. Wu; J.J. Carroll, Eds.; John Wiley & Sons, Inc.: Hoboken, NJ, USA, pp. xxi-xxxiv, (2011).
4. S. Bachu; W. Gunter *Acid gas injection: A study of existing operations (phase 1: Interim report)*; PH4/15, IEA Greenhouse Gas R&D Programme; Alberta, Canada, p 64, (2003).
5. S. Bachu; W.D. Gunter In *Overview of acid-gas injection operations in western Canada*, The 7th international conference on greenhouse gas control technologies, 2005; Elsevier: pp 443–448, (2005).
6. Q. Li; X. Li; L. Du; G. Liu; X. Liu; N. Wei. Potential sites and early opportunities of acid gas re-injection in China. In *Sour gas and related technologies*, Y. Wu; J.J. Carroll; W. Zhu, Eds.; Wiley Scrivener: New York, pp. 131–140, (2012).
7. B.E.Buschkuehle; K. Michael *Subsurface characterization of acid-gas injection operations in northeastern british columbia*; EUB/AGS Earth Sciences Report 2006-05; Alberta Energy and Utilities Board: Edmonton, Alberta, Canada, p 142, (2006).
8. K. Michael; B.E.M. Buschkuehle. Acid-gas injection at west stoddart, british columbia: An analogue for the detailed hydrogeological characterization of a CO_2. *J Geochem Explor* 89, (1-3), 280–283, (2006).
9. E. Campbell-Stone; R. Lynds; C. Frost; T.P. Becker; B. Diem. The wyoming carbon underground storage project: Geologic characterization of the moxa arch and rock springs uplift. *Energy Procedia* 4, 4656–4663, (2011).
10. J.J. Carroll Acid gas injection -the next generation. http://www.gasliquids.com/papers/EGPA%20May%202009.pdf
11. J. Stopa; J. Lubas; S. Rychlicki In *Underground storage of acid gas in poland - experiences and forecasts*, 23rd World Gas Conference, Amsterdam, 2006; Amsterdam, (2006).

12. J. Lubaś; W. Szott. 15-year experience of acid gas storage in the natural gas structure of borzęcin–poland. *Nafta Gaz* 66, (5), 333–338, (2010).
13. K. Michael; M. Buschkuehle *Subsurface characterization of acid-gas injection operations in the peace river arch area*; ERCB/AGS Special Report 090; Energy Resources Conservation Board: Edmonton, Alberta, Canada, p 186, (2008).
14. K. Michael; M. Buschkuehle *Subsurface characterization of acid-gas injection operations in the provost area*; ERCB/AGS Special Report 091; Energy Resources Conservation Board: Edmonton, Alberta, Canada, p 143, (2008).
15. S. Bachu; M. Buschkuehle; K. Haug; K. Michael *Subsurface characterization of the edmonton-area acid-gas injection operations*; ERCB/AGS Special Report 092; Energy Resources Conservation Board: Edmonton, Alberta, Canada, p 134, (2008).
16. S. Bachu; M. Buschkuehle; K. Haug; K. Michael *Subsurface characterization of the pembina-wabamun acid-gas injection area*; ERCB/AGS Special Report 093; Energy Resources Conservation Board: Edmonton, Alberta, Canada, p 60, (2008).
17. S. Bachu; K. Haug; K. Michael *Stress regime at acid-gas injection operations in western Canada*; ERCB/AGS Special Report 094; Energy Resources Conservation Board: Edmonton, Alberta, Canada, p 42, (2008).
18. S. Bachu; M. Buschkuehle; K. Michael *Subsurface characterization of the brazeau nisku q pool reservoir for acid gas injection*; ERCB/AGS Special Report 095; Energy Resources Conservation Board: Edmonton, Alberta, Canada, p 62, (2008).
19. Tumbler Ridge News Hydrogen sulphide leak northwest of tumbler ridge. http://tumblerridgenews.com/?p=7935
20. M.Pooladi-Darvish; S.Bachu; H.Hong. Pressure build-up and decay in acid gas injection operations in reefs in the zama field, Canada, and inplications for CO_2 storage. *Oil Gas Sci Technol* 66, (1), (2011).
21. S. Bachu; M. Pooladi-Darvish; H. Hong In *Causes of pressure buildup and decay in acid gas injection operattions in reefs in the zama field, Canada, and implications for CO_2 storage*, International Conference on Deep Saline Aquifers for Geological Storage of CO_2 and Energy, IFP, France, 2009; IFP, France, (2009).
22. S. Bachu; T.L. Watson. Review of failures for wells used for CO_2 and acid gas injection in Alberta, Canada. *Energy Procedia* 1, (1), 3531-3537, (2009).

14

Control and Prevention of Hydrate Formation and Accumulation in Acid Gas Injection Systems During Transient Pressure/Temperature Conditions

Alberto A. Gutierrez and James C. Hunter

Geolex, Incorporated, Albuquerque, NM, USA

Abstract

Dry acid gas injection (AGI) systems are typically comprised of compression/dehydration facilities which compress treated acid gas (TAG), primarily CO_2 and H_2S into dedicated AGI well(s). During normal operations, the pressure and temperature (P/T) of the TAG are maintained within the TAG's liquid or supercritical phase, well outside the field in which hydrates may form. However, during startup, upset conditions or power failures, transient conditions often occur allowing hydrates to form and accumulate downstream of the compressors, blocking the TAG flow, causing unacceptable pressures, temporarily rendering the well inoperative and potentially damaging compression or well equipment. Using equilibrium models and field experience, Geolex, Inc.˚ (Geolex) has developed best management practices and procedures (BMPs) to minimize potential hydrate formation in these situations and the safe removal of hydrates in AGI systems. This paper details the scientific bases for those BMPs and their application to several AGI systems which have experienced hydrate problems.

14.1 General Agi System Considerations

The design, construction and operation of safe and efficient AGI systems require careful evaluation of the TAG's physical, chemical and thermodynamic properties. These include the anticipated ranges of composition, pressure and temperatures likely to occur during the operational lifetime of the system. This evaluation should begin with modeling of the real-world operating conditions to identify where in the operating phase envelope the TAG might enter the hydrate-forming region. Careful planning, intelligent design and implementation of these BMPs can minimize the time that the TAG enters the hydrate-forming region, and design and operating procedures can also reduce the H_2O fraction at the TAG during each compressor stage. These BMPs also include the introduction of additives (e.g., methanol) to the TAG stream after compression through engineered systems to depress the hydrate formation temperature and inhibit the formation of hydrates during unstable P/T conditions often encountered during start-up or upset conditions that result in rapid changes in P/T conditions within the system The AGI system must be designed to allow the prompt and safe blowdown of the well, piping and compression facilities in the event that hydrates do form within the system or in the event of mechanical failures that may require a workover or testing of the well or surface facilities. In addition, Geolex has developed BMPs that prevent the formation of hydrates and highly corrosive conditions within AGI wells during start-up or resumption of injection after upsets or mechanical failures that caused rapid changes in P/T conditions within the piping leading to the well or within the AGI well itself.

14.2 Composition And Properties Of Treated Acid Gases

The composition of the TAG streams generated by "sweetening" processes at "sour" natural gas processing facilities are primarily driven by the concentrations and ratios of H_2S and CO_2 present in the field gas mixture entering the plant. Furthermore, these properties may change over time as different wells and fields are added or removed from the gathering system feeding the processing facility. The average daily volume of the TAG is determined by the composition and the gross inlet volume to the plant. The total TAG flow in the three case studies presented in this paper ranges from approximately 14,000 to 450,000 cubic meters per day (0.5 MMCFD to 16 MMCFD).

The data utilized in this paper is derived from work at natural gas plants throughout the US and overseas. TAG stream compositions from sour natural gas processing facilities typically contain 1% to 60% H_2S and 40% to 99% CO_2 and less than 1% residual hydrocarbons (C1 - C7). At the end of the sweetening process, the low pressure, gaseous phase TAG is generally saturated with H_2O vapor, comprising approximately 4% by mole at 45°C (113°F).

For the purpose of this paper we consider the properties and behavior of two "end members" of typical compressed and dehydrated TAG streams: a "CO_2-rich stream" with 90% CO_2, 10% H_2S and a "H_2S-rich stream" with 90% H_2S, 10% CO_2. Calculations using the CSMGem program (1) demonstrate that the CO_2-rich TAG stream has a critical point of 33 °C (91°F) and 7.465 MPa (1083 psig), with a density of approximately 460 kg/m³. For the H_2S-rich TAG stream the critical point is 91.8 °C (197 °F) and 9.15 MPa (1327 psig), with a density of 310 kg/m³ (Figures 14.1a and 14.1b).

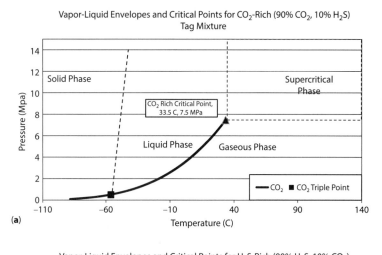

(a)

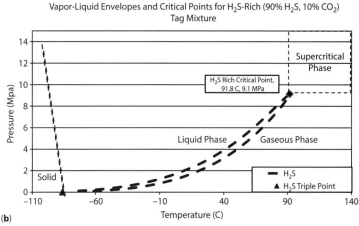

(b)

Figure 14.1 (a) Vapor-Liquid Envelopes and Critical Points for CO_2-Rich (90% CO_2, 10% H_2S) TAG Mixture (b) Vapor-Liquid Envelopes and Critical Points for H_2S-Rich (90% H_2S, 10% CO_2) TAG Mixture

As a result of progressive compression of the TAG from the amine unit through five stages, the TAG stream becomes largely dehydrated; however, the small amount of remaining residual water can result in hydrate formation under transient P/T conditions often encountered during start-up operations. Under stable operating conditions, AGI well-head injection temperatures and pressures are typically 35 to 45 °C and 10 to 25 MPa, placing the compressed CO_2-rich TAG streams well into the supercritical phase field. At these temperatures and pressures, the supercritical TAG has densities ranging from 500 to 830 kg/m³.

H_2S-rich TAG streams require very high TAG temperatures [(over 92 °C) (198 °F) to maintain supercritical conditions. In these conditions, the TAG density ranges from 540 to 590 kg/m³. For this reason, these systems often inject TAG in the liquid phase rather than the supercritical phase. In addition, due to the lower density of the TAG, these systems often require more elevated surface injection pressures to achieve the same downhole pressure necessary to maintain injection.

14.3 Regulatory And Technical Restraints On Injection Pressures

The initial permitted surface maximum allowable operating pressure (MAOP) for AGI wells is typically governed by state or provincial oil and gas regulatory agencies, based on various formulas for assuring that the bottom hole injection pressure remains below the reservoir rock's parting pressure at the proposed injection depth. The parting pressure is calculated as:

$$P_p = P_{res} + \upsilon(P_{ob} - P_{res})/(1-\upsilon) \qquad (14.1)$$

where: P_p = Parting pressure
P_{ob} = Overburden pressure
P_{res} = Reservoir pressure
υ = Poisson's ratio (dimensionless)

Since Poisson's ratio is dimensionless, parting pressures can be calculated in any consistent units. Under typical conditions the parting pressure can range from 62% to 72% of the lithostatic pressure, depending on the values of Poisson's ratio (ranging from 0.25 for clastic rocks to 0.35 for carbonates). Ultimately the permitted surface MAOP is based on the parting pressure minus the hydrostatic pressure of the TAG in the well (determined by the average TAG density, the depth of injection and reservoir pressure conditions).

Many regulatory agencies recommend or require a step-rate injection test be performed in the reservoir prior to final approval to inject to evaluate the true formation parting pressure (2). Step-rate tests can directly determine the parting pressure by observing a decline in the pressure to injection rate curve at the pressure corresponding to the effective parting pressure. The results of the step rate test which determines the observed parting pressure can be used to confirm, reduce or increase the final MAOP in order to maintain safe and effective injection conditions.

Depending on the depth of the injection zone (typically 2000 to 3500 meters), MAOPs may range from 10 to 25 MPa, well above the critical pressure for CO_2-rich TAG streams at typical operating compressed TAG temperatures of 30 to 45 °C (86 to 113°F). These supercritical streams have densities of approximately 500-830 kg/m³. Supercritical conditions for H_2S-rich TAG streams can only be achieved if compressed TAG temperatures are kept above 92 °C (198°F). In these conditions, the supercritical TAG has a density of 607 kg/m³ [(at 20 MPa and 95 °C) (2901 psig and 203°F)].

14.4 Phase Equilibria, Hydrate Formation Boundaries And Prevention Of Hydrate Formation In Agi Systems

14.4.1 Hydrate Formation Conditions in AGI Compression Facilities

The formation of hydrates requires three conditions: 1) a hydrate guest molecule, 2) water, and 3) a certain range of temperature and pressure (generally low temperatures and high pressures). Figure 14.2 is a phase diagram showing the hydrate-formation boundaries for the CO_2-rich and H_2S-rich TAG streams discussed above. This figure clearly shows that all three conditions may exist during start-up of AGI wells. Specifically, hydrates may form in CO_2-rich systems below 20 °C (68 °F), and in H_2S-rich systems below 30 °C (86 °F) throughout the pressure ranges found in wellhead and downhole locations.

If the TAG's pressure and temperature are maintained above the hydrate boundaries, hydrates do not form. However, the TAG stream may pass into the hydrate-formation field, particularly if temperatures drop during compressor start-up, shutdown, or sudden temperature control system

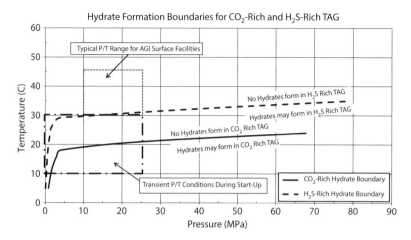

Figure 14.2 Hydrate Formation Boundaries for CO2-Rich and H_2S-Rich TAG.

or compressor failure. In these cases, abrupt drops in TAG pressure or temperature can result in uncontrolled hydrate formation.

Figures 14.3a and 14.3b show the cooling rates of the CO_2-rich and H_2S-rich TAG streams under uncontrolled decompression from an initial pressure of 25 MPa (3626 psig) and a temperature of 45 °C (113 °F). During decompression cooling, both TAG streams pass quickly into the liquid phase, and may cool into the solid phase as decompression progresses. Uncontrolled decompression can form water ices, hydrates, solid CO_2 and H_2S, and the abrupt temperature drops can compromise the strength of many alloys. Clearly, control systems and procedures must be applied to prevent these conditions.

14.4.2 Hydrate Controls in AGI Compression Facilities

BMPs for hydrate control include awareness that hydrates can form throughout the surface and subsurface parts of the system, and that prevention measures must recognize and address each component of the system. The five major control areas are:

1) consistent and stable temperature control from the compressors to the well head,
2) reduction of water in the TAG stream,
3) the introduction of inhibitors to reduce the freezing point of the hydrates,

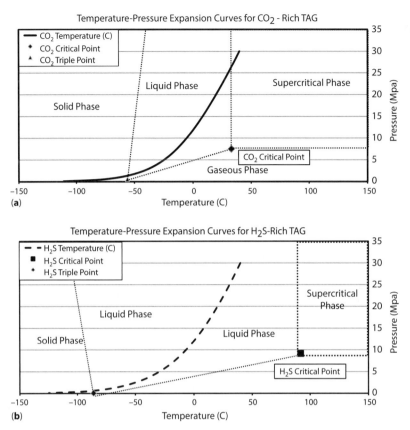

Figure 14.3 (a) Temperature-Pressure Expansion Curves for CO_2 – Rich TAG (b) Temperature-Pressure Expansion Curves for H_2S-Rich TAG.

4) reduction or elimination of nucleation sites where hydrates are preferentially formed, and
5) engineered systems to safely vent, clear and purge the piping from surface compression facilities to the reservoir (including well tree and downhole equipment) for maintenance, repairs, or mitigation of hydrate accumulations.

At the compressor, it is important to continuously monitor pressure and temperature at each compression stage, and to alarm the operators if these parameters exceed the acceptable ranges. The pipeline between the compressors and the well head must be insulated and heated, if warranted by anticipated ambient weather conditions. Temperature and pressure along the pipeline should also be monitored, with appropriate alarms for unacceptable P/T variations that may result in the formation of hydrates within the system.

The TAG from the gas plant amine unit is generally saturated in water, with a concentration of approximately 0.05 kg/m³ at 40 °C (104 °F). Thus an uncompressed gaseous TAG stream of 50,000 m³/day (approximately 2MMCFD) would contain about 2,500 kgs of water. A significant fraction of this water is removed at each compression stage, but some water will remain in the final stage. Compressor operations should be optimized to insure that only a minimum amount of water reaches the final stage.

Inhibitors such as methanol are commonly used in the natural gas pipeline industry to reduce the probability of hydrate formation in lines. Geolex has used standard engineering calculations (e.g., the Hammerschmidt Equation (3)) to determine the ratio of methanol to residual water in the compressed TAG stream to achieve a desired depression of the freezing curve of the hydrates. An example of this method is shown in Figure 14.4 and the table below using the CO_2-rich TAG stream. An addition of methanol representing 10% by weight of the residual water in the TAG stream depresses the freezing point by approximately 5 °C (9 °F), and the addition of methanol representing 30% by weight of the residual water in the TAG stream will depress the freezing point by approximately 20 °C (36 °F). Methanol or similar additives may not be necessary during normal operations if proper temperature and pressure control is achieved; however, it is needed to address variable temperature and pressure conditions during startup, unplanned upsets or shutdowns where transient pressure and temperature conditions exist.

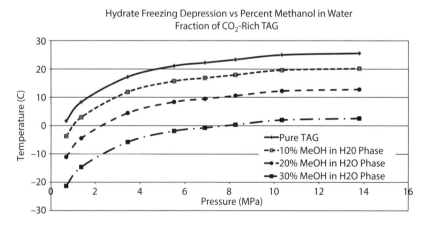

Figure 14.4 Hydrate Freezing Depression vs Percent Methanol in Water Fraction of CO_2-Rich TAG

TAG Temperature(°C)	Liters per 10,000 m³ TAG for 10% MeOH	Liters per 10,000 m³ TAG for 20% MeOH	Liters per 10,000 m³ TAG for 30% MeOH
0	4.8	9.6	14.4
10	9.4	18.9	28.3
20	17.3	34.5	51.8
30	30.4	60.8	91.1
40	51.2	102.3	153.5
50	83.0	166.0	248.9
60	130.0	260.0	390.0
Methanol Dosing Rate (in Liters) per 10,000 m³ CO_2 - Rich TAG to Prevent Hydrates			

Hydrates tend to form in portions of the system where nucleation is favored, such as joints and threads, valves, meters, fittings, and bends in lines and changes in line diameter. BMPs include piping and well designs that minimize potential nucleation sites.

To prevent hydrate formation during either planned or emergency compressor or system shutdowns, plant operators are now designing and implementing equipment and procedures to allow safe, controlled venting of TAG from the surface facilities to flares or other acceptable disposal units, while maintaining the TAG's P/T conditions outside the hydrate formation range. These systems include detailed checklists and semiautomatic control networks for compressor start-up, normal running, planned and emergency shutdowns, purging cycles, and re-starting. Venting equipment includes dedicated piping with choke valves to allow careful reduction of gas pressures to maintain safe temperatures and pressures while going to flare.

14.5 Formation, Remediation And Prevention Of Hydrate Formation During Unstable Injection Conditions – Three Case Studies

14.5.1 Case 1: CO_2 – rich TAG (90% CO_2, 10% H_2S) Injection into a 2,000 m Deep Clastic Reservoir

This AGI well was designed by Geolex and permitted to receive up to 140,000 m³/day (5 MMCFD) of compressed TAG from a natural gas processing plant located adjacent to the well site. The TAG composed of approximately 90% CO_2 and 10% of H_2S, and was formerly treated using a Claus-process sulfur reduction unit (SRU). Economic, environmental and operational problems with the SRU were the key factors in selecting an AGI solution at this gas processing facility. A significant economic issue was the need to blend natural gas back into the TAG stream in order to maintain combustion in the initial stage of the SRU.

The AGI well for this facility was drilled and completed in August 2010, and was completed at a total depth of 2,018 m in a Mesozoic sandstone reservoir, capped below by dense shales, and above by carbonates and mudstones. A total of 42 m of reservoir was perforated between 1,936 and 1,978 m, based on analyses of geophysical logs, warm-back tests, and conventional cores. After completion the well's static shut-in pressure was approximately 2.7 MPa (391 psig).

The original MAOP was 13.64 MPa (1979 psig), but following the completion and analysis of a step-rate test in November 2011, an increase of the MAOP to 14.85 MPa (2154 psig) was requested and approved in February 2011.

The surface compression facilities for this AGI system included three identical 5-stage compressors, each rated for a maximum operating pressure of 14.8 MPa (2147 psig). Each compressor set was capable of injecting approximately 86,000 m^3/day (3.04 MMCFD). Thus any two compressors operating together would have the capacity of 172,000 m^3/day (6.08 MMCFD), or 123% of the planned maximum injection rate of 140,000 m^3/day (5 MMCFD). This capacity allows the plant to significantly increase their throughput without installing additional compressor capacity.

The compressor system is connected to the AGI wellhead by a 75 mm stainless steel pipeline extending approximately 75 m from the compression facility which was completed in May 2011, after which the compressors were tested and calibrated prior to injection. The control logic for the compressors involved go/no go decision trees based on the acceptable temperatures and pressures in and between each of the five stages. Considerable problems were encountered during the calibration process, and over several days pressures fluctuated from 0 to 10 MPa (0 to 1450 psig) on time scales minutes to hours. These transients caused wide swings of temperature and pressure in the TAG stream from the compressors to the reservoir.

After several days of calibration, the system appeared to be stable, and injection was stabilized at 8.9 MPa (1291 psig) after three hours of steady increase. After approximately 11 hours, the pressure abruptly increased to 10.7 MPa (1552 psig) over approximately 20 minutes after which a mechanical failure shut down the compressor (Figure 14.5). The well head was shut in, but over the next few weeks the well head pressure remained at approximately 8.9 MPa (1291 psig) (the average pressure during the 11-hour injection pressure) instead of quickly falling down to the reservoir pressure of 2.7 MPa (391 psig). The persistence of this elevated pressure indicated that some obstruction had occurred in the well's injection tubing, below the well head.

As seen in Figure 14.2, at a pressure of 8.9 MPa (1291 psig), hydrates will begin to form at temperatures below approximately 19 °C (66 °F). Although temperature was not recorded at the well head during this testing and startup, compressor temperatures were generally over 35 °C. A review of a warm-back study, performed during initial injection testing in October 2010 however showed that downhole temperatures as low as 20 °C (68 °F) were encountered in the upper 100 m of the borehole with higher temperatures below that level. Thus it is possible that hydrate formation conditions may have occurred in the upper portion of the well bore during the initial start-up period.

In consultation with the plant operator, Geolex developed and implemented a method to remediate the suspected hydrate blockages or other causes for the elevated shut-in pressure in the well, and re-establish the injectability of the well. The method included:

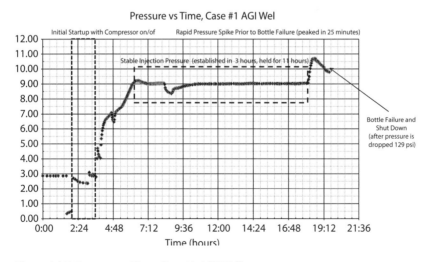

Figure 14.5 Pressure vs Time, Case #1 AGI Well

- Injection of approximately 800 liters (211 US gal) of methanol into the well head through the Christmas tree head, at rates of 11 to 19 liters per minute (2.9 to 5.02 gpm), and pressures of 8.9 to 9.6 MPa (1291 to 1392 psig) (current well head pressures), using plant equipment and personnel. Methanol rapidly degrades hydrates into free liquids and gasses.
- Mobilizing a well service contractor to perform a step test at the well, using pressures as high as 21 MPa (3046 psig) in an effort to force the methanol down through the tubing into the formation, and physically displacing the remaining hydrates,
- Recording step-rate data to determine if the well's injectivity had returned to the behavior seen in the November 16, 2010 step rate test, and
- Monitoring the post-test well pressure, to confirm that the blockage had been removed allowing the pressure to return to the original shut-in pressure of approximately 2.7 MPa (391 psig).

A tailgate safety meeting was held, a work permit was prepared and approved, and the methanol injection was initiated. The well head was isolated from any surface line, the well head cap was bled, and the methanol pump was connected to the well. At that time, the well head static pressure was 8.9 MPa (1291 psig). After approximately 2 hours, a total of 833 liters (220 US gal) of methanol were injected, at gauge pressures of 9.3 to 9.9 MPa (1342 to 1436 psig). After injection, the well head pressure stabilized at 8.8 MPa (1276 psig). This pressure change did not indicate any significant change in the well.

After removing the methanol equipment, the pumping contractor rigged up their pumping rig and 4 water trucks with a total capacity of approximately 54,000 liters (14265 US gal) of fresh water. After attaching the pumping rig to the well and providing a safety meeting for all involved, the step test began at 238 liters per minute. After 20 minutes, the pumping rate was increased to 398 liters per minute (105 gpm). The following steps were spaced at 20 minutes, until a final maximum rate of 715 liters per minutes (189 gpm) was reached. The complete clearing and testing program injected 51,200 liters (13526 US gal) of water after the methanol injection. Water flow and well-head pressure were continuously monitoring during the test, and the final well head pressure was monitored for 15 minutes after the end of the test and checked again approximately 12 hours later. Following the rig down of the pumping equipment, well pressure was monitored with the existing digital gauges at the well head.

Following the end of the step rate test, the well head pressure decreased to 7.5 MPa (1088 psig) after 15 minutes. Approximately one hour after the test, the well head pressure hat declined to 5.5 MPa (798 psig). The next morning, approximately 12 hours after the test, the well head pressure was 2.7 MPa (392 psig). This is essentially the original shut-in pressure observed during the initial development and testing of the well in November 2010.

The well blockage was the result of hydrate formation as indicated by the circumstances of the events, the persistence of the blockage, and the response of the blockage to a treatment plan appropriate for hydrate remediation strongly indicate that hydrates were the cause of the problem. Major modifications to the compressors and their controls, has allowed the operators to closely maintain optimum TAG pressures and temperatures in the surface systems. Since then, the well's operation has been completely successful and injection has not been interrupted except for scheduled maintenance.

14.5.2 Case 2: CO_2-Rich TAG (75% CO2, 25% H2S) Injected Into a 3050 m Deep Carbonate Reservoir

This AGI well was also designed by Geolex and was permitted to inject up to approximately 55,000 m³/day (2MMCFD) of TAG from the adjacent natural gas processing facility. The TAG consists of approximately 85% CO_2 and 15% H_2S. The TAG was formerly burned in a flare stack. Environmental issues, including the emissions of sulfur, and the operational costs of adding natural gas to the TAG stream to maintain combustion in the flare, were the primary drivers in the decision to drill an AGI well. The flare will not be removed and will remain active to handle any upsets that may occur at the plant or the AGI well.

The AGI well was spudded in March 2012 and drilling was completed in June 2012. Geolex completed and tested the well beginning on June 13, 2012 and ending October 25, 2012. The reservoir target was a Permian Wolfcamp submarine debris fan formed on the shelf-slope facies of the Delaware Basin. Due to limited well control in this area and depth, Geolex employed 3D seismic interpretation to identify and characterize a promising target immediately southeast of the plant. The debris fan is isolated vertically and laterally by surrounding deep-water shales and muddy carbonates.

Following drilling, well testing included geophysical logging, side-wall coring and warm back studies. Six porous and permeable zones were identified between 2,920 and 3,088 m and a total of 60 m of section was perforated and acid-treated. A step-rate test was then performed and analyzed. A packer and tubing were then installed at 2,881 m and the well completed with a subsurface safety valve (SSV) at 75 m, and the well completed with a corrosion-resistant "Christmas tree".

The original permitted MAOP was 20.4 MPa. Analyses of the step rate test verified this pressure, and no request was made to increase the original MAOP.

The compressor facility, including two identical compressors rated at 39,640 m3/day (1.4 MMCFD) at a running pressure of 9.65 MPa, is connected to the well head via an approximately 145 m, 75 mm insulated stainless steel line. As originally built, there were no provisions for controlled blow down of the line or the well head.

Following installation and preliminary testing of the compressors, initial injection began in February 2013, lasting for 47 hours before pressure increases lead to automatic shut down by the compressor. Initial start-up included approximately 12 hours operating at 0.24 MPa (35 psig) and 1.0 °C (33.8°F). Pressure and temperature were then increased, over approximately 6 hours to the targeted operational pressure of 9.5 MPa (1378 psig). Pressure was stable for 12 hours, then dropped to 5.6 MPa (812 psig) for one hour. Pressure then returned to 9.5 MPa (1378 psig) for 4 hours before rapidly increasing to 11.6 MPa (1682 psig) for 11 hours before overpressure shut down the compressor.

As shown in Figure 14.6, the TAG temperature remained below the hydrate formation temperature for this TAG composition of 21.2 °C (70°F) over much of the initial startup. During this period, hydrates gradually accumulated in the surface piping, the Christmas tree and the upper tubing of the well. After consultation with Geolex and the compressor vendor, the pressure was briefly raised to 12.4 MPa (1799 psig) after increasing the compressor cut out to that level. This pressure was reached in less than 20 minutes. After shut down, the well head pressure remained constant at approximately 12 MPa (1740 psig), a very similar behavior to the conditions observed in the Case 1 well. Blockages were also observed in the surface pipeline from the compressors to the well.

In consultation with the plant operator, Geolex and Parsons-Brinkerhoff Energy Storage Services, Inc. (PB) developed a remedial plan to remove the hydrates. This plan included:

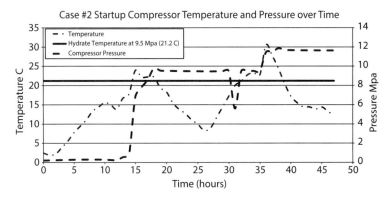

Figure 14.6 Case #2 Startup Compressor Temperature and Pressure over Time

- Design and fabricate necessary tubing to allow slow and careful blow down of the TAG in the surface pipeline, directing the vented TAG to the existing flare
- Close all wellhead valves and bleed off the residual TAG in the space above the uppermost section of the Christmas tree, using supplied-air personal protection
- Open the top of the Christmas tree and connect temporary piping to a choke manifold and then to a rental 3-phase separator
- Fabricate and install a connection into the flare line
- Connect the separator to the flare line, using approximately 120 m of temporary piping
- Open the well valves and, using the choke and separator, gently bleed down the well until static well pressure (atmospheric) was achieved,
- Connect and remove the choke, separator and connections to the flare
- Rig up and connect a pumping truck and pumped approximately 14,000 liters of brine, monitoring the pressure and rate,
- Pump an additional 350 liters (92 US gal) of methanol
- Rig down the pump truck and monitor well head pressure until atmospheric pressure was observed.

In March 2013 the remedial strategy was implemented over 3 days, with a successful removal of the hydrates and the return of the well to normal pressure. After modifying the start-up procedures in detail, and designing and installing a permanent blow down system, the well will be returned to service in late summer 2013.

14.5.3 Case 3: CO2-Rich TAG (82% CO_2, 18% H_2S) Injected Into a 2950 m Deep Carbonate/Clastic Reservoir

This AGI well was designed and permitted by Geolex and installed under our supervision. It is permitted to inject up to approximately 200,000 m³/day (7 MMCFD) of TAG consisting of approximately 82% CO_2 18% H_2S from an adjacent natural gas processing facility. The well has been operating since 2009 and over the first 18 months of operation experienced significant difficulties with maintaining adequate temperature control of the TAG stream. As a result of these fluctuations, phase changes occurred within the tubing that allowed for the condensation of free water in the tubing. As a result of this condition, the basal 100 foot portion of the tubing experienced significant corrosion and resulted in a tubing leak. This leak was detected through the careful analysis of annular pressure fluctuation data and the response of the annular fluid to the preparations for conducting a regularly scheduled MIT test. After the leak was discovered and reported to the appropriate regulatory agencies, the well was worked over and the tubing leak repaired. In addition, the ultimate cause of the temperature control problems was resolved by modifying the temperature control modules and the location on the compressor skid. Subsequent to the well workover, the operator has successfully improved the controls on the compression system to assure that a more reliable and consistent P/T regime is maintained during injection preventing the phase changes that allowed for condensation to occur within the tubing in the TAG stream.

However, the nature of the electrical supply to the compressors and other mechanical issues at the plant continue to result in periodic short term spikes in injection flow rate, pressure and temperature variations. While these events now are much better controlled and often resolved within a matter of hours, it has been necessary to implement the BMPs described earlier in this paper to prevent the development of hydrates during these unstable or transient P/T conditions in the TAG stream and the well bore. A recent example that is well demonstrated in the available data indicated a pressure spike during startup which was a result of hydrate formation during a restart of the AGI system after a four day shutdown of the natural gas processing plant for scheduled maintenance. Figures 14.7a and 14.7b shows the behavior of the well during the month in which this shut down occurred.

The maintenance shutdown occurred from May 6th to May 10th 2013. Figure 14.7a shows a generally stable TAG injection pressure of approximately 10.3 MPa (1500 psig) and temperature of 50°C (122°F) leading up to the shutdown on the 6th. Early on the morning of the 10th when the

CONTROL AND PREVENTION OF HYDRATE FORMATION   271

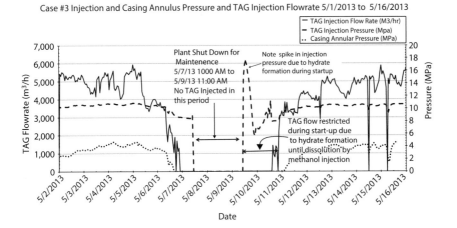

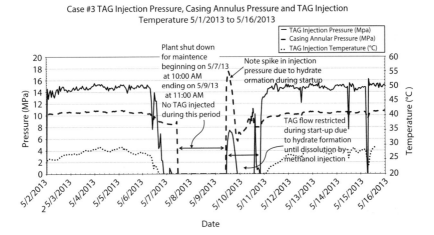

Figure 14.7 (a) Case #3 Injection and Casing Annulus Pressure and TAG Injection Flowrate (b) Case #3 TAG Injection Pressure, Casing Annulus Pressure and TAG Injection Temperature

AGI facility began to receive flow again the injection pressure spiked to approximately 17.2 MPa (2500 psig) as hydrates formed in a regime where the unstable TAG temperature fluctuated between 15 to 40°C (60-100°F) causing a shutdown of the compressors. Methanol was then injected using the feed system into the TAG line immediately upon restarting the compressors. For the next 16-20 hours the methanol injection was continued while the temperature of the TAG could be stabilized to the normal operating temperature of approximately 50°C (122°F). The immediate and dramatic effect of the methanol's depression of the hydrate formation curve can be seen in the rapid pressure decline and stabilization observed on Figure 14.7b during the day following the initiation of methanol injection and the removal of hydrates from the system.

This example shows the immediate and dramatic effect of hydrate formation in the wellbore in response to temperature fluctuations during unstable startup conditions in an AGI well. While in this instance the immediate action of the operator and the built in pressure safety systems prevented any damage to the injection equipment or the well components, the situation would have been prevented by initial injection of a methanol pad prior to resumption of injection and a constant feed rate of methanol based on the volume of TAG being injected during startup.

14.6 Discussion And Conclusions

The prevention of hydrates during transient P/T conditions in AGI wells is crucial to the safe and efficient operation of these systems and to prevent damage to well or compression equipment. This paper analyzes the physical chemistry of hydrate formation under conditions often experienced during the cold startup of AGI systems and as a result of upset conditions that cause outages of compression or other interruptions in normal injection operations. As a result of the review and analysis of phase equilibrium and hydrate formation boundaries in typical AGI systems and through the field experience gained from various instances of hydrate formation within AGI systems, Geolex has developed a series of BMPs that will assure that operators prevent the conditions which will result in hydrate blockages, pressure spikes and potential damage to compression, surface piping and well equipment.

These BMPs include:

- design and construction of systems that will permit the addition of methanol in front of the TAG stream during startup or resumption of operations after long shutdowns (cold start-up)
- design and construction of systems that will allow for metered injection of methanol based on TAG volume and composition until discharge P/T conditions stabilize and TAG is safely out of the hydrate formation zones
- careful monitoring of P/T conditions throughout the injection process and implementation of process alarms that will alert operators of potential for hydrate formation and

allow for mitigation measures to be implemented prior to well blockage or equipment damage, and
- the use of trained professionals during start-up experienced with AGI operations and the control and prevention of hydrates in AGI systems.

Typically, AGI systems are associated with larger gas processing operations which include automated plant controls and programmable logic controllers (PLCs) that permit the continuous monitoring of injection conditions as part of integrated plant operations. The use of trained professionals to monitor and adjust process conditions such that hydrate formation is prevented during AGI operations will assure that critical systems will not fail or be damaged resulting in costly plant shutdowns, gas delivery interruptions and excess emission events due to H2S flaring (which may result in fines or other regulatory actions).

References

1. E. D. Sloan and Carolyn Koh, *CSMGem Version 1.10*, Center for Hydrate Research, Dept. of Chemical and Petroleum Refining, Colorado School of Mines, Golden, Colorado, Release date January 1, 2007.2. Calculation of New Fracture Parting
2. *www.pe.tamu.edu/schechter/baervan/Annual_5/chapter1_2.pdf*
3. K. C. Covington, J. T. Collie III and S. D. Beherns, "Selection of Hydrate Suppression Methods in Gas Streams", in *Proceedings of the Seventy-Eight GPA Annual Convention, Nashville, TN: Gas Processors Association*, pp. 46-52, 1999.

15
Review of Mechanical Properties Related Problems for Acid Gas Injection

Qi Li[1], Xuehao Liu[1], Lei Du[2], and Xiaying Li[1]

[1] *State Key Laboratory of Geomechanics and Geotechnical Engineering (SKLGME), Institute of Rock and Soil Mechanics (IRSM), Chinese Academy of Sciences, Wuhan 430071, China*
[2] *China Petroleum Engineering Southwest Company, Chengdu 610017, China*

Abstract

In recent years, acid gas injection (AGI) has drawn widespread attention around the world due to the increasingly stringent sulfur emission regulations and continuously improved economic feasibility of AGI technology. However, there are some fundamentals of AGI that still need to be figured out comprehensively before large-commercial projects are carried out.

This chapter mainly focuses on mechanical properties-related problems of an AGI process. In general, fluid migration within a reservoir rock and long-term evolution in integrity of caprock are, to a large extent, affected by H_2S/CO_2-brine–rock mineral interactions, which take place in the formation coupling thermal-hydro-mechanical-chemical-biological (THMCB). For a reservoir rock, multi-component fluids containing injected acid gas and formation brine exist in mixed phase behaviors (such as liquid, supercritical, and gas phase), the fluids will migrate through porous media and interact with rock minerals. In the meanwhile, these chemical reactions will accompany with a large amount of mineral dissolution and precipitation, such as carbonates dissolution, hydrate formation, and server sulfur deposition. Both of these phenomena could change rock micropore structure, and also affect fluid transport characteristics, new precipitated minerals could even plug injection wells. For caprock, injected acid gas could impair caprock integrity, which embodies in two aspects. One is rock mechanics, its properties (such as fracturing pressure) will be influenced over long-term geochemical reaction, rock stress will change due to pressure accumulation in large-scale injection process and pressure drop during acid gas dissolution. The other one is caprock sealing efficiency, it will vary substantially according to new channels formed by mineral dissolution as well as the evolution of breakthrough pressure through micro-pores during geochemical reactions.

In the light of the above discussion, we strongly recommend that it is necessary to conduct the experimental research before accurate numerical simulation, which could answer how the long-term geochemical reactions affect the rock mechanics during acid gas injection.

15.1 Introduction

Acid gas injection (AGI) is regarded as an effective technology in reducing atmospheric emissions of noxious substance (H_2S) and greenhouse gas (mainly CO_2)[1]. It is a process of injecting acid gas into a reservoir for geological storage, which the acid gas is usually captured from the process of natural gas production [2]. The expense also has been provedless costly than recovery of sulphur in the gas by the Claus process [3, 4], which is still widely used in China [5].

From the first AGI project taken place at the Chevron Acheson on Edmonton, Alberta in 1989. So far, there are near eighty projects around the world [6], but there is still no one project yet in China [7, 8]. Leakage of acid gas through caprock or wellbore into ground surfaces or valuable aquifers could have detrimental effects. An AGI project might also be suspended if the reservoir pressure accumulating over the approved value, or if the selected site is estimated not suitable any more.

It seems that the mechanical factors in AGI do not obtain enough attention as it deserve. Though pressure build-up and decay [9-12], potential leakage from wellbores [13, 14] and caprock [15] were studied to a certain extent, as well as lot of works about numerical simulation [16]. Nevertheless, much more research need undertake to address mechanical problems [17], especially other factors like chemical reactions affecting mechanical properties, stress change or failure mode. Moreover, long-term effects from thermal, mechanical, hydraulic, geochemical, and biological factors due to H_2S/CO_2-brine-rock interactions demand a more improved and comprehensive understanding.

15.2 Impact Elements

Acid gas injection might lead to various risks occurring in wellbore, in reservoir, and in caprock (Figure 15.1).

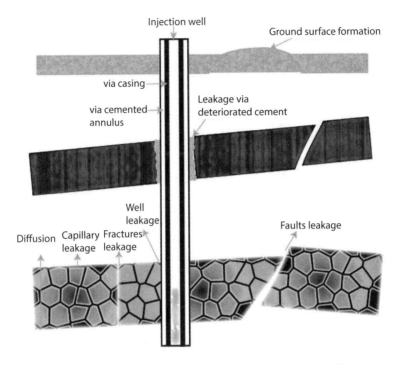

Figure 15.1: Key problems for acid gas injection through caprock and wellbore

15.2.1 Well

What problems addressed here happen at wells are 1) phase behavior and physical properties for acid gas mixtures, and 2) integrity of wellbores.

15.2.1.1 *Phase Behavior and Physical Properties*

The complexity of phase behaviors and uncertain properties of acid gas bring about certain gaps transporting in wellbore, besides flowing on reservoirs.

Phase transitions might lead to considerable pressure change when supercritical fluids turn into gas or liquid state through well injection, and severe hydrate plug when fluids are provided with conditions of enough water content, high pressure and low temperature. Specifically, the supercritical point of CO_2 is 31°C, 7.38MPa, and H_2S at 100.4°C, 9MPa, which is shown in Figure 15.2. As a vital data in controlling phase behaviors, the supercritical point of acid gas, however, varies along with its contents of H_2S acutely [18] and components like mercaptans [19] elusively. For another part, it is hard to prevent any hydrate formed from mixed fluids in a changeable condition of pressure and temperature. What need to be clear here is the hydrate of H_2S dissociating at 273k, 98.6kPa and at 283K, 275kPa [20].

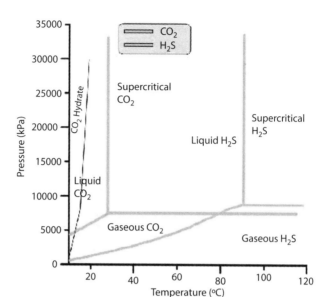

Figure 15.2: Phase behavior for CO2 and H2S

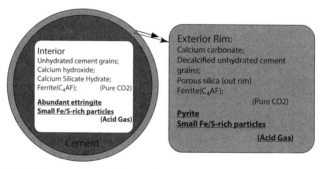

Figure 15.3: Mainly mineralogical changes between pureCO2 and acid gas (21 mol% H2S and CO2) exposed cement after a period of 28 days at a temperature of 50°C and a pressure of 15MPa [32].

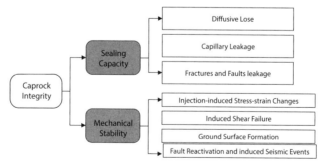

Figure 15.4: Risks Concluded in Caprock

For acid gas mixtures, as have been mentioned above about supercritical point, hydrate formation condition, and water content measurements, other important P-T dependent physical properties include bulk compressibility (the property of CO_2 typically an order of magnitude higher than that of water), viscosity (typically 10 times lower than of water (μCO_2=10-4Pa.s at 10MPa and 7°C)) ,density [21], solubility [22], relative permeability [23], dew point and bubble point [24, 25]. Thus, these physical properties for acid gas mixtures should be measured experimentally along changeable P-T condition.

15.2.1.2 Integrity of Wells

In a typical injection well, leakage could take place 1) in the annulus, 2) along the casing, 3) along cement matrix within pore space and fractures, and 4) along or through plugs where the interfaces between cement and formation rock, cement and casing, and casing and plugs [26], which are partly shown in Figure 15.1.

Watson and Bachu [27] reviewed 315,000 oil, gas and injection wells in Alberta, especially 79 wells used for CO_2 injection and acid gas disposal [28]. Mechanical, thermal, hydraulic and chemical factors which affecting the potential for well leakage are analyzed here respectively.

Mechanical causes impacted on well integrity during well completion and injection include mud displacement, gas migration during cement setting, and cement sheath stress failure. Specifically speaking, a hydraulic fracturing treatment and a loading from far-field formation stresses, such as tectonic stress, subsidence and formation creep, may cause a pressure decrease, and then lead to the failure of the wellbore structure. Cyclic pressure fluctuation may also cause fatigue fractures. For casing failure, uncemented casing is thought the main factor [27]. The failure of cement seal is caused primarily by poor bonding of cement to the formation rock. Cyclic pressure and temperature variations also could lead to debonding, tensile failure or stress crushing of the cement.

Temperature variation could be caused by different absolute values between the injecting fluids and formation rock, and thermal gradients cross the wellbore structure and different thermal expansion coefficients between these materials like casing steel, cement, fluids and formation rock, and phase transition of injecting fluids [29]. This variation could induce dilation or shrink of the casing and cement, and then cause a longitudinal shearing stress at the cement/casing interface or radial fracturing in the cement sheath. Moreover, since the thermally-induced expansion is cubical, so induced stress cracks are tensional rather than compressional [26].

As a hydraulic cause, well leakage could also occur through fractures induced by cement shrinkage. Since cement shrinkage leads to circumferential fractures propagate, thus gradually cause leakage along cases and the abandoned well. Small radial strains brought by shrinkage will cause relaxation of the radial stress and an increase in the tangential stress between cement and rock. Once the condition under which pore pressure is greater than the radial stress is reached, the bond between cement and rock maybe lost [26].

Leakage via deteriorated cement, caused by geochemical reactions in acid gas-brine-cement [30, 31], is also an issue for the long-term safety of acid gas geological storage. Complex geochemical reactions occurring under reservoir conditions are recognized gradually. The conversion of ferrites to pyrite after H_2S reactivity was observed by Jacquemet [31] with cement at pressures (50 MPa) and temperatures (120, 200°C). Furthermore, it is concluded that the addition of H_2S to the CO_2 sequestration system resulted in two main mineralogical differences in Portland cement [32]:(1) the precipitation of significant amounts of ettringite (evidenced in the interior region of the cement), and (2) the precipitation of pyrite in the carbonated rim of the cement, The mineralogical changes and alteration front subsequent to the hardening of cement can lead to strength loss and degradation, even mechanical damage.

The mineralogical changes of cement induced by geochemical reactions can also change the physical properties of the material (texture, porosity, permeability, diffusivity and mechanical resistance) [31]. Not only for cement and rock, the chemical reactions also cause steel corrosion to happen [33]. Hence, several pathways are evolved and then produced for the H_2S–CO_2 fluids to leakage.

15.2.2 Reservoir

As we know, the acid gas injected into reservoirs exists as three specific forms: 1) in a supercritical state gathering below caprock and distributing in capillary pores, 2) in solution state dissolving into formation water as icons, and 3) in mineral state precipitating into solid after chemical reactions.

15.2.2.1 Fluid Migration Mechanisms

After injection into reservoir, large amount of supercritical acid gas mixtures accumulate around the injection zone below reservoir-caprock interface as a supercritical state. Then partly of acid gas starts to dissolve in formation water as a solution state, as well as process of diffusion and chemical reaction, thus lead to the density increasing (2%-3% for fluid dissolution of pure CO_2) [34]. The balance of fluid interface between supercritical acid gas mixtures and brine could also be disturbed by factors from mechanics (differential pressure caused by injection), thermal (differential temperature) and chemistry (mineral dissolution and precipitation) [35], thereby taking place finger phenomenon. This process is controlled by Damkohler/Rayleigh² [36], and work like an avalanche to promote fluid convection and immigration.

15.2.2.2 Geochemical Reactions

Acid gas mixtures dissolved into brine result in an increase in H+, and a decrease in pH, which lower than pure CO_2 dissolution in water yields a pH≈3 at reservoir conditions. Through certain geochemical reactions, some mixtures storage as minerals like carbonates, silicates and sulphate [37]. Specifically, acid gas could induce formation dissolution, fines migration, precipitation and scale potential, oil or condensate banking and plugging, asphaltene and elemental sulphur deposition, and hydrate plug. In the common case, dolomite, calcite and pyrite are highly sensitive to the injected acid gas mixtures. Given enough time with hydrogen sulfide, rock minerals like carbonates, silicates and pyrite will totally dissolve [38, 39], thereby precipitating of anhydrite, sulfur solid and amorphous silica.

15.2.2.3 Influence Injection Rate

Those geochemical reactions mentioned above could do harmful to continuous acid gas injection. The factors including minerals precipitation, fines migration and deposition, sulphur deposition and hydrate formation, would result in porosity and permeability decreasing near injected zone, thereby plug the injection well or pipelines.

This reaction displays a way of sulfur deposition mainly taking place at injection well.

$$H_2S + O_2 \rightarrow H_2O + S_8 \downarrow + SO_{2,aq} \tag{15.1}$$

This reaction mainly happens in reservoir condition.

$$H_2O + SO_{2,aq} \rightarrow H^+ + SO_4^{2-} + S_8 \downarrow \tag{15.2}$$

15.2.3 Caprock

Geological storage projects should prevent the injected fluids (acid gas for AGI, and CO_2 for CCS) leakage into a ground surface or overlying formation over hundreds of years. Its viability and long-term safety depends largely on sealing capacity and mechanical stability of cap rock. In terms of sealing capacity, leakage mechanisms are classified as follows: 1) on a pore scale, diffusive lose and capillary leakage; 2) on a fracture scale, fractures and fault leakage. In terms of mechanical stability, it includes

15.2.3.1 Diffusion

The process of diffusion takes place through the whole life of AGI project, and could become a dominant factor in regard to reactions and long term sealing. Usually, diffusive loss takes into considerate only if the break through pressure is not exceeded, and its leakage quantity is small enough to be neglected [40].

15.2.3.2 Capillary Leakage

Capillary leakage is taken placed if the breakthrough pressure is exceeded, and its amount much larger than diffusion. The breakthrough or displacement pressure is defined as the minimum pressure required to establish a connected filament through the largest interconnected water-saturated pore throats of the rock. It is controlled by interfacial tension (IFT) [41] and wettability (θ), which means the sealing capacity of a given caprock is lower with respect to pure CO_2 than traditional hydrocarbons, not regarding to the acid gas mixtures. Thus, it is not correct to conclude that because it has sealed hydrocarbons for a geological time, the caprock would also provide a good seal for pure CO_2 or acid gas mixtures. Properties for H_2S and detailed acid gas mixtures should be measured and understood.

$$P_c = \frac{2 \text{IFT} \cdot \theta}{r} \quad (15.3)$$

Leakage through faults or fracture networks can be rapid and catastrophic. Fluids could leakage through existing hydraulic tunnel, through the weak strata where main cap-rock pinch out naturally, through faults, and through induced fractures by geochemistry reactions or mechanical failure.

Mechanical stability of caprock could affect the sealing capacity, and also need to be pointed out. Large amount of acid gas injection induced stress-strain and pore pressure change, thus might lead to shear or tensile failure of the caprock, even causing ground surface formation, faults reactivation, or induced seismic events.

15.2.3.3 Injection-induced Stress-strain and Pore pressure changes

Acid gas injection increase the pore pressure, thus induce stress field changes around the injection zones (Figure 15.5. Condition①→②: the

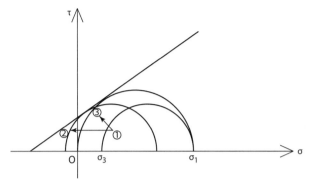

Figure 15.5: Failure mechanisms of rock induced by injection in Mohr-Coulomb

total effective stress decreases along with the increasing of pore pressure, thus tend to meet the critical line). Specifically, the vertical effective stress remains nearly constant to the over-burden, whereas the horizontal stress increasing significantly with acid gas injection. This effect could be explained that the rock-mass expend much free vertically but is confined in lateral expansion [42]. In this case shear and tensile failure of cap-rock might be induced when the increasing horizontal stress meeting the Mohr-Coulomb criteria (Figure 15.5, condition ①→③: the vertical effective stress keeps nearly constant, while the horizontal effective stress increasing along with fluids injection.). To understand the mechanism of this pore pressure change is also beneficial to determine the maximum approved well-head injection pressure and total approved injection volume for an AGI project [43]. Moreover, the in situ stress field will evolve along with the evolution of the pore pressure during continuous acid gas injection, and it tends to be one of key problems for long-term safety and stability to clarify the maximum pore pressure that rock can bear before fracturing or yield.

15.2.3.4 Induced Shear Failure of the Caprock

Shear or tensile failure tends to take place at the reservoir-caprock interface [42], because horizontal stress increases with fluids accumulating below caprock, which of immigrating driving by differences in density, in icons concentration, and in temperature. For details in thermal factor, the injected acid gas is much cooler than the reservoir rock, thus cooling effects (shrinkage) could occur and reduce the compressive stress, producing a tensile stress. The greater differential temperature between the injected fluids and reservoir rock is, the more extensive scope of the tensile stress or fractures is. The factors that elevate this failure mechanism conclude below [44]:

- High reservoir compressibility;
- Stiff caprock;
- Large pressure changes;
- Low strength caprock, especially containing weak bedding planes;
- Shallow depths (as the normal stress acting cross the reservoir-caprock interface will be small);
- Domed or anticlinal reservoirs

15.2.3.5 Ground Surface Formation

Continuous gas injection is regard as a process of accumulating pore pressure in reservoir rock. Furthermore, this continuously increasing pore pressure would lead to stress expansion in a reservoir rock based on elastic-plastic deformation or fracturing, and this volumetric expansion is shown macroscopically cap-rock formation or even ground surface formation in selected areas. For example, the ground surface formation in the Krechba field, a CO_2 geological project in Salah, Algeria, have already showed an accumulate uplift of 2.5 cm at a rate of 5 mm per year monitored by InSAR [8, 45]. Distinctively contrast to a ground subsidence due to pressure drop in the nearby depleted gas fields.

15.2.3.6 Fault Reactivation and Induced Seismic Events

Fault reactivation will occur when the maximum shear stress acting in the fault plane exceeds the shear strength of the fault, thereby could provide a possible path for leakage, and even might induce seismic events. Inactive faults could turn to reactivate: 1) due to pore pressure increase with a reduction in effective stress in the fault plane, 2) due to in situ stress changes induced by reservoir pressure changes, 3) due to reservoir compaction in the overburden [44], and 4) due to its shear strength deduced by fluids lubricating or geochemical reactions. At Weyburn, a CO_2 storage project in Australia, approximately 100 micro-seismic events with magnitudes ranging from -1 to -3 have been detected since 2004[46]. Another example in the Rocky Mountain Arsenal, a number of earthquakes occurred in the Denver areas as a result of nerve gas injection into a 3,700 m deep fractured reservoir in the 1960s [47].

Inclusion, integrity of caprock includes sealing capacity which should defend physical properties evolution caused by long-term H_2S/CO_2-brine-rock interaction, and mechanical stability which defend the variation of stress field.

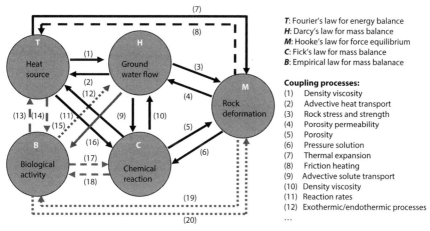

Figure 15.6: Schematic of an Unified THMCB System for Acid Gas Mixtures Flowing

15.3 Coupled Processes

Acid gas injection could induce a range of coupled thermal, hydraulic, mechanical, chemical, and biological processes (THMCB), including heat transfer, multiphase fluids flow in porous media, mechanical responses, geochemical reactions, and biological activities [48-50] (Figure 15.6). In this unified THMCB system [51-53], energy balance in using the Fourier' law for thermal, force balance in using the Hooke' law for mechanics, fluid mass balance in using the Darcy' law for hydraulic, icons mass balance in using the Fick' law for chemistry, and organism mass balance in using an empirical law for biological activities. Specifically, heat transfer will affect reaction rates. These induced chemical reactions could change porosity and pressure solution in rock mechanics, thereby effect fluid flow by change in permeability, stress and strength.

It is reasonable to expect that caprock permeability and mechanical properties will sometimes be affected by geochemical factors. To understand the prior coupled process, chemical factor affect on mechanics take placed by at least three forms: (1) the dissolution or precipitation of minerals leading to increased porosity and permeability, directly impacting the pore pressure field and indirectly affecting the stress fields; (2) the H_2S/CO_2-brine-rock interaction weakens the mechanical properties of the rock framework, thus could also affect the state of stress; and (3) the aperture change of the fractures induced by chemical reactions, which is regarded as an equivalent strain in calculating the stress.

15.4 Failure Criteria

The constitutive models usually contain elastic, elastoplastic, rheological, and damage. Only a small number of models, including poroelastic constitutive models and dual porosity models, have been applied to the geomechanical models for CO_2 storage [54]. An innovative model of rock failure mechanism in AGI is looking forward to building up, which could highly improve the accuracy in the long-term, such as considering mechanical properties evolved by chemical reactions and thermal transfer. Before this model could be developed, a significant number of experiments should be taken firstly, such as H_2S flooding experiment and failure process of the rock which interacted by acid gas mixtures.

15.5 Conclusions

The key mechanical-related problems for acid gas injection might be concluded as follows:

1) Phase behaviors and physical properties for acid gas mixtures, and integrity of injection wells;
2) Fluid migration mechanisms and geochemical reactions in reservoir rock;
3) Diffusive lose, capillary leakage, fractures and fault leakage in sealing capacity of caprock;
4) Injection-induced stress-strain changes, induced shear failure, ground surface deformation, and fault reactivation and induced seismic events.
5) THMCB processes coupled is also a vital problem for an AGI project, especially in what certain the coupled factors affect the mechanical properties in a long-term view.
6) Before an innovative and more suitable failure criteria developed for an AGI project, experiments on H_2S flooding and failure process of the rock which interacted by acid gas mixtures should be taken first.

15.6 Acknowledgements

Q.L. thanks the hundred talent program of Chinese Academy of Sciences and NSFC (Grant No. 41274111).We also thank J.J. Carroll for very good suggestions and comments to raise the great improvements of this manuscript.

References

1. A. Chakma.Acid gas re-injection - A practical way to eliminate CO_2 emissions from gas processing plants.*Energy Conversion and Management*, Vol. 38, p S205-S209,(1997).
2. J. S. Eow.Recovery of sulfur from sour acid gas: A review of the technology. *Environmental Progress*, Vol. 21, p 143–162,(2002).
3. L. Whatley.Acid-gas injection proves economic for West Texas gas plant.*Oil & Gas Journal*, Vol. 98, p 58–62,(2000).
4. R. Sikora; S. Wong; W. D. Gunter; et al. Economic and Emission Accounting for Acid-Gas Injection Projects - An Example from Keyspan Brazeau River, Alberta, Canada, in *Seventh International Conference on Greenhouse Gas Control Technologies (GHGT-7)*: Vancouver, Canada. p. 1–9, (2004)
5. Q. Li; X. Liu; L. Du; et al.Economics of Acid Gas Injection with Comparison to Sulfur Recovery in China.*Energy Procedia*, Vol. 37, p. 2505–2510.(2013).
6. S. Bachu; K. Haug; K. Michael; et al., "Deep Injection of Acid Gas in Western Canada", in C.F. Tsang and J.A. Apps, Eds., *Developments in Water Science*, Elsevier, p. 623–635, (2005).
7. Q. Li; X. Li; L. Du; et al., "Potential Sites and Early Opportunities of Acid Gas Re-injection in China", in Y. Wu, J.J. Carroll, and W. Zhu, Eds., *Sour Gas and Related Technologies*, Wiley Scrivener, p. 131–140, (2012).
8. J. P. Morris; Y. Hao; W. Foxall; et al., A study of injection-induced mechanical deformation at the In Salah CO_2 storage project.*International Journal of Greenhouse Gas Control*, Vol. 5, p. 270–280,(2011).
9. M. Pooladi-Darvish; S. Bachu; H. Hong. Pressure Build-up and Decay in Acid Gas Injection Operations in Reefs in the Zama Field, Canada, and Implications for CO_2 Storage.*Oil & Gas Science and Technology-Revue D Ifp Energies Nouvelles*, Vol. 66, p.67–80,(2011).
10. S. Sayegh; S. Huang; Y. P. Zhang; et al.Effect of H_2S and Pressure Depletion on the CO_2 MMP of Zama Oils.*Journal of Canadian Petroleum Technology*, Vol. 46, p. (2007).
11. F. Biscay; A. Ghoufi; V. Lachet; et al., Monte Carlo Simulations of the Pressure Dependence of the Water-Acid Gas Interfacial Tensions.*Journal of Physical Chemistry B*, Vol. 113, p.14277–14290,(2009).

12. S. Bachu; K. Haug; K. Michael. Stress Regime at Acid-Gas Injection Operations in Western Canada, ERCB/AGS, Alberta, p. 49, (2008).
13. D. M. LeNeveu.Analysis of potential acid gas leakage from wellbores in Alberta, Canada.*International Journal of Greenhouse Gas Control*, Vol. 5, p. 862–879,(2011).
14. C. Diller. Field/WellIntegrity Issues, Well-Abandonment Planning, and Workover Operations on an Inadequately Abandoned Well: Peace River, Alberta, Case Study.*SPE Drilling & Completion*, Vol. 26, p. 540, (2011).
15. S. A. Smith; P. McLellan; C. Hawkes; et al. "Geomechanical testing and modeling of reservoir and cap rock integrity in an acid gas EOR/sequestration project, Zama, Alberta, Canada", in J. Gale, H. Herzog, and J. Braitsch, Eds., *Greenhouse Gas Control Technologies 9*, p. 2169–2176, (2009).
16. P. Samier; A. Onaisi; S. d. Gennaro.A Practical Iterative Scheme for Coupling Geomechanics With Reservoir Simulation.*Spe Reservoir Evaluation & Engineering*, Vol. 11, p. 892–901, (2008).
17. X. Li; H, Koide; T. Ohsumi. CO2 Aquifer Storage and the Related Rock Mechanics Issues.*Chinese Journal of Rock Mechanics and Engineering*, Vol. 22, p.989–994,(2003).(*in Chinese*)
18. D. B. Bennion; F. B. Thomas; B. E. Schulmeister, et al. The Phase Behaviour of Acid Disposal Gases and the Potential Adverse Impact on Injection or Disposal Operations.*Journal of Canadian Petroleum Technology*, Vol. 43, p24-29,(2004).
19. F. Man;J. J. Carroll. "Limitations And Challenges Associated With The Disposal Of Mercaptan-Rich Acid Gas Streams By Injection - A Case Study", *Acid Gas Injection and Related Technologies*, John Wiley & Sons, Inc., p. 129–141, (2011).
20. J. J. Carroll and A. E. Mather.An examination of the vapor-liquid equilibrium in the system propane-hydrogen sulfide.*Fluid phase equilibria*, Vol. 81, p.187–204,(1992).
21. J. J. Carroll and D. W. Lui, Density, phase behavior keys to acid gas injection. *Oil & Gas Journal*, Vol. 95, p. 63–72,(1997).
22. Z. H. Duan; R. Sun; J. W. Hu. A model for calculating the solubility of gases (CO2, H2S.)used for the sequestration of global warming gases. *15th Annual V M Goldschmidt Conference*, Moscow, ID, Pergamon-Elsevier Science Ltd, 2005.
23. B. Bennion and S. Bachu, Drainage and Imbibition Relative Permeability Relationships for Supercritical CO2/Brine and H2S/Brine Systems in Intergranular Sandstone, Carbonate, Shale, and Anhydrite Rocks.*SPE Reservoir Evaluation & Engineering*, Vol. 11, p. 487–496, (2008).
24. J. J. Carroll, Phase equilibria relevant to acid gas injection, part 1 - Non-aqueous phase behaviour.*Journal of Canadian Petroleum Technology*, Vol. 41, p. 25–31,(2002).
25. J. J. Carroll, Phase equilibria relevant to acid gas injection: Part 2 - Aqueous phase behaviour.*Journal of Canadian Petroleum Technology*, Vol. 41, p. 39–43,(2002).

26. M. Zhang and S. Bachu. Review of integrity of existing wells in relation to CO2 geological storage: What do we know? *International Journal of Greenhouse Gas Control*, Vol. 5, p. 826–840,(2011).
27. T. L. Watson and S. Bachu. Evaluation of the Potential for Gas and CO2 Leakage Along Wellbores.*SPE Drilling & Completion*, Vol. 24, p. 115–126, (2009).
28. S. Bachu and T. L. Watson. Review of failures for wells used for CO_2 and acid gas injection in Alberta, Canada.*Energy Procedia*, Vol. 1, p. 3531–3537,(2009).
29. M. Loizzo; O. A. Akemu; L. Jammes, et al. Quantifying the Risk of CO2 Leakage Through Wellbores.*SPE Drilling & Completion*, Vol. 26, p. 324–331, (2011).
30. J. Pironon, N. Jacquemet, T. Lhomme, et al., Fluid inclusions as micro-samplers in batch experiments: A study of the system C-O-H-S-cement for the potential geological storage of industrial acid gas.*Chemical Geology*, Vol. 237, p. 264–273,(2007).
31. N. Jacquemet, J. Pironon, and J. Saint-Marc, Mineralogical changes of a well cement in various H_2S-CO_2(-brine) fluids at high pressure and temperature. *Environmental Science & Technology*, Vol. 42, p. 282–288,(2008).
32. B. G. Kutchko; B. R. Strazisar; S. B. Hawthorne, et al., H2S-CO2 reaction with hydrated Class H well cement: Acid-gas injection and CO2 Co-sequestration. *International Journal of Greenhouse Gas Control*, Vol. 5, p. 880–888,(2011).
33. S. Nesic, Key issues related to modelling of internal corrosion of oil and gas pipelines – A review.*Corrosion Science*, Vol. 49, p. 4308–4338,(2007).
34. S. Bachu, J. M. Nordbotten, and M. A. Celia, "Evaluation of the spread of acid-gas plumes injected in deep saline aquifers in western canada as an analogue for CO_2 injection into continental sedimentary basins", in E.S. Rubin, D.W. Keith, and C.F. Gilboy, Eds., *Greenhouse Gas Control Technologies, Volume I*, Elsevier, pp. 479–487, (2005).
35. S. B. Hawthorne, D. J. Miller, Y. Holubnyak, et al., "Experimental investigations of the effects of acid gas (H2S/CO2) exposure under geological sequestration conditions", in J. Gale, C. Hendriks, and W. Turkenberg, Eds., *10th International Conference on Greenhouse Gas Control Technologies*, pp. 5259–5266, (2011).
36. J. T. H. Andres and S. S. Cardoso, Onset of convection in a porous medium in the presence of chemical reaction.*Physical Review E*, Vol. 83,(2011).
37. L. Richard, N. Neuville, J. Sterpenich, et al., Thermodynamic analysis of organic/inorganic reactions involving sulfur: Implications for the sequestration of H_2S in carbonate reservoirs.*Oil & Gas Science and Technology-Revue De L Institut Francais Du Petrole*, Vol. 60, p. 275–285,(2005).
38. T. F. Xu, Y. Ontoy, P. Molling, et al., Reactive transport modeling of injection well scaling and acidizing at Tiwi field, Philippines.*Geothermics*, Vol. 33, p. 477–491,(2004).
39. T. Xu, K. Pruess, J. A. Apps, et al., Mineral Sequestration of CO_2 mixed with H_2S and SO_2 in Sandstone-Shale Formation.*EOS, Transactions American Geophysical Union*, Vol. 85, p. 319–346,(2004).

40. M. Fleury, P. Berne, and P. Bachaud, Diffusion of dissolved CO2 in caprock. *Energy Procedia*, Vol. 1, p. 3461–3468, (2009).
41. V. Shah, D. Broseta, G. Mouronval, et al., Water/acid gas interfacial tensions and their impact on acid gas geological storage.*International Journal of Greenhouse Gas Control*, Vol. 2, p. 594–604,(2008).
42. J. Rutqvist, J. T. Birkholzer, and C.F. Tsang, Coupled reservoir–geomechanical analysis of the potential for tensile and shear failure associated with CO2 injection in multilayered reservoir–caprock systems.*International Journal of Rock Mechanics and Mining Sciences*, Vol. 45, p.132–143,(2008).
43. British Columbia Geological Survey, Acid Gas Injection: A Study of Existing Operations Phase I: Final Report, Ministry of Energy, Mines and Petroleum Resources, British Columbia Geological Survey, p. 71, (2003).
44. C. D. Hawkes, S. Bachu, and P. J. McLellan, Geomechanical factors affecting geological storage of CO_2 in depleted oil and gas reservoirs.*Journal of Canadian Petroleum Technology*, Vol. 44, p. 52–61,(2005).
45. J. Rutqvist, D. W. Vasco, and L. Myer, Coupled reservoir-geomechanical analysis of CO_2 injection and ground deformations at In Salah, Algeria. *International Journal of Greenhouse Gas Control*, Vol. 4, p. 225–230,(2010).
46. J. B. Riding, "THE IEA WEYBURN CO2 MONITORING AND STORAGE PROJECT", in S. Lombardi, L.K. Altunina, and S.E. Beaubien, Eds., *Advances in the Geological Storage of Carbon Dioxide*, Springer Netherlands, p. 221–230, (2006).
47. P. A. Hsieh and J. D. Bredehoeft, A reservoir analysis of the Denver earthquakes: A case of induced seismicity.*Journal of Geophysical Research*, Vol. 86, p. 903–920,(1981).
48. H. Yasuhara, Thermo-Hydro-Mechano-Chemical Couplings that Define the Evolution of Permeability in Rock Fractures, in *College of Earth and Mineral Sciences, Department of Energy and Geo-Environmental Engineering*, The Pennsylvania State University, (2005).
49. http://isrm.net/fotos/gca/1163783695yasuhara_summary.pdf
50. D. Elsworth and H. Yasuhara, Short-Timescale Chemo-Mechanical Effects and their Influence on the Transport Properties of Fractured Rock.*Pure and Applied Geophysics*, Vol. 163, p. 2501–2570,(2006).
51. J. S. Liu, J. C. Sheng, A. Polak, et al., A fully-coupled hydrological-mechanical-chemical model for fracture sealing and preferential opening.*International Journal of Rock Mechanics and Mining Sciences*, Vol. 43, p. 23–36,(2006).
52. Q. Li and K. Ito, "Numerical Analysis and Modeling of Coupled Thermo-Hydro-Mechanical (THM) Phenomena in Double Porous Media", in R.H. Laughton, Ed. *Aquifers: Formation, Transport and Pollution*, Nova Science Publishers, p. 403-413, (2010).
53. Q. Li. Coupled T-H-M-C-B effects on long-term performance assessment of HLW disposal in Japan.*The 18th Symposium on Environmental Chemistry*, Tsukuba, Japan, Japan Society for Environmental Chemistry (JEC), (2009).

54. Q. Li; K, Ito. An Integrated Thermal-Hydraulic-Mechanical-Chemical-Biological (THMCB) Multiscale and Multiphysics Coupled System with Applications to Geological Disposal Problems.*AIST Symposium: The Annual Conference of GREEN*, (2008).
55. Q. Li, Z. S. Wu, X.L. Lei, et al., Experimental and numerical study on the fracture of rocks during injection of CO_2-saturated water.*Environmental Geology*, Vol. 51, p. 1157–1164,(2007).

16
Comparison of CO_2 Storage Potential in Pyrolysed Coal Char of different Coal Ranks

Pavan Pramod Sripada[1], MM Khan[1], Shanmuganathan Ramasamy[1], VajraTeji Kanneganti[2], Japan Trivedi[2], and Rajender Gupta[1]

[1]*Canadian Centre for Clean Coal, Department of Chemical and Materials Engineering, University of Alberta, Edmonton, AB: Canada.*
[2]*School of Mining and Petroleum, Department of Civil and Environmental Engineering, University of Alberta, Edmonton, AB: Canada.*

Abstract

In recent years, post-underground gasification (UCG) sites act as a potential sink for carbon dioxide storage due to their enhanced storage capacities. However, the magnitude of sequestration possible is not well understood, particularly, the influence of coal properties. In the present study, the adsorption capacity was determined for different coal rank samples (sub-bituminous and bituminous). Furthermore, a comparison is drawn between the adsorption capacities of the virgin coal and the pyrolyzed coal chars. The chars were pyrolyzed at a temperature of 950°C with a constant heating rate of 10°C / minute. The adsorption capacity of the samples was determined using a volumetric adsorption setup at operating pressures of 600 and 1000 psig. The experiments were performed with gas temperatures of 35 and 45°C. The result shows that adsorption capacity of coal chars is much higher than the virgin coal samples. It was observed from the pore analysis study that the macro-porous surface areas of coal char are much greater than the case of virgin coals, leading to better sorption capacities.

16.1 Introduction

Carbon dioxide being one of the most discharged greenhouse gas (GHG) by the industry, which is contributing in a significant magnitude to the global climate change (1). Many 'green' technologies are being developed to either reduce carbon dioxide emission or to capture carbon dioxide and identify avenues for sequestration (2). Among the available technologies, Geological storage of GHG is practised due to its unique advantages like Coal Bed Methane recovery (CBM), Enhanced Oil Recovery (EOR), enormous storage capacity, and agreeable economics(1).

In Canada, the coal reserves in Alberta province are mostly too deep to be mined by traditional methods (3,4,5). Within the coal technologies, Under Ground Gasification (UCG) technique has paved way to access these deep mines. In specific, the attractive economics and reduced carbon emission has emerged the technique viable (6,7). Further, the UCG white paper led by Sherritt International Corporation will lead to more UCG demonstration projects in Alberta, Canada (4).UCG technology was implemented in many countries like Russia, USA, Australia, Canada, South Africa, India, and Japan (8,9,10). Alberta province has one of the deepest UCG operations (i.e., 1400 m) in the world (11).

Figure 16.1 illustrates the outline of vertical linked well UCG process. In this process, the injection line and a production line are vertically drilled into the coal seam at a set-distance from each other. Initially, the coal seam was dried and ignited with the help of oxidant. Further, a mixture composing of air, oxygen, and steam are feed through the injection line in a particular ratio to initiate the gasification process. In this process, a synthetic natural gas was produced, which primarily consists of carbon monoxide, hydrogen, methane, hydrocarbons, carbon dioxide, and moisture (8,10,12).

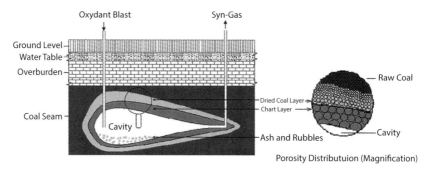

Figure 16.1 Schematic of the UCG process and the post burn cavity

The magnified image in Figure 16.1 represents the four different strata in post UCG site such as rubble, gasified layer, pyrolyzed layer, and virgin coal. The rubble that is formed as a result of the gasification settles at the bottom of the cavity. Gasified layer is the one that is left after gasification process has taken place. Pyrolyzed char layer is that layer of coal which has been exposed to high temperatures in a nearly inert environment or that is said to have undergone pyrolysis. The third layer is the raw coal that has not experienced any change at all. The layers (see Figure 16.1) of the post UCG site will have enhanced porosity in comparison to the virgin coal[13], thus it become a potential viable and a beneficial option in terms of reduced costs and better storage capacity(14,15,16,17). However, the nature of the resulting porosity in post UCG process was not well understood till date for the development of Carbon Capture and Storage (CCS) technique. In particular, a fundamental knowledge about the coal properties is required to assess the rate of CO_2 adsorption and storage capacity.

In this work, the adsorption capacities of carbon dioxide on virgin coals was estimated and compared with pyrolyzed coal sample. In addition, the influence of coal rank was also studied and compared. Further, a surface area analysis study was performed for the studied sample.

16.2 Apparatus, Methods, & Materials

The adsorption capacity measurements were carried using a volumetric adsorption method. This method was chosen so that larger samples can be analysed and the effects due to heterogeneity in coal samples can be considered during experiments. Figure 16.2 shows the laboratory scale experimental setup that was designed.

Core of the experimental setup consists of the reference and the sample cells, where both are placed in a Thermo-Scientific water bath to maintain isothermal conditions during experiments. Pressure Transducers and thermocouples are placed close to these cells to measure the gas pressure and the gas temperature at all times. The temperature readings are obtained by using a NI-DAQ card and a read-out from the lab view software. The pressure readings were translated using TRH central software from Omega. A high pressure syringe pump (*ISCO 500D*) was used to pressurize the gas into the cells.

In this study, the void volume of the sample was determined by helium displacement method. In this method, a known volume of Helium is injected into the sample cell, the Helium molecules would penetrate

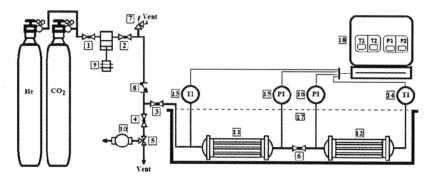

Figure 16.2 Experimental set-up for volumetric adsorption test in an isothermal environment. [1, 2, 3, 4, 5, & 6: Valves; 7: Vent; 8: Check valve; 9: High pressure syringe pump; 10: Vacuum Pump; 11: Reference cell; 12: Sample cell; 13 & 14: Temperature sensor; 15 & 16: Pressure transducer; 17: Water bath; 18: Data acquisition system]

through the studied samples' pores. The volume occupied by the Helium molecules represents the void volume of the sample cell. Prior to the start of any experiment, the system was evacuated using a vacuum pump to ensure 0 atm. Afterwards, the Helium gas was injected into the reference cell at a high pressure of 1200 psi. After stabilizing, the gas was equilibrated between the sample and the reference cell. The final pressures were noted to calculate the helium void volume of the sample. Bulk volume of the coal char samples was also estimated using a glass bead displacement to determine the effective porosity of the sample.

Before the start of any adsorption capacity determination experiment, a vacuum is created within the setup with the help of a vacuum pump. The gas is pressurized in the syringe pump to any desired pressure and injected into the reference cell. After stabilizing in the reference cell for an hour, the gas is let into the sample cell where the sample is placed by opening the valve present in between them for equilibration. The time for equilibration was maintained at 2 hours for any run. The pressures from the transducers are noted before and after equilibration process. The process described above is repeated again. After the final pressure, the gas is then released through the vent and vacuum conditions are brought again with the vacuum pump.

A mass balance is used to calculate the gas that is adsorbed on a solid. For the system at the initial conditions, the amount of gas present in the reference cell can be written as

$$n_{inital} = \rho_1 \times V_R \qquad (16.1)$$

Where, ρ_1 is the molar density of the gas at the set temperature and pressure. V_R is the volume of the reference cell (inclusive of the valves and the tube fittings). The Helium and the Carbon dioxide molar density data have been obtained from NIST web book (18).

The equation below represents the mass balance

$$\rho_1 \times V_R = \rho_i \times V_R + \rho_i \times (V_{cell} - V_{solid} - V_a) + \rho_a \times V_a \qquad (16.2)$$

V_{cell} is the total volume of the sample cell (inclusive of the valves and the tube fittings). V_{solid} is the skeletal volume of the solid (Does not include the volume in the pores of the solid). Naturally, the term $V_{cell} - V_{solid}$ is the void volume of the system. For the present study, Helium was used for the determination of the void volume of the samples.

The terms with the suffix 'a' in (16.2) refers to the properties of the adsorbed phase and cannot be experimentally determined. Hence, we define Gibbs Surface Excess adsorption as shown below,

$$n_{excess} = V_a(\rho_a - \rho_i) \qquad (16.3)$$

The surface excess is the difference between the actual amount adsorbed at a density of r_a and the amount that would be adsorbed if the density of the gas in the adsorbed phase is same as the bulk gas phase (19,20).

16.2.1 Sample Characterization

In this study, two coals of different ranks were used for analysis. Proximate and the Ultimate analysis for the coal samples are provided in Table 16.1. Coal A is the coal with a higher rank and the Coal B is a low-rank Coal. In addition, Coal B is a coking coal while Coal A is a sub-bituminous coal. Particle size used for the adsorption varied from 22–32 mm.

The coal char samples were pyrolyzed using LECO TGA710 instrument at a temperature of 950°C with a constant heating rate of 10°C / minute. The coal chars in this study are referred to as Coal A- 950 and Coal B-950.

Figure 16.3 shows BET surface areas for the studied sample using Micromeritics ASAP 2020 with nitrogen as the probe molecule at 77 K. Results clearly shows that the char sample has more surface area in comparison to the virgin coal samples for both the coal ranks. Surface area and micropore analyses experiments were carried on ~ 400 mg samples. The samples were degassed at 250°C till they reached vacuum set point of 0.5 Pa. The degassing time was at least 4 hours per sample.

Table 16.1 Proximate and Ultimate Analysis of the Coals (21)

	Proximate analysis (wt. %)			
	Moisture (ad)	Ash (ad)	VM (daf)	FC (ad)
Coal A	5.01	10.39	31	58.39
Coal B	1.26	14.07	27	61.85
	Ultimate analysis (wt. %; dry basis)				
	S	C	H	N	O
Coal A	0.56	66.8	3.86	0.78	8.99
Coal B	0.61	79.4	4.48	1.46	3.83

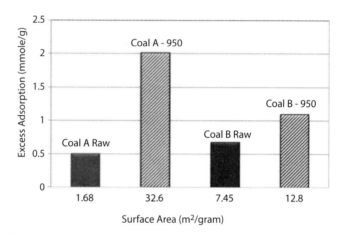

Figure 16.3 Surface Analysis of the virgin coal and coal char at 950°C

16.3 Results And Discussion

16.3.1 Repeatability of adsorption experiments

Figure 16.4 describes the sorption isotherms of Coal A. Multiple runs were performed at a temperature of 45.5°C and were tested at similar pressures of (± 4 psi). It is evident from the figure that there is an increase in the excess sorption till the pressure of 40 bar, similar to a Langmuir trend, as the pressure closes on the critical pressure of carbon dioxide a sharp minimum is observed and a rapid increase thereafter. The maximum sorption capacity for the coal was 2.2, 2, and 1.8 mmole/g observed to be at a pressure close to 120 bar for Run 1,2, and 3 respectively.

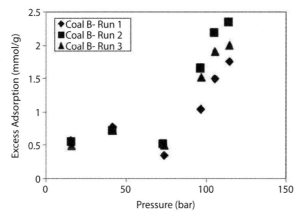

Figure 16.4 Repeatability of the adsorption measurements in the experimental setup

The excess adsorption data trend obtained from the multiple run resembles each other. However, the standard deviation of the data obtained above 80 bar is higher in comparison to the low pressure data. In specific, the mean standard deviation for the sorption capacities for all the pressures is found to be 0.190 mmole/g.

16.3.2 Adsorption capacities of coal

Many carbon dioxide-coal adsorption studies were performed by research groups all over the globe (22, 23, 24, 25, and 26). The data from this work (see Figure 16.5) compares well with the sorption studies by Krroos et al.(2002). It can be seen that the three coals have a similar trend of increasing excess adsorption till about 50 bar and then a sharp dip in the range of 75 and 100 bar for the coal in present study and literature data respectively. The maximum sorption capacity of coal A and coal B is around 1 and 2 mmole/g respectively. Coal B has a greater sorption capacity than coal A for all the pressures. The difference in the sorption capacities of the coals can be attributed to the surface area of the pores. From the BET surface area analysis, coal A has a greater surface area when compared to coal B. This difference would reflect that there is lesser number of micropores available for the gas to penetrate and get adsorbed. The sorption capacity can also be linked with the coal property. The rank, vitrinite, ash and volatile matter content are three of the parameters that can be considered for addressing this behavior. Many researchers have suggested that the sorption capacity

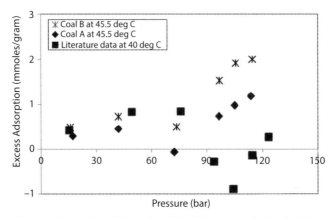

Figure 16.5 Excess Adsorption of Coal A and Coal B compared with the literature data (22).

increases with the decreasing ash, volatile matter content and increases with the vitrinite content in the raw coal (27, 28, 29, 30, 31).

Vitrinite content in Coal A is 32.2% and for Coal B is 44.8%. Since vitrinite is a carbon based maceral, more of vitrinite is going to mean more carbon available for carbon dioxide adsorption. Also, from the proximate analysis, it can be noticed that the volatile matter is lesser for coal B than coal A. The effect of coal property is consistent with expected results from other researches. However, it is only a general trend that has been observed for the coal property effect and quantitative analysis on the effect of each of these parameters with respect to the excess adsorption of carbon dioxide is yet to be done.

16.3.3 Adsorption capacities of coal chars

Figure 16.6 describes the adsorption capacity on coal chars. For coal chars it is observed that the dip in the excess sorption curves is not as sharp as in the case of the raw coal. The adsorption capacity of the coal chars is at least 2 times higher in the case of coal A and at least 1.5 times in the case of coal A. The sorption capacity of coal A -950 is greater than the coal B 950. The surface area analysis supports this outcome. The surface area of coal char A was 32 m²/g and coal char B is 12 m²/g. It should also be noted that the increase in the surface area due to pyrolysis at 950 degrees in coal A is approximately 19 times the raw coal whereas for coal B-950 it is just 1.7 times. Such a large difference in the increase in surface area is the action of heat on the kind of coal. Coal B being a coking coal, the tar that did not

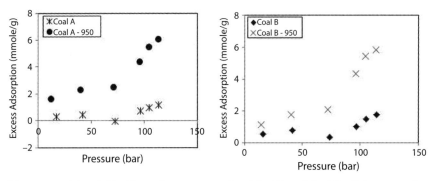

Figure 16.6 (a) and (b): Adsorption amounts of coal A -950 and Coal B – 950 compared with their respective raw coals

come out of the coal during the pyrolysis blocks the pores of the material, thus decreasing the surface area available for adsorption. It can also be argued that the large increase in surface area is not reflected in the amount adsorbed as the quantity of gas adsorbed in both the case were around 6 mmole/g. The nature of macerals present in the coal has to be considered in order to explain such behaviour.

16.3.4 Effect of temperature on blank test

Figure 16.7 shows the comparative plot of mass balance residue against pressure for blank runs at two different isotherms at 35.5°C & 45.5°C. The mass balance residue is the difference final moles and the initial mole in the system.

Blank test experiments were performed to understand the system induced errors such as the behaviour of gas, pressure sensor error, equilibration times, and the fluctuation in the water bath temperature. In this study (see Figure 16.7), temperature is the only significantly varied parameter hence it can be deduced that the effects presented are solely due to the behaviour of gas as a function of temperature. Figure 16.7 clearly show that there a significant reduction in the mass balance residue due to the influence of temperature. However, this effect became abnormal at critical pressure of 73 bar. This can be attributed by the error of the pressure transducer which is about 0.14% of the full scale range (i.e., ±3.5 PSI). Pressure transducer accuracy becomes important because the CO_2 molar density is very sensitive near the critical pressure and this can potentially lead to large errors. From our observation, we found that in addition to the accuracy of the pressure transducer and volume ratio between the reference and sample cell plays a vital role in reducing the magnitude of error.

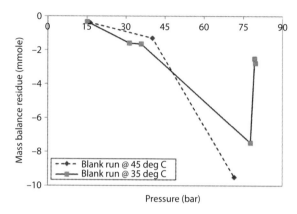

Figure 16.7 Comparison of Blank Tests conducted at different gas temperatures

16.4 Conclusion

In this paper, the CO_2 storage potential of pyrolyzed coal sample produced at 950°C was compared with virgin coal sample using volumetric set-up. The current results suggest that the adsorption capacity of pyrolyzed sample is comparatively higher than virgin sample. In addition, adsorption capacity varies with coal rank and operating pressure. From adsorption point of view, the sample porosity increases the surface area and enhances the storage potential. The effect of pore volume plays a vital role in determining the total storage capacity, in particular for pyrolyzed sample. As such, the CCS system subsequent to UCG will be a preferable option to provide a cleaner energy with a smaller carbon footprint. The knowledge obtained from this study will be beneficial for both undergoing and any future projects in underground coal gasification with carbon capture system. From the BET surface area analysis, it is observed that the surface area of coal chars is greater than the surface area of virgin coals. The sorption capacity experiments indicate that the carbon dioxide adsorption is greater in coal chars. Higher rank coals are found to adsorb more CO_2.

References

1. IPCC, Special Report on Carbon Dioxide, Capture and Storage, Cambridge University Press, Cambridge, UK and New York, USA, 2005.
2. European Commission-Cummunity Research, Directorate-General for Research Sustainable Energy Systems, Report on "CO2 capture and storage projects", Luxembourg, 2007.

3. Richardson R.J., "Alberta's 2 trillion tonnes of 'unrecognized' coal", Alberta Innovates Energy and Environment solutions, August, 2010.
4. Richarson R.J.H., Singh.S., "Prospects for underground coal gasification in Alberta, Canada", Proceedings of the Institution of Civil Engineers, Energy, Vol.165 Issue EN3, pp. 125–136, 2012.
5. Pana, C. Review of underground coal gasification with reference to Alberta's potential; Energy Resources Conservation Board, ERCB/AGS, Open File Report 2009–10, 56 p, 2009.
6. Blinderman MS, Anderson B "Underground coal gasification for power generation:
7. Efficiency and CO2-emissions". Proc. 12th International Conference on Coal Science, Cairns, Australia,; Paper No. 13C1,2003.
8. Liu S-Q., Liu J-H., Yu.L., "Environmental benefits of Underground Coal Gasification" , Journal of Environmental Sciences, Vol. 14, No.2, pp. 284–288, 2002.
9. Gregg D.W., Edgar T.F., Underground Coal Gasification, AIChE Journal, Vol. 24 , No. 5, pp. 753–781, 1978
10. Klimenko A.Y., "Early Ideas in Underground Coal Gasification and Their Evolution", Energies, 2, pp.456–476, 2009.
11. Shafirovich.E.,Varma A.," Underground Coal Gasification: A Brief Review of Current Status", Ind. Eng. Chem. Res., 48, pp7865–7875, 2009,.
12. Swanhillssynfuels-Alberta Innovates – Enery and Environment solutions, "Swan hills in-situ coal gasification technology development – Final Outcomes Report", 2012.
13. Eskom Holdings SOC limited, UCG Technology, website: http://www.eskom.co.za/c/article/86/ucg-technology/, 2013.
14. Mahajan O.P., Walker P.L., Jr., "Porosity of coals and coal products", Pennsylvania.
15. Roddy R.J., Younger P.L., "Underground coal gasification with CCS: a pathway to decarbonising industry", Energy Environ. Sci., 3, 400–407, 2010.
16. Friedmann S.J., Ravi Upadhye.R., Kong F.M., Prospects for underground coal gasification in carbon-constrained world, Energy Procedia 1,pp. 4551–4557, 2009.
17. Self S.J., Reddy B.V., Rosen M.A., Review of underground coal gasification technologies and carbon capture., International Journal of Energy and Environmental Engineering, 3:16, 2012.
18. Reichle D., Houghton J., Kane B., Ekmann J., Benson S., Clarke J., Dahlman R., Hendry G.,Herzog H., Hunter-Cevera J., Jacobs G., Judkins R., Ogden J., Palmisano A., SocolowR.,Stringer J., Surles T., Wolsky A., Woodward N., York M. Carbon Sequestration Research and Development, U.S. Department of Energy Report DOE/SC/FE-1, (1999).
19. Burgess. D.R., "Thermochemical Data" in NIST Chemistry WebBook, NIST Standard Reference Database Number 69, Eds. P.J. Linstrom and W.G. Mallard, National Institute of Standards and Technology, Gaithersburg MD, 20899, http://webbook.nist.gov, 2013.

20. Sircar S., Excess Properties and Thermodynamics of Multicomponent Gas Adsorption, J. Chem. Soc., Faraday Trans. I,81, pp. 1527–1540, , 1985.
21. Sircar S.,Gibbsian Surface Excess for Gas Adsorptions- Revisited Ind. Eng. Chem. Res., 38, pp. 3670–3682, 1999.
22. Tian S., "Fragmentation of Large Coal Particles at High Temperature in a Drop. Tube Furnace", MSc Thesis, University of Alberta, 2011.
23. Krooss B.M., Van Bergen F., Gensterblum Y., Siemons N., Pagnier H.J.M., David P., High-pressure methane and carbon dioxide adsorption on dry and moisture-equilibrated Pennsylvanian coals, International Journal of Coal Geology 51, pp. 69– 92, 2002.
24. Ozdemir E., Morsi B.I., Schroeder K., Importance of Volume Effects to Adsorption Isotherms of Carbon Dioxide on Coals, Langmuir, 19, pp. 9764–9773, 2003.
25. Jian X., Guan P., Zhang W., Carbon dioxide sorption and diffusion in coals: Experimental investigation and modeling, Science China, Earth Sciences,Vol.55 No.4: 633–643, 2012.
26. Bae J.S., Bhatia S.K., High-Pressure Adsorption of Methane and Carbon Dioxide on Coal,Energy& Fuels, 20, pp. 2599–2607, 2006.
27. Yee, D., Seidle, J.P., Hanson, W.B.,. Gas sorption on coal and measurement of gas content. In: Law, B.E., Rice, D.D. (Eds.), Hydrocarbons from Coal. AAPG Studies in Geology, p.38, 1993.
28. Clarkson C.R., Bustin R.M., Binary gas adsorptionrdesorption isotherms: effect of moisture and coal composition upon carbon dioxide selectivity over methane International Journal of Coal Geology Vol. 42, pp 241–271, 2000.
29. Mastalerz M., Gluskoterb H., John Rupp J., Carbon dioxide and methane sorption in high volatile bituminous coals from Indiana, USA International Journal of Coal Geology 60 pp. 43– 55, 2004.
30. Crosdale P.J., Beamish B.B, Marjorie Valix M., Coalbed methane sorption related to coal composition International Journal of Coal Geology 35,pp. 147–158, .2004
31. Day S., Duffy G., Sakurovs R., Wei S., Effect of coal properties on CO2 sorption capacity under supercritical conditions International Journal of Greenhouse gas control. Vol 12. pp 342– 352,2008.
32. Carroll, R. E., and Pashin, J. C. "Relationship of sorption capacity to coal quality: CO2 sequestration potential of coalbed methane reservoirs in the Black Warrior basin," International Coalbed Methane Symposium Proceedings, Tuscaloosa, Alabama, University of Alabama College of Continuing Studies, Paper 0317, 11 p., 2003.

17
Capture of CO_2 and Storage in Depleted Gas Reservoirs in Alberta as Gas Hydrate

Duo Sun[1], Nagu Daraboina[1], John Ripmeester[2], and Peter Englezos[1]

[1]Department of Chemical and Biological Engineering The University of British Columbia, Vancouver, BC, Canada
[2]Steacie Institute for Molecular Sciences, National Research Council Canada, Ottawa, ON, Canada

Abstract

Storage of CO_2 in depleted natural gas reservoirs is considered an attractive option to mitigate climate change concerns arising from the emissions of CO_2 from sources such as thermal power plants. A number of depleted gas reservoirs located in Northern Alberta with the pressure and temperature conditions of 2–5 MPa and 1–10°C, respectively, are identified as potential sites for CO_2 storage in gas hydrate form. CO_2 hydrate is ice-like crystalline compound consisting of hydrogen-bonded water molecules and CO_2 molecules. The water molecules build a network with hydrogen bonds to stabilize CO_2 molecules in it. Equilibrium pressures of CO_2 hydrate corresponding to 1 and 10°C are 1.41 and 4.29 MPa, respectively. Therefore, the pressure and temperature conditions of depleted gas reservoirs in Northern Alberta are within CO_2 hydrate formation region. There are several computer simulations that have demonstrated the possibility of storing CO_2 as solid hydrate in these depleted natural gas pools. However, there has not any laboratory demonstration of CO_2 injection and storage in a simulated depleted gas reservoir.

We are investigating the injection of CO_2 in a bed of silica sand partially saturated with water (wet sand) at pressure and temperature conditions similar to those in a typical depleted gas reservoir in Northern Alberta. Through the pressure drop measurements, the CO_2 gas consumption is determined and also the amount of water converted to hydate is calculated. The results indicate that most CO_2 gas is consumed in a *Semi-batch gas injection* experiment, and 35–40%,

40–45%, and 45–50% of water formed hydrate in *Batch*, *Series-batch*, and *Semi-batch gas injection* experiments, respectively.

17.1 Experimental

The apparatus is a modification of the apparatus described by Linga et al., [1]. A high pressure vessel also can be called crystallizer (CR) will be used. A temperature-controlled water bath was employed to have a required experimental temperature. Four thermocouples were located to measure temperature inside the crystallizer at different positions. The crystallizer pressure was regulated by a PID controller and a control valve. A DAQ (data acquisition system) was employed to record the pressure and temperature data into a computer and transfer order to the control valve.

Table 17.1 shows the experimental conditions for gas cap mode CO_2 injection. A sand bed which simulated conditions of depleted gas reservoir in Northern Alberta was prepared by placing 1650 g of silica sand in the crystallizer. The volume of water required to fill the void space of sand to make water saturation of 0.25 was calculated to be 90 mL. Silica sand and water (sand bed) filled all the inside volume of crystallizer but left a gas phase area (gas cap) above sand bed and a gas phase area among silica sand due to 0.75 gas saturation. A thermocouple was set to measure the temperature of gas cap and another three were set into the sand bed to measure the temperatures of sand bed in top, middle and bottom position of sand bed. Then the crystallizer was closed and located into a temperature controlled water bath of 274.15 K (1 °C). When the crystallizer temperatures measured by four thermocouples reached a temperature of 274.15 K, the crystallizer was pressurized with CO_2 to 1.5 MPa and depressurized

Table 17.1 Experimental conditions for gas cap mode CO2 injection

Experiment	T_{ini} (K)	P_{ini} (KPa)	Sand (g)	Water (mL)
Batch gas injection	274.15	3200	1650	90
Series-batch gas injection	274.15	3200	1650	90
Semi-batch gas injection	274.15	3200	1650	90

to atmospheric pressure for three times to eliminated the presence of air in the experiment system. Subsequently, the CO_2 injection procedure was conducted according to the following three procedures.

(a) Batch gas injection: The crystallizer was pressurized with CO_2 to 3200 KPa and without any further injection of gas. Due to hydrate formation the pressure dropped and the experiment was stopped after 24 hours.
(b) Series-batch gas injection: The crystallizer was pressurized with CO_2 to 3200 KPa. Because of hydrate formation the pressure in the crystallizer decreased. After 6 and 20 hours of operation the crystallizer was pressurized with CO_2 to restore the pressure to 3200 KPa. The experiment was stopped after 24 hours.
(c) Semi-batch gas injection: The crystallizer was pressurized with CO_2 to 3200 KPa then the pressure in the crystallizer was maintained constant of 3200 KPa by continuously injecting CO_2 with the help of the PID controller. The experiment was stopped after 24 hours.

Pressure and temperature of crystallizer data was logged every 20 seconds.

17.2 Results And Discussion

Figure 17.1 shows pressure and temperature profiles of the crystallizer in gas cap mode CO_2 injection experiments and Figure 17.2 shows the profiles during the first 50 minutes. Pressure decreased due to CO_2 in the crystallizer consumed to form hydrate and to dissolve in water. The amount dissolved in water is very small (0.05 mol) compared to that to form hydrate. In (a) *Batch gas injection* experiment, pressure dropped quickly in the first 3 hours and then exhibited a moderate decrease. Hydrate crystal formation is an exothermic process. A temperature rise is seen to occur at the 2-3 minute mark and it is attributed to hydrate nucleation. This period of time until nucleation occurs marks the induction time. Rapid pressure decrease can also be observed in (b) *Series-batch gas injection* experiment at first 3 hours. Temperature rise was also observed at the 1-3 minute mark. When the crystallizer was pressurized to 3200 KPa again, the pressure dropped quickly in 20 minutes and then exhibited a moderate decrease. However the pressure decreased moderately after pressurized the crystallizer at the

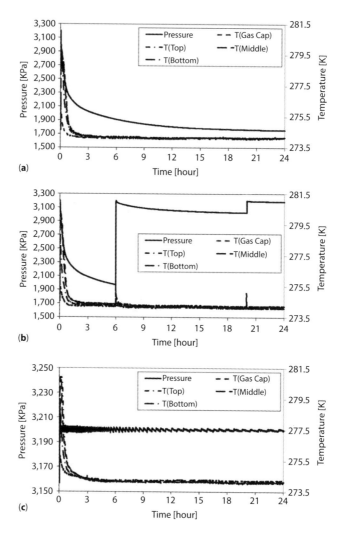

Figure 17.1 Pressure and temperature profiles in crystallizer in gas cap mode CO2 injection experiments. (a) Batch gas injection experiment, (b) Series-batch gas injection experiment, and (c) Semi-batch gas injection experiment.

20 hour mark because there was less hydrate formation potential (less water available). In (c) *Semi-batch gas injection* experiment, constant pressure was maintained at 3200 ± 3 KPa. Temperature dropped down to 274.15 K in two hours and then no significant change was observed. Temperature rise was observed at the 3-5 minute mark as it was seen during the *Batch* and *Series-batch gas injection* experiments.

Table 17.2 shows the number of moles of CO_2 in the hydrate phase (CO_2 stored in hydrate), Percent of water formed hydrate (hydrate formation

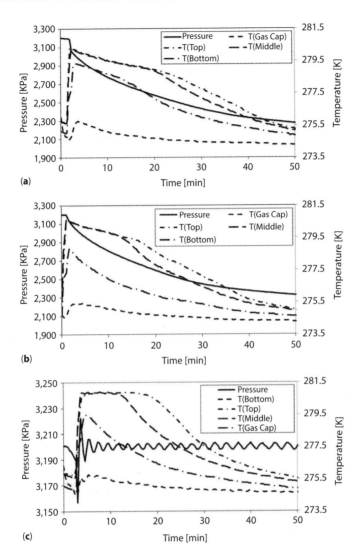

Figure 17.2 Pressure and temperature profiles in crystallizer during the first 50 minutes. (a) Batch gas injection experiment, (b) Series-batch gas injection experiment, and (c) Semi-batch gas injection experiment.

ratio), and the CO_2 hydrate saturation for all three gas cap mode CO_2 injection experiments. The data indicate that the *Semi-batch gas injection* experiment results in more CO_2 stored which is 0.41 mol compared to the other two injection experiments. The percent of original water converted to hydrate in *Batch, Series-batch* and *Semi-batch gas injection* experiments was 39 %, 42% and 47% respectively. After 24 hours, the CO_2 hydrate

Table 17.2 The number of moles of CO2 stored in hydrate (nCO2,stored), water conversion ratio (Rw), and the CO2 hydrate saturation (Sh) for gas cap mode CO2 injection experiments.

Experiment	$n_{CO2,stored}$ (mol)	R_w (%)	S_h
Batch gas injection	0.338	39	0.107
Series-batch gas injection	0.361	42	0.114
Semi-batch gas injection	0.413	47	0.130

saturation reaches 0.11 to 0.13 in the experiments. The rest of the volume in pore space of sand bed is liquid water and gaseous CO_2.

17.3 Conclusions

CO_2 gas can be stored in hydrate form in a laboratory-scale fixed bed of sand particles which simulates a depleted natural gas reservoir. It was found that 38-47 % of the amount of water can form hydrate with CO_2 after 24 hours of a gas injection experiment. The hydrate saturation can reach 0.11 to 0.13 during this period of time. Under *Semi-batch gas injection* more CO_2 gas can be consumed to form gas hydrate. CO_2 storage capacity in hydrate form is much more than in vapor form at the same P-T conditions. In this case CO_2 injection into depleted gas reservoirs in Northern Alberta to store CO_2 in hydrate form can be considered as a potential method to reduce GHG.

Reference

1. Linga, P., Haligva, C., Nam, S. C., Ripmeester, J. A., Englezos, P. Gas hydrate formation in a variable volume bed of silica sand particles. *Energy & Fuels.* **2009**, *23*(11), 5496–5507.

18

Geological Storage of CO_2 as Hydrate in a McMurray Depleted Gas Reservoir

Olga Ye. Zatsepina[1], Hassan Hassanzadeh[1], and Mehran Pooladi-Darvish[2]

[1] *Department of Chemical and Petroleum Engineering, University of Calgary, Calgary, AB, Canada*
[2] *Fekete Associates Inc., Calgary, AB, Canada*

Abstract

The growing concern about climate change causes the public, industry, and government to pay more attention to reducing CO_2 emissions. Geological storage of CO_2 is considered as a promising method, with a major concern of a possible gas leakage. Mineral trapping converts the CO_2 into a solid compound. However, the time scale of such reactions varies from centuries to millennia. In contrast, the kinetics of CO_2 hydrate formation—leading to trapping of CO_2 in the solid form—is quite fast, providing the opportunity for long-term storage of CO_2. In this chapter, a particular geological setting suitable for the formation of CO_2 hydrate is investigated. We study the storage of CO_2 in a depleted gas pool of Fort McMurray area to provide a local option where significant CO_2 emissions are associated with production of oil sands and bitumen. While the storage capacity of a single pool may look unimpressive, the overall capacity is expected to be large because of their number in the area. The storage of CO_2 in many small gas pools distributed over the area will provide more options to the industry, in addition to knowledge of the reservoirs and caprocks associated with the production history.

In this chapter, we provide preliminary sub-surface engineering design examining effects of injection rate and number of wells. Sensitivity to the initial saturation of water and the effect of heat removal are also considered. Furthermore, we simulate a shut-in case as the most realistic condition in field scale CO_2 sequestration.

Simulation results presented in this chapter show that when CO_2 is injected into this reservoir, pressure initially rises to meet conditions appropriate for hydrate formation. As hydrate forms, the reservoir temperature increases following hydrate three-phase equilibrium. Formation of hydrate occurs at rates controlled by heat transfer between the reservoir that has warmed up due to hydrate formation and cold cap/base rocks. The volumetric CO_2 storage capacity of the pool is found to be more than five times greater than the initial gas-in-place at the start of injection (IGIP).

18.1 Introduction

Carbon dioxide is a greenhouse gas, so its capture and storage to avoid accumulation in the atmosphere are important components of climate change mitigation. Since CO_2 released from biological, igneous, or chemical activities has been stored in the upper crust of the Earth for millions of years, options for its storage may involve geological settings including sedimentary basins or saline aquifers. However, safety issues are vital in choosing a geological formation to store CO_2. Therefore, a low probability of CO_2 leakage and an effective trapping mechanism are of great significance in choosing a storage facility. Requirements for the secure storage are met in such mechanisms as mineral and solubility trapping. However, these processes are slow since it may take centuries for CO_2 to naturally dissolve into the aquifer water (Hassanzadeh et al., 2009) or to form stable carbonates with calcium and magnesium (Gunter et al., 2004).

Contrary to that, formation of CO_2 hydrate that occurs when CO_2 is dissolved in water at low temperatures and high pressures is mostly fast. Hydrates are solids composed of the framework of water molecules with gas molecules trapped in "cages" within the framework. Storage of CO_2 in the form of hydrate is attractive because of a high density of the solid structure and fast kinetics that depends on thermodynamic conditions.

This paper focuses on storing CO_2 in the form of hydrate in depleted reservoirs of natural gas. It continues and builds upon our earlier studies of CO_2 injection in a hypothetical depleted methane reservoir (e.g., Zatsepina and Pooladi-Darvish, 2012). In the mentioned papers, we discover that hydrate formation during period of eight months is primarily controlled by the sensible heat of the reservoir between the initial temperature and equilibrium temperature at maximum pressure. Over longer periods of time, dissipation of heat of hydrate formation causes the reservoir cooling and leads to additional formation of hydrate, with rate controlled by this heat conduction. In the current paper, a real depleted gas reservoir from Fort McMurray area is investigated. Similar to the previous studies, the thermal simulator CMG STARS [2008] is employed here.

We start the paper with a brief presentation of theoretical background and assumptions discussed in details in Zatsepina and Pooladi-Darvish (2012). Then, the geological model of the depleted gas reservoir and Base Case of the study are introduced. This is followed by sensitivity studies on injection conditions, properties of the reservoir, and the effect of heat removal. Lastly, a 70-year shut-in period after the reservoir pressure has reached maximum is examined.

18.2 Fundamentals

In this section, hydrate phase equilibriums along with a discussion of some of the modeling assumptions are briefly presented.

18.2.1 Gas Flow

When CO_2 is injected in the reservoir, it pushes the in-situ natural gas away from the injection well. Since CO_2 is heavier than methane, it settles at the bottom, while methane moves toward the reservoir top. Mixing between the two gases occurs at the leading edge of the CO_2 front and is enhanced if mechanism of dispersion or reservoir heterogeneity is involved. In our model, the gas mixing is mainly due to numerical dispersion, with the length scale comparable to that of the gas diffusion over a typical simulation run time (e.g., 30 years).

18.2.2 Hydrate Phase Equilibrium

To predict hydrate phase equilibriums, a multi-phase Gibbs energy minimization program CSMGem (Ballard and Sloan, 2004) and an empirical expression that relates pressure, temperature, and composition of the vapor phase at three-phase equilibrium developed by (Adisasmito et al., 1991) have been used. (In STARS, formation of mixed hydrates is modeled using a reaction formulation coupled with equilibrium relations specified using table of k-values that mimics the hydrate phase boundaries.)

18.2.3 Assumptions

The vapor phase composition affects not only P-T conditions of hydrate stability, but also molar fraction of gases in hydrate which is assumed to be equal to 1 (Zastepina and Pooladi-Darvish, 2012). This means that the vapor phase with composition, for example, 60% CO_2 and 40% CH_4 is in equilibrium with CH_4-CO_2 hydrate of the same composition (i.e., 60% CO_2 and 40% CH_4).

All hydrate is assumed to be in equilibrium with gas; the previously formed hydrate changes to reflect new gas composition. As a result, the vapor and hydrate phases in the modeling have the same compositions. Consequently, the CO_2 fraction in the hydrate phase would increase with injection time as CO_2 molecules substitute methane in the structure.

We have not found estimations of kinetic constant in porous medium and use intrinsic rate constants for CO_2-hydrate decomposition and formation reported by Clarke and Bishnoi (2004; 2005). Our calculations indicate that the reported constants are large enough for P-T conditions of CO_2-hydrate formation and decomposition to follow the equilibrium curve. We did not consider additional resistances to mass transfer in porous media, so that the pressure and temperature of the hydrate formation and decomposition follow the equilibrium conditions. If kinetic constants in porous medium are smaller than those reported by Clarke and Bishnoi (2004; 2005) by orders of magnitude, then a deviation from the equilibrium curve may occur. A change in permeability to the hydrate saturation is incorporated through the use of a model where changes in fluid saturation affect gas mobility.

We neglect the solubility of methane in water, while the solubility of CO_2 is included in the modeling. The influence of salinity on hydrate stability is neglected. These assumptions are incorporated in STARS for simulation studies reported next.

18.3 Reservoir

18.3.1 Geological Model

This study considers a depleted gas reservoir of Fort McMurray area. The reservoir is at a depth of 217 meters; it includes a gas cap and a water leg (Figure 18.1). Interpretation of geophysical well logs in the pool and its immediate surroundings provide maps of the structure, net pay, and porosity. The structure is composed of a main basal flow unit with an inclusion of shale on its top. The shale inclusion is overlaid by sand which is indicated by

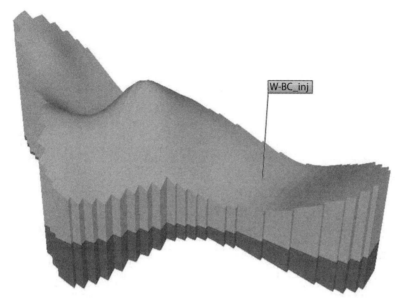

Figure 18.1 Geological model of depleted gas reservoir to study.

a hill in the center of the reservoir. Average value of the reservoir porosity is 32%, while permeability is assumed to be 500 mD. The initial saturations of gas and water are 75% and 25%, respectively. The reservoir temperature is 7°C, pressure after depletion is 0.7 MPa, and the remaining volume of the gas-in-place (methane) is 178×10^6 sm³[1]. (Pressure / production history match provide with the reservoir pressure prior to production of 1560 kPa and the original volume of gas-in-place of 405×10^6 sm³.)

We model a fully penetrated vertical well. The CO_2 gas is injected at a constant rate of 0.3×10^6 sm³/day, with the reservoir pressure restricted to no greater than 4 MPa. In practice, this maximum pressure is determined by operating or safety requirements, typically 10-20% below the formation fracture pressure. Most of the shallow reservoirs in Northeastern Alberta are significantly under-pressurized (to as much as half of hydrostatic pressure before any production occurred). Therefore, fracture pressure in these reservoirs would be significantly larger than the pressure before any production occurred. The gas is injected at a temperature of 10°C. This temperature is chosen so that no hydrate forms in the vicinity of the wellbore (since the

---

[1] Standard pressure and temperature in this paper are defined as P=1atm (14.7 psi) and T=15.56 °C (60 °F), respectively.

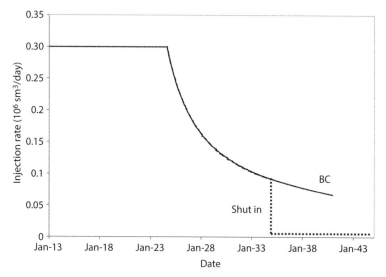

Figure 18.2 CO2 injection rate in Base Case (solid line) and Shut in (dashed line).

three-phase equilibrium pressure at 10° C is 4.4 MPa). We have previously shown that by controlling the temperature of the injected CO_2, hydrate formation in the vicinity of the wellbore where formation plugging may occur could be avoided. Formation of hydrate away from the wellbore and at saturations observed in our studies is not expected to affect injectivity.

Figure 18.2 shows injection rate as a function of time in Base Cases and Shut-in Case. In Base Case, CO_2 gas is initially injected at a constant rate of 0.3×10^6 sm³/day (~0.2 Mt/year[2]). From year 2024 to 2041, when the pressure approaches 4 MPa, the rate decreases from 0.3×10^6 m³/day to 0.07×10^6 m³/day. In Shut-in Case, the injection is stopped at year 2035.

The reservoir initial conditions, properties, and operating constraints are given in Table 18.1. The heat loss at the cap and base rock borders is solved using semi-analytical infinite-overburden heat loss model included in STARS.

18.3.2 Base Case

Figure 18.3 shows changes in average values of pressure, temperature, and hydrate saturation with time. During the first 7 years, pressure increases with injection, while no hydrate forms. At an average pressure of 3 MPa,

---

[2] CO2 density at standard condition is considered to be 1.857 kg/m³.

Table 18.1 Reservoir parameters

PARAMETERS	
Reservoir mean depth (m)	217
Initial temperature (°C)	7
Initial (start of injection) pressure (MPa)	0.7
CAP/BASE ROCKS	
Volumetric heat capacity (J/m³-°C)	3.76×10^6
Total thermal conductivity (J/m-day-°C)	1.5×10^5
RESERVOIR	
Volumetric heat capacity (J/m³-°C)	2.6×10^6
Hydrate heat capacity (J/gmol-°C)	203.7
Rock thermal conductivity (J/m-day-°C)	2.468×10^5
Hydrate thermal conductivity (J/m-day-°C)	3.396×10^4
Water thermal conductivity (J/m-day-°C)	5.53×10^4
Saturation of water	0.25
Saturation of gas	0.75
Average Porosity (%)	32
Permeability (mD)	500
Diffusion coefficient (m²/day)	0

average hydrate saturation is 0.003. As hydrate forms gas is consumed, so the rate of pressure increase lessens. Formation of hydrate is also signified by an increase in temperature, which continues to rise with increasing hydrate saturation. After 20 years and injection of more than 5 times the initial gas-in-place, the average reservoir pressure and temperature have increased to 3.88 MPa and 8.8° C, respectively.

Horizontal distributions of temperature and hydrate saturation in the middle of the reservoir (crossing the injector) at years 2018, 2023, 2033 (or 5, 10, 20 years of injection) are shown in Figures 18.4 and 18.5, respectively. After 5 years of the injection, pressure increases from the initial value of 700 kPa to 2500 kPa (below hydrate three-phase equilibrium at 7°C) to represent a situation prior to hydrate formation. After 10 and 20 years of the injection, pressure rises to 3450, and 3880 kPa, respectively. These values correspond to situations when hydrate is stable.

318   GAS INJECTION FOR DISPOSAL AND ENHANCED RECOVERY

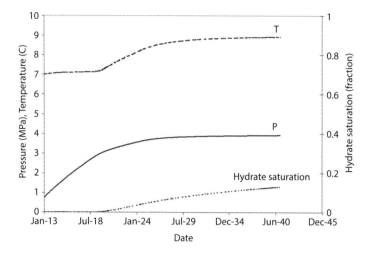

Figure 18.3 Average pressure, temperature, and hydrate saturation as functions of time.

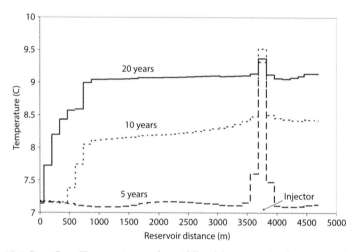

Figure 18.4 Base Case: Temperature in the middle of the reservoir after 5, 10, and 20 years of injection.

The distributions of temperature reveal a spike around the injector. They show that the reservoir temperature has increased by 1°C and 1.9°C after 5 and 10 years of hydrate formation. The corresponding distributions of hydrate show absence of hydrate around the injector as well as increasing with time saturations of hydrate (hydrate is absent at year 2018). The hydrate distributions reveal an additional feature which is represented by two peaks.

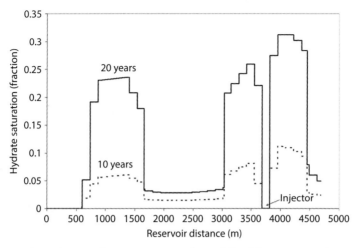

Figure 18.5 Base Case: Hydrate saturation in the middle of the reservoir after 10 and 20 years of injection.

Maps of temperature and hydrate saturation in the gas cap at years 2023 and 2033 are shown in Figures 18.6 and 18.7, respectively. The two temperature maps exhibit similar features. They are mostly affected by the temperature of injected CO_2, initial temperature, and heat of hydrate formation. Three main regions could be distinguished there: the vicinity of the wellbore, the outskirts, and in-between. Prior to hydrate formation, temperature changes from 10°C of injected CO_2 to the initial reservoir temperature of 7°C. Once hydrate forms releasing heat, an "in-between" region develops. In this region, the methane is fully displaced by CO_2, with P-T conditions governed by hydrate phase equilibrium. From the "in-between" region to the outskirts, temperature decreases with the distance from the injector. (Since the hydrate reservoir temperature is increased, heat is conducted along the temperature gradients, with lower temperatures found at borders with the cap rock and water leg.)

The maps of hydrate saturation mostly reflect the same features discussed above. Hydrate is absent around the wellbore and on the outskirts. (Hydrate is absent in the vicinity of the wellbore because of the high temperature, although hydrate forms closer to the wellbore as pressure increases with injection time.) An "in-between" region shows the mentioned above additional feature: two areas with an increased saturation of hydrate.

Figure 18.8 presents a vertical slab of the reservoir and brings insight on the areas of increased hydrate saturation. It reveals an increased saturation of hydrate at borders with the rock and water. These saturations are

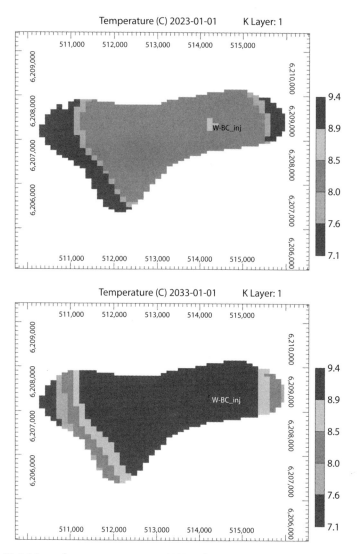

Figure 18.6 Maps of temperature at years 2023 and 2033.

higher in two regions with a decreased thickness due to faster heat removal noticed in hydrate maps. Figure 18.8 shows that spread of the hydrate zone at the bottom of the reservoir exceeds that at the top. This is caused by settling of CO_2, which promotes formation of hydrate.

Methane is displaced by injected CO_2, so that pure CO_2 hydrate forms in most of the reservoir. The mixed hydrates contribute slightly to the hydrate saturation at the rear end. The mixed hydrate composition reflects

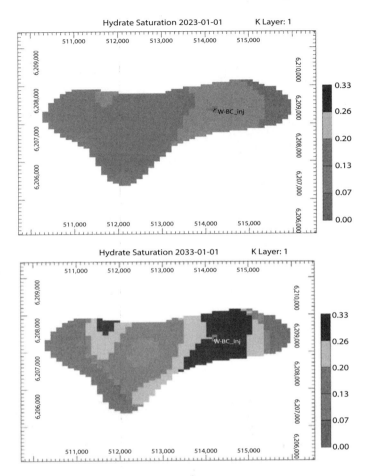

Figure 18.7 Maps of hydrate at years 2023 and 2033.

composition of the vapor phase in the mixing zone. As the fraction of CO_2 in the vapor phase decreases with distance away from the front, the equilibrium pressure increases. At the distance where equilibrium pressure exceeds the reservoir pressure, hydrate ceases to form.

Figure 18.9 shows volumes of CO_2 and methane stored in the form of hydrate as a function of time. At 2041, volume of CO_2 stored in hydrate is 920×10^6 sm³ (1.71 Mt), while volume of methane in hydrate is 35 times smaller, 26×10^6 sm³. After injection of approximately 2000×10^6 sm³, the hydrates have taken up 946×10^6 sm³ of the injected CO_2, with the rest in the free gas phase. A comparison between the initial gas-in-place and the volume of CO_2 stored in hydrate suggests that the later is 5.3 times larger than the initial gas-in-place (178×10^6 sm³).

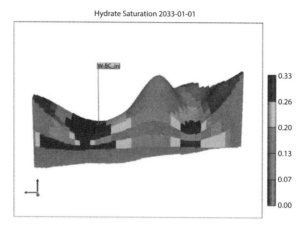

Figure 18.8 Vertical slab of hydrate saturation at year 2033.

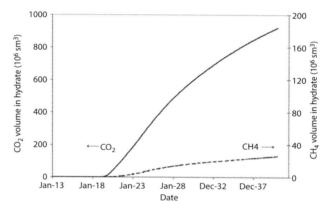

Figure 18.9 Carbon dioxide and methane stored in hydrate form.

18.4 Sensitivity Studies

A number of simulations were conducted to understand factors that affect hydrate formation when CO_2 is injected at a constant rate of 0.3×10^6 sm³/day (~0.2 Mt/year). Table 18.2 shows the list of cases where we examine the effects of the injection rate, number of injectors, initial water saturation, and removal of heat. Results are discussed in the following.

18.4.1 Effect of the Injection Rate

Effects of the injection rate are investigated by comparing results obtained in Case 1, Case 2, and the Base Case. Injection rates in Cases 1 and 2 are

Table 18.2 List of Cases (A no-entry in the table means the Base-Case property)

Case	Description	Injection rate Million sm³/day (~Mt/year)	Number of wells	Sw	Heat removal
BC	Base Case (BC)	0.3 (0.2)	1	0.25	Present
Case 1	Injection rate	0.15 (0.1)			
Case 2	Injection rate	0.6 (0.4)			
Case 3	No. of injection wells	0.3 (0.2) per well	2		
Case 4	Sw & IGIP adjusted to BC IGIP			0.45	
Case 5	Sw (not adjusted)			0.45	
Case 6	Heat removal				Absent

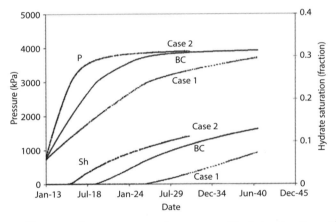

Figure 18.10 Changes in the average reservoir pressure and hydrate saturation with time — effect of the injection rate.

0.15e6 and 0.6e6 sm³/day, respectively. Figure 18.10 shows average reservoir pressure and hydrate saturation, while Figure 18.11 – injection rate and standard volume of CO_2 in hydrate as functions of time. Figure 18.10 reveals that the injection rate is responsible for the pressure at which hydrate formation is first initiated. The higher the injection rate, the faster is the onset of hydrate formation. Consequently, Case 2 with the highest rate is associated with the longest period of hydrate formation and, correspondingly, the greatest amount of CO_2 formed prior to the reservoir pressure reaching 4 MPa.

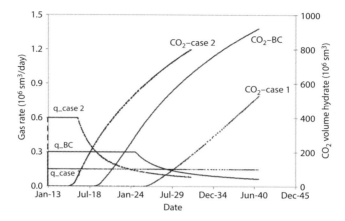

Figure 18.11 Changes in the injection gas rate and volume of CO2 stored in hydrate with time — effect of the injection rate.

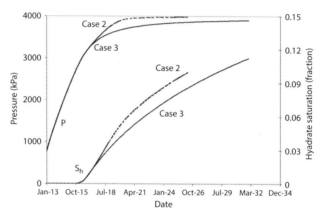

Figure 18.12 Changes in the average reservoir pressure and hydrate saturation with time — effect of the number of wells.

18.4.2 Effect of the number of wells

Effect of the number of injectors is investigated by comparing results obtained in Case 3 and Case 2. There are 2 wells with an injection rate of 0.3e6 sm³/day each in Case 3 and one well with an injection rate of 0.6e6 sm³/day in Case 2. Figure 18.12 shows the average reservoir pressure and hydrate saturation. It reveals that increasing the number of wells does not result in substantial change in hydrate saturation when compared with a corresponding increase in the injection rate, especially if cost of an additional well is considered.

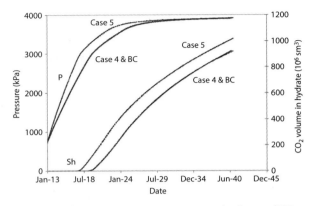

Figure 18.13 Changes in the average reservoir pressure and volume of CO_2 stored in hydrate with time — effect of the initial water saturation.

18.4.3 Effect of the initial saturation of water

Effects of the initial saturation of water are investigated by comparing results obtained in Case 4, Case 5, and the Base Case. The initial saturation of water in Cases 4 and 5 is 0.45. Change in the initial saturation of water from 0.25 to 0.45 affects the reservoir initial gas-in-place (IGIP). An adjustment in the reservoir porosity in order to fit IGIP of the Base Case is made in Case 4. Consequently, the IGIP in Cases 4 and 5 differ: IGIP in Case 5 is almost twice smaller. Figure 18.13 represents average reservoir pressure and volume of CO_2 stored in hydrate as functions of time. (Hydrate saturations are not shown because of the IGIP difference.) Figure 18.13 shows that the increased saturation of water in Case 4 has no influence on the amount of CO_2 stored in hydrate. In Case 5, a decreased IGIP causes the pressure at which hydrate formation is first initiated to be reached faster. Earlier onset of hydrate formation results in a longer period of hydrate formation and greater amount of CO_2 stored.

18.4.4 Effect of the heat removal

Effect of removal of the hydrate formation heat is investigated by comparing results obtained in Case 6 and the Base Case. In Case 6, the heat removal at the border with cap/base rocks is absent. Figure 18.14 shows the average reservoir pressure, temperature, and hydrate saturation as functions of time. Prior to hydrate formation, temperature differs by 0.15 C causing a small variation in the formation onset for the two cases. With time, a difference in average hydrate saturations increases because

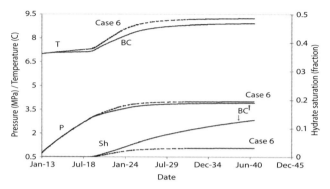

Figure 18.14 Changes in the average reservoir pressure, temperature, and hydrate saturation with time — effect of heat removal.

lower rates of thermal flow result in reduced rates of hydrate formation. Although amounts of heat transferred from the reservoir to cap/base rocks in the Base Case and Case 6 are significantly different, variations in average temperatures are small. Figure 18.14 confirms that heat transfer is the mechanism which controls formation of hydrate in reservoir. When heat removal is absent, the amount of CO_2 stored prior to reaching 4 MPa is 4.2 times less than in the Base Case.

18.5 Long-term storage

Shut-in Case represents situation when CO_2 injection stops after 23 years when the reservoir pressure has reached a value of 3.9 MPa. By the time when the pressure reaches 3.9 MPa, the reservoir temperature has increased from 7° to 8.9°C due to hydrate formation, while most of the parts of cap and base rocks are at the initial temperature. As heat of hydrate formation is transferred to the rocks, the reservoir temperature decreases providing a driving force to form hydrate. When hydrate forms, the reservoir temperature follows the hydrate phase equilibrium reflecting changes in pressure. In this case, the increased temperature in the reservoir dissipates, the reservoir cools, and additional hydrate forms, with the rate controlled by the conduction of heat.

Figure 18.15 shows that, during a 71-year shut in period, reservoir pressure and temperature decrease by 0.6 MPa and 1.1°C, respectively, while the average saturation of hydrate increases from 0.1 to 0.14. The additional hydrate formed during the shut-in period stores 247×10^6 sm^3 (~0.5 Mt) of CO_2, which corresponds to the additional volumetric storage capacity of 1.6 IGIP.

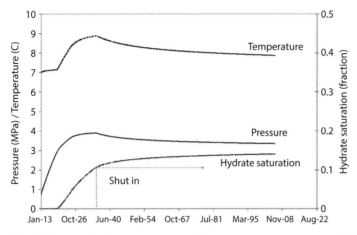

Figure 18.15 Changes in the average reservoir pressure, temperature, and hydrate saturation with time — 23-year injection with 71-year shut in.

18.6 Summary and conclusions

We have studied hydrate formation in a depleted gas reservoir of Fort McMurray area. Injection of CO_2 occurs over long periods of time (e.g., 20 years), so that the reservoir temperature increased due to hydrate formation dissipates and the reservoir cooling leads to formation of more hydrate and pressure reduction. In this case, the rate of hydrate formation is controlled by conduction of heat. When CO_2 is injected in the reservoir, the hydrate formation occurs as soon as three-phase equilibrium conditions are met. As pressure increases with the injection, the hydrate keeps forming and temperature increases following the equilibrium.

Our simulation results indicate that more hydrate forms at top / bottom, where heat of hydrate formation dissipates to the cap / base (via water leg) rocks. In all cases studied, the volume of CO_2 that turned into hydrates was greater than the initial gas-in-place. Although the numbers are not relatively small (e.g., 5 times), the overall storage in the Fort McMurray area is expected to be large because of the number of depleted gas pools there.

As CO_2 is injected in the methane gas reservoir, a zone where the two gases mix appears. At corresponding values of pressure, temperature, and gas composition, CO_2-CH_4 mixed hydrates form to affect the storage capacity. Molecular diffusion of CO_2-CH_4 is not included in the modeling. Numerical truncation errors result in numerical dispersion, which leads to mixing of CO_2 and CH_4 at the gas front. Our results show that the volume of methane stored in hydrate is 35 times smaller than that of CO_2. Heterogeneity of the reservoir rock is not included in the current modeling but could contribute to dispersion of the gas front. As a result, the mixed hydrates could have a greater contribution compared to that of this work.

Heating of the reservoir due to the exothermic reaction of hydrate formation during CO_2 injection, has also been studied. Our results show that the reservoir temperature increases by 2 °C when pressure reaches 4 MPa. Cooling of the reservoir after the end of the injection in shut-in case, has been examined. Our results predict that more of the CO_2 will turn into hydrate, with a decline in the average reservoir pressure leading to an increase in security of the storage.

In the previous studies, it was found that by controlling the temperature of the injected CO_2, hydrate formation in the vicinity of the wellbore could be avoided. Formation of hydrate away from the wellbore and at saturations observed in the study was not expected to affect injectivity. Main conclusions of this study could be summarized as follows.

(i) Decrease in the injection rate results in longer times to reach three-phase equilibrium pressure of hydrate formation, so that the total amount of formed hydrate decreases.

(ii) Profits from increase in the number of injectors and corresponding increase in the injection rate are comparable.

(iii) Effect of uncertainty in the initial saturation of water is connected with changes in the volume of gas in reservoir. Decrease in the gas volume results in shorter times to reach three-phase equilibrium pressure of hydrate formation, so that the total amount of formed hydrate increases. Otherwise, water does not limit hydrate formation.

(iv) Removal of heat from the reservoir controls formation of hydrate in reservoir, while the initial temperature, saturation of water, and maximum pressure affect hydrate saturation.

(v) 3.8 Mt of CO_2 was injected into the depleted gas reservoir during 28 years, with 1.85 Mt turning into hydrate and 1.95 Mt remaining as free gas in the reservoir.

(vi) A CO_2 volume more than 5.8 times greater than the initial gas-in-place volume could be stored as hydrate in the depleted gas reservoir. Although this number is not impressive, this pool is expected to be one of many similar storage sites in the area.

18.7 ACKNOWLEDGEMENTS

We would like to acknowledge Carbon Management Canada Inc. (CMC-NCE) for providing financial support for this work.

REFERENCES

1. Adisasmito, S., Frank, R.J., and Sloan, E.D. Hydrates of carbon dioxide and methane mixtures. *J. Chem. Eng. Data.* (1991). 36: 68–71.
2. Ballard, A.L., and Sloan Jr., E.D., The next generation of hydrate prediction: Part III. Gibbs energy minimization formalism. *Fluid Phase Equilibria* (2004). 218: 15–31.
3. Clarke M. and Bishnoi, P.R., Determination of the intrinsic rate constant and activation energy of CO_2 gas hydrate decomposition using in-situ particle size analysis. *Chemical Engineering Science* (2004). 59: 2983–2993.
4. Clarke, M. and Bishnoi, P.R., Determination of the intrinsic kinetics of CO_2 gas hydrate formation using in-situ particle size analysis. *Chemical Engineering Science* (2005). 60: 695–709.
5. CMG STARS. Advanced process and thermal reservoir simulator. (2008). Computer Modelling Group Ltd., Calgary, AB, Canada.
6. Gunter, W., Bachu, S., and Benson, S. The role of hydrogeological and geochemical trapping in sedimentary basins for secure geological storage of carbon dioxide, Geological Society, London, Special Publications. (2004). 233: 129–145
7. Hassanzadeh, H., Pooladi-Darvish, M., and Keith, D.W. Accelerating CO_2 dissolution in saline aquifers for geological storage – mechanistic and sensitivity studies. *Energy & Fluids.* (2009). 23: 3328–3336.
8. Zatsepina, O.Y. and Pooladi-Darvish, M., CO2 storage as hydrate in depleted gas reservoirs, *SPE Reservoir Evaluation and Engineering – Reservoir Engineering*, (2012). 15 (1), 98–108.

19
A Modified Calculation Method for the Water Coning Simulation Mode in Oil Reservoirs with Bottom Water Drive

Weiyao Zhu, Xiaohe Huang, and Ming Yue

Civil and Environmental Engineering School, University of Science Technology Beijing, Beijing, China

Abstract
The model of highness and shape of water cone were established for the oil reservoirs that were driven by bottom water. Because the water cones, the top and bottom pressures are equal, the dynamic water cone interface equation was built on the condition that the oil well produced at constant production rate. According to the boundary and initial conditions, the shape of water cones of vertical wells was obtained. The model quantitatively describes the influence of oil production, reservoir thickness, perforating degree and permeability, and so on. Based on the model, the time of bottom water breakthrough and the section forms of water coning at different time can be determined.

19.1 Introduction

There exists cone-like bottom water-oil interface during oil production. As production rate increases, the height (or depth) of water-coning increases until the water coning becomes unstable and breaks through into the well. This phenomenon is referred to two phase interface coning.

From the shape of water cone profile, the side-inclined planes are curves rather than straight lines. Previous researches [1-5] presented several models of water coning shape. Some assumed that the side-inclined is a curve, while others simply consider the water cone profile as the circular cone.

19.2 Mathematical Model

For oil reservoir with good permeability, large reservoir thickness and the perforated degree less than 2/3, the oil phase flow near the perforations was considered as a spherical centripetal flow.

Initial oil-water interface was considered as a horizontal plane, bottom water flow to wells after oil well production. Basic assumption: oil reservoir is homogeneous reservoir; water-drive oil is performed with piston like displacement; the capillary pressure is neglected; both oil and water flow obey Darcy's law.

The oil-water interface was expressed in $z = \xi(r,t)$:

$$F(r,z,t) = z - \xi(r,t) = 0 \tag{19.1}$$

Consider an abrupt interface between two incompressible liquids: oil and water (Figure 19.1) in a homogeneous isotropic reservoir. We take the material derivative of function F,

$$\phi \frac{\partial \xi}{\partial t} = V \bullet \nabla[z - \xi(r,t)] \tag{19.2}$$

In (19.2), V is velocity. The velocity potentials φ may be defined for each sub region:

$$\varphi_1(r,z,t) = \frac{K_1}{\mu_o}(p + \gamma_o z) \tag{19.3}$$

$$\varphi_2(r,z,t) = \frac{K_2}{\mu_w}(p + \gamma_w z) \tag{19.4}$$

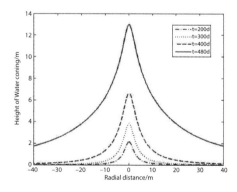

Figure 19.1: Section shape of water cone at different time

By combining (19.3) and (19.4), we obtain (19.5):

$$\xi = \frac{\mu_w}{K\Delta\gamma}\varphi_2(r,z,t)\Big|_{z=\xi} - \frac{\mu_o}{K\Delta\gamma}\varphi_1(r,z,t)\Big|_{z=\xi} \tag{19.5}$$

where: $\Delta\gamma = \gamma_w - \gamma_o$. Substitution of $V = -\nabla\varphi$ into (19.2), obtained boundary conditions on the oil water interface:

$$\phi\frac{\partial\xi}{\partial t} = \nabla\phi_1 \cdot \nabla[z-\xi(r,t)] = 0 \tag{19.6}$$

$$\phi\frac{\partial\xi}{\partial t} = \nabla\phi_2 \cdot \nabla[z-\xi(r,t)] = 0 \tag{19.7}$$

Substitution of (19.5) into (19.6) and (19.7), φ_1 and φ_2 at the interface satisfies the conditions:

$$\phi\left(\frac{\mu_o}{K\Delta\gamma}\frac{\partial\varphi_1}{\partial t} - \frac{\mu_w}{K\Delta\gamma}\frac{\partial\varphi_2}{\partial t}\right) - \frac{\mu_o}{K\Delta\gamma}(\nabla\varphi_1)^2 + \frac{\mu_w}{K\Delta\gamma}(\nabla\varphi_1 \cdot \nabla\varphi_2) - \frac{\partial\varphi_1}{\partial z}$$
$$= 0 \quad (z=\xi) \tag{19.8}$$

$$\phi\left(\frac{\mu_o}{K\Delta\gamma}\frac{\partial\varphi_1}{\partial t} - \frac{\mu_w}{K\Delta\gamma}\frac{\partial\varphi_2}{\partial t}\right) + \frac{\mu_w}{K\Delta\gamma}(\nabla\varphi_2)^2 + \frac{\mu_o}{K\Delta\gamma}(\nabla\varphi_1 \cdot \nabla\varphi_2) - \frac{\partial\varphi_2}{\partial z}$$
$$= 0 \quad (z=\xi) \tag{19.9}$$

The initial condition is:

$$\frac{\mu_w}{K\Delta\gamma}\varphi_2(r,z,t) - \frac{\mu_o}{K\Delta\gamma}\varphi_1(r,z,t) = 0 \quad (t=0, z=0) \tag{19.10}$$

and φ_1 and φ_2 have to satisfy the following boundary conditions:

$$\frac{\partial\varphi_1}{\partial z}\Big|_{z=h} = 0 \tag{19.11}$$

$$\frac{\partial\varphi_2}{\partial z}\Big|_{z=-b} = 0 \tag{19.12}$$

$$\lim_{R\to 0}\left(R^2\frac{\partial\varphi_1}{\partial R}\right) = \frac{Q}{2\pi} \quad R^2 = r^2 + (z-z_w)^2 \tag{19.13}$$

Thus our problem is to determine (φ_1, φ_2) each in its respective sub region, by satisfying (19.4), the boundary conditions on the external boundaries, and (19.8) and (19.9) on the interface separating the two regions (or liquids). Once (φ_1, φ_2) are known, (19.5) may be used to determine the shape of the interface ($\xi = \xi(r,t)$).

19.3 Solution

An exact solution by present available mathematical tools is most complicated because of the nonlinearity of (19.8) and (19. 9). $z_n = \xi(r, t_n)$ in each time step at a constant production rate can be calculated by numerical calculation. Or we can use the method of small perturbations to get approximate solution.

By using the method of small perturbations (19.8) and (19.9) are linearized, and we got the new boundary conditions on the oil water interface:

$$\frac{\phi\mu_o}{K\Delta\gamma}\frac{\partial\varphi_1}{\partial t} - \frac{\phi\mu_w}{K\Delta\gamma}\frac{\partial\varphi_2}{\partial t} - \frac{\partial\varphi_1}{\partial z} = 0 \qquad (z=\xi) \qquad (19.14)$$

$$\frac{\phi\mu_o}{K\Delta\gamma}\frac{\partial\varphi_1}{\partial t} - \frac{\phi\mu_w}{K\Delta\gamma}\frac{\partial\varphi_2}{\partial t} - \frac{\partial\varphi_2}{\partial z} = 0 \qquad (z=\xi) \qquad (19.15)$$

Because of the singularity introduced by the drain, let the required solution for φ_1 be decomposed into two parts, φ_{11} and φ_{12} (both satisfying the Laplace equation):

$$\varphi_1(r,z,t) = \varphi_{11}(r,z) + \varphi_{12}(r,z,t) \qquad (19.16)$$

and such that φ_{11} represents the singular character of φ_1 at $r = 0$, $z = z_w$ whereas φ_{12} is an unsteady state solution which is regular everywhere in oil reservoir. Hence:

$$\varphi_{11}(r,z) = \frac{Q}{4\pi h}\frac{1}{\sqrt{r^2 + (z+z_w)^2}} - \frac{Q}{4\pi h}\frac{1}{\sqrt{r^2 + (z-z_w)^2}} \qquad (19.17)$$

The boundary conditions for φ_{12} and φ_2 are:

$$z = h, \quad \frac{\partial\varphi_{12}}{\partial z} = \frac{Q}{4\pi h}\left\{\frac{h+z_w}{\left[r^2+(z+z_w)^2\right]^{3/2}} - \frac{h-z_w}{\left[r^2+(z-z_w)^2\right]^{3/2}}\right\} \qquad (19.18)$$

$$z = -b, \quad \frac{\partial\varphi_2}{\partial z} = 0 \qquad (19.19)$$

When $z = 0$, $\varphi_{11} = 0$, by combining (19.14) and (19.15), we obtain:

$$\frac{\phi\mu_o}{K\Delta\gamma}\frac{\partial\varphi_{12}}{\partial t} - \frac{\phi\mu_w}{K\Delta\gamma}\frac{\partial\varphi_2}{\partial t} - \frac{\partial\varphi_{12}}{\partial z} = -\frac{Qz_w}{2\pi h\left[r^2 + (z+z_w)^2\right]^{3/2}} \quad (z=0) \quad (19.20)$$

$$\frac{\phi\mu_o}{K\Delta\gamma}\frac{\partial\varphi_{12}}{\partial t} - \frac{\phi\mu_w}{K\Delta\gamma}\frac{\partial\varphi_2}{\partial t} - \frac{\partial\varphi_2}{\partial z} = 0 \quad (z=0) \quad (19.21)$$

The initial conditions become:

$$\frac{\mu_w}{K\Delta\gamma}\varphi_2(r,z,t) - \frac{\mu_o}{K\Delta\gamma}\varphi_{12}(r,z,t) = 0 \quad (t=0, z=0) \quad (19.22)$$

Now we have to solve (19.18) through (19.22).

Let us now assume, since φ_{12} and φ_2 are everywhere regular, that they may be represented by Fourier integral transforms enabling separation of variables.

Finally we get shape of the interface:

$$\xi = \frac{\mu_o Q}{2\pi hK\Delta\gamma}\int_0^\infty \frac{1}{\lambda}\frac{\cosh[\lambda(h-z_w)]}{\sinh(\lambda h)} \cdot \left\{1-\exp\left[\frac{-\lambda t}{\frac{\phi\mu_o}{K\Delta\gamma}\text{ctgh}(\lambda h) + \frac{\phi\mu_w}{K\Delta\gamma}\text{ctgh}(\lambda b)}\right]\right\}\cos(\lambda r)d\lambda \quad (19.23)$$

19.4 Results and Discussion

The basic data adopted by the numerical calculation are as follows: thickness of bottom aquifer 55 m, thickness of reservoir 16 m, the perforated thickness 3 m, permeability 10 mD, porosity 0.32, formation water viscosity 0.7 mPa·s, formation water density 1.10 g/cm³, oil viscosity 10 mPa·s, oil density 0.9 g/cm³, flow of production well 16 m³/d.

Through calculation, water breakthrough time is 480 days. Figure 19.1 shows the profile of water cone at different production time. The figure indicates that the height of water coning increases with radial distance decreasing and the peak is located in the wellbore.

The relationship between maximum height of water coning and production time are shown in Figure 19.2. The peak of water coning is derived from (19.30) by setting $r = r_w$. The figure indicates that, rate of rise of water coning peak increases over production time. There exists a critical

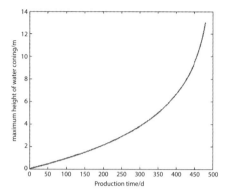

Figure 19.2: The relationship between maximum height of water coning and production time

production time after which the rate increases dramatically over time until the breakthrough of water coning.

19.5 Conclusions

For oil reservoir with good permeability, large reservoir thickness, and the perforated degree less than 2/3, the oil phase flow near the perforations was considered as a spherical centripetal flow.

Consequently a mathematical model for calculating the height and profile of water coning is established. This model quantitatively describes the relationship between the water coning height and influences of oil production, reservoir thickness, perforating degree and permeability etc. Based on the derived model, bottom-water breakthrough time and the profile of water coning at different time can be determined.

19.6 Nomenclature

b	thickness of water layer, m
K	reservoir permeability, m²
h	thickness of oil reservoir, m
p	pressure, MPa
Q	vertical well production, m³/d
r	radial distance, m
z_w	distance between vertical well and border along z axis, m
φ	porosity

γ_o oil specific weight, N/m³
γ_w oil specific weight, N/m³
μ_o oil viscosity, mPa·s
μ_w water viscosity, mPa·s

References

1. Zhu Zhongqian, Cheng LinsongSome. Dynamic Problems on Bottom Water Coning During High Water Cut Stage in Sandstone Bottom-Water Reservoir[J]. Xinjiang Petroleum Geology, 2001, 22(3):235–237.
2. Li Chuanliang. An Analysis on Water Coning Shape-a discussion with Mr. ZHU Sheng-ju[J]. Xinjiang Petroleum Geology, 2002,23(1):74–75.
3. Boyun Guo, J-E. Molinard, and R.L. Lee. A General Solution of Gas/Water Coning Problem for Horizontal Wells. SPE25050
4. Li Chuanliang. Determination of Perforation Intervals of Wells in Bottom-Water Reservoirs with Barriers [J]. Xinjiang Petroleum Geology, 2004, 25(2):199–201.
5. Zhu Shengju. A Study for Predicting Water Breakthrough Time of Oil-Well in Low Permeability Bottom-Water Reservoir with no Barrier [J]. Xinjiang Petroleum Geology, 2001,22(2):153–154.

20

Prediction Method on the Multi-scale Flow Patterns and the Productivity of a Fracturing Well in Shale Gas Reservoir

Weiyao Zhu, Jia Deng and M.A. Qian

School of Civil and Environmental Engineering, University of Science and Technology Beijing, Beijing, China

Abstract

The flow regime was described by combining the mechanics of continuous media and molecular kinematics method for the shale gas reservoir with nano–micro scale pore. The fluid flow state was judged by Knudsen number. The different regional flow mechanism and flow state characteristics were illustrated. Then the flow state chart was drawn. Based on this, the variation between permeability adjustment factor and Knudsen number was calculated. In this chapter, the multi-scale seepage model considering of diffusion, slippage, and desorption effect was established. Using conformal transformation and equivalent flow resistance method, the productivity formula of vertical and horizontal fracturing well in consideration of diffusion, slip, and desorption absorption was obtained. The conclusions show that three flow states exist in tight shale gas reservoir, including transition, slip, and continuous flow. Under the formation pressure for the shale gas reservoir with nano–micro scale pore, Knudsen number is greater than 0.1, and the Knudsen adjustment factor shows bigger deviation between Darcy formula and the multi-scale seepage model. The result shows that Darcy formula cannot be applied in shale gas reservoir with nano pore. The multi-scale vertical and horizontal fracturing well productivity was analyzed by numerical calculation. Compared with the field data, this method was verified effectively and practically. The flux of total gas increases with fracture conductivity. When the fracture conductivity increases to 2 $\mu m^2 \cdot cm$, the gas production basically remains steady. With the increase of fracture penetration ratio, gas production rate gradually increased. The flux of free gas and desorption gas increased with formation pressure, when the pressure increased to a certain degree, the flux of total gas slowly increases.

20.1 Introduction

Core analyses show that the porosity and pore volume in Shale gas reservoirs are nano-scale. The flow in very low permeability shale gas reservoirs is non-linear, which undergoes a transition from a Darcy regime to other regimes where molecular collisions with the pore walls have a significant effect on transport. So the Darcy formula can't be applied in these shale gas reservoirs. Structure characteristics analyzed results showed that, the sizes of the main nanopores are in a range of 5~200 nm, and the permeability is in a range of 1×10^{-9}~1×10^{-3} μm². The flow in tight shale gas reservoirs which not only includes seepage, but also diffusion, slip and desorption absorption is different from conventional reservoirs obviously. Therefore, it is necessary to establish a new seepage theory which can describe the flow law in nanopores and the multi-scale pores coupled flow for shale gas reservoirs' effective development.

Javadpour [1] put forward the conclusion that gas flow in nanopores is different from the Darcy flow, and had a test on the mean free path of gas and Knudsen number.

Wang and Reed [2] showed free gas flow can be a non-Darcy type in matrices, but a Darcy type in natural and hydraulic fractures.

Freeman [3] indicated pore throat diameters on the order of molecular mean free path lengths will create non-Darcy flow conditions, where permeability becomes a strong function of pressure.

Michel [4] developed a model to describe the transport of gas in tight nanoporous media by modifying the original Beskok and Karniadakis equation [5] through Knudsen number. In the paper, the model was built based on the premise that the slip coefficient b is assigned a theoretical value of negative one corresponding to full-slip under slip regime conditions. And this model can't apply for the entire range of flow regimes.

In our paper, a new model was found based on Beskok and Karniadakis equation in consideration of different slip coefficient b. Then the multi-scale seepage model considering of diffusion, slippage and desorption effect was established. By use of conformal transformation and equivalent flow resistance method, the productivity formula of vertical and horizontal fracturing well in consideration of diffusion, slip and desorption absorption was obtained, which can describe the potential of productive well, and enhance the productivity.

20.2 Multi-scale flow state analyses of the shale gas reservoirs

The flow in the reservoir was described macroscopically by a continuous assumption, For example, molecular Interaction was not considered in commonly used gas seepage equation. Since the tight shale gas reservoir has micro- and nano-scale porous, with the reduction of the pore scale, the continuous flow assumption is no longer fully valid. Then, the flow regime was described by combining the mechanics of continuous media and molecular kinematics method. Gas flow state in porous media vary depending on petrophysical properties of the media and the average free path of gas molecules (Civan, [6]), and summarized the research of Liepmann [8], Stahl [9] and Kaviany [10], using Knudsen Number to divide gas flow region was put forward, and the gas flow was divided into 3 region:(1) continuous flow;(2) slip flow;(3) transition flow. At a low Knudsen number (Kn<0.001) the no-slip boundary condition in the continuous flow regime is valid. The Darcy equation is a suitable equation; At a high Knudsen number regime (0.001<Kn< 0.1), the slip boundary condition in the continuous flow regime is valid. The Knudsen equation is a suitable equation; At a higher Knudsen number regime (0.1<Kn<10), the gas flow was transition flow, the continuous approach breaks down, the slip boundary condition was valid, The Burnett equation is a suitable equation ;When Kn>10, the gas flow was free molecule flow ,and it was in the free-molecule regime, The Fick equation is a suitable equation.

The flow mechanism and characteristic of shale gas in different regions was different, the flow state can be described by Knudsen Number and the flow equation. The author basing Knudsen dimensionless, used the Knudsen Number to judge the gas flow state in different scales of porous media, drew a flow chart and analyzed the flow state.

In 1934, Knudsen dimensionless numbers, Kn, was defined by Knudsen:

$$K_n = \frac{\overline{\lambda}}{r} \quad (20.1)$$

$$\lambda = \frac{K_B T}{\sqrt{2\pi}\delta^2 P} \quad (20.2)$$

where $\overline{\lambda}$ is the gas phase molecular mean free path in m; r is the pore throat diameter in m; K_B is the Boltzmann constant (1.3805 × 10-23J/K),δ is the collision diameter of the gas molecule; P is pressure and T is temperature.

Table 20.1 presents the gas properties of a typical shale gas sediment. Figure 20.1 presents the Knudsen number as a function of pressure for different pore sizes ranging from 10 nm to 50 μm. The flow state was different in different pore and pressure condition, which was transition flow, slip flow and continuous flow was the conventional seepage model, in Nano-pore the gas flow was mainly continuum flow and slip flow, with the pressure increased, part of it transformed to continuous flow. When r>50μm, the flow was continuous flow. For example, in the shale gas reservoir, when

Table 20.1 Gas molecule collision diameter of different components

Gas Components	Mole (%)	Collision Diameter(δ,nm)	Molar Mass (kg/kmol)
CH_4	87.4	0.4	16
C_2H_6	0.12	0.52	30
CO_2	12.48	0.45	44

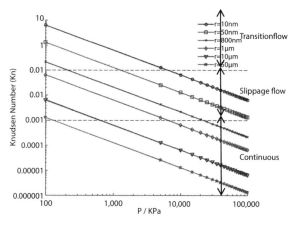

Figure 20.1 Relationship between Knudsen number and the pressure under different pore diameter.

the pressure was 10~20 MPa, the pore was 10~30 nm, the gas flow was slip flow. So for the shale gas reservoir, the flow in the pore was slip flow. For instance, the following cited the Long maxi reservoir, which the nanopore was mainly 2~40 nm, accounting for 88.39% of the total pore volume, 98.85% of the specific surface area; the 2~50 nm mesopore provides the main pore volume space, the micropore and mesopore less than 50nm provides the main specific surface area.

The simulation results were shown in Figure 20.2, the shale gas flow mechanism can be seen under the combination of different pore size. Under the condition of 10~20 MPa, in the sentiments combined with pores of 70 nm pore diameter which account for 80%, and pores of 3μm pore diameter which account for 20%,the flow was continuous flow. While in the sentiments combined with pores of 10 nm pore diameter which account for 70%, and pores of 1μm pore diameter which account for 30%, the flow was continuous flow. So for the different combination of pore size, the flow mechanism of shale gas was different.

20.3 Multi-scale seepage non-linear model in shale gas reservoir

20.3.1 Non-linear model considering on diffusion and slippage effect

Seepage model: Slippage effect and the collisions with the pore wall are not considered in the Darcy flow model. While the flow in very low permeability shale gas reservoirs is non-linear, which undergoes a transition from a Darcy regime to other regimes where molecular collisions with the pore walls have a significant effect on transport. Beskok and Karniadakis equation shows a relational expression between flow velocities and pressure gradient:

Beskok and Karniadakis equation:

$$v = -\frac{K_0}{\mu}(1+a\,K_n)\left(1+\frac{4K_n}{1-bK_n}\right)\left(\frac{dP}{dx}\right) \quad (20.3)$$

Darcy's law describes the flow velocities by:

$$v = -\frac{K}{\mu}\frac{dP}{dx} \quad (20.4)$$

By comparing (20.3) and (20.4), permeability adjustment factor can be defined as:

$$K = K_0 \varsigma \quad (20.5)$$

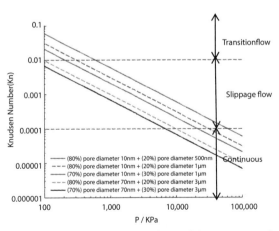

Figure 20.2 Relationship between Knudsen number and the pressure under different pore diameter combination.

$$\varsigma = (1 + a K_n)\left(1 + \frac{4K_n}{1 - bK_n}\right) \tag{20.6}$$

where Kn is the Knudsen number given by: Kn = λ /r, a is the rarefication coefficient, which is the correction of the bulk viscosity μ, b is the slip coefficient; and λ is the mean free-path of a molecule covered between the molecular collisions.

The rarefication coefficient by Beskok and Karniadakis [5] is:

$$a = \frac{128}{15\pi^2} \tan^{-1}\left(4K_n^{0.4}\right) \tag{20.7}$$

Permeability adjustment factor gained by (20.6) is:

$$\varsigma = 1 + a K_n + \frac{4K_n}{1 - bK_n} + \frac{4a K_n^2}{1 - bK_n} \tag{20.8}$$

Here, the first two terms represent the first order correction to no-slip flow in the Beskok-Karniadakis model. In this paper, slippage effect is considered.

When (20.7) is incorporated into (20.8) and simplified, the model is shown as follows:

$$\begin{aligned} v &= -\frac{K_0}{\mu}(1 + a K_n)\left(1 + \frac{4K_n}{1 - bK_n}\right)\left(\frac{dP}{dx}\right) \\ &= -\frac{K_0(1 + 4K_n + 4bK_n^2)}{\mu}\left(\frac{dP}{dx}\right) \end{aligned} \tag{20.9}$$

Figure 20.3 shows the relationship between permeability adjustment factor and Knudsen number under different slip factor. When the Knudsen number is closed to zero, the effect of collisions with the pore walls is small, thus the slippage effect can be ignored. When the Knudsen number is greater, the slippage factor have a great effect on the permeability adjustment factor.

Volumetric flux equation: we use the definition of the mean free-path of a molecule from Guggenheim [7]. Knudsen diffusivity is given by Civan:

$$\lambda = \sqrt{\frac{\pi z RT}{2M_w}} \frac{\mu}{P} \tag{20.10}$$

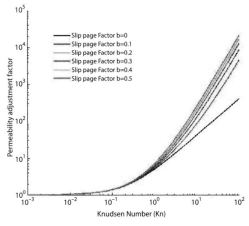

Figure 20.3 Relationship between permeability adjustment factor and Knudsen number under different slip factor.

$$D_K = \frac{4r}{3}\sqrt{\frac{2zRT}{\pi M_w}} \tag{20.11}$$

where R is universal gas constant, μ is the viscosity of the gas, T is the temperature, M_w is molecular weight, z is gas compressibility factor, λ is the mean free-path of a molecule, D_k is the Knudsen diffusivity.

When (20.11) is substituted into (20.10), the mean free-path of a molecule is shown as follows:

$$\lambda = \frac{3\pi}{8r} \cdot \frac{\mu}{p} \cdot D_K \tag{20.12}$$

When (20.12) is substituted into (20.1), the Knudsen number is shown as follows:

$$K_n = \frac{\lambda}{r} = \frac{3\pi}{8r^2} \cdot \frac{\mu}{p} \cdot D_K \tag{20.13}$$

Then the seepage model is formulated by combining (20.13) and (20.3):

$$\begin{aligned} v &= -\frac{K_0(1+4K_n+4bK_n^{\,2})}{\mu}\left(\frac{dP}{dx}\right) \\ &= -\frac{K_0}{\mu}\left(1+\frac{3\pi}{2}\frac{\mu}{r^2}D_K\frac{1}{p}+\frac{b}{4}\left(\frac{3\pi}{2}\frac{\mu}{r^2}D_K\frac{1}{p}\right)^2\right)\left(\frac{dP}{dx}\right) \end{aligned} \tag{20.14}$$

Here

$$K_0 = \frac{r^2}{8}$$

So (20.14) is rearranged as:

$$v = -\frac{K_0(1+4K_n+4bK_n^2)}{\mu}\left(\frac{dP}{dx}\right)$$

$$= -\frac{K_0}{\mu}\left(1+\frac{3\pi}{16K_0}\cdot\frac{\mu D_K}{p}+\frac{b}{4}\left(\frac{3\pi}{16K_0}\cdot\frac{\mu D_K}{p}\right)^2\right)\left(\frac{dP}{dx}\right) \quad (20.15)$$

According to the gas state equation, the density of gas under the isothermal condition is

$$\rho_g = \frac{T_{sc}Z_{sc}\rho_{gsc}}{P_{sc}}\cdot\frac{p}{TZ}$$

Equation (20.15) is multiplied by the density of gas, and thus the mass flux is obtained

$$\Gamma = -\frac{\rho K_0}{\mu}\varsigma\left(\frac{dP}{dx}\right) \quad (20.16)$$

The mass flux is defined as:

$$q = v \cdot A$$

Where, for radial fluid flow, $A = 2\pi r h$.
Thus, mass flux is obtained as follows:

$$q_m = \frac{K_0}{\mu}[1+\frac{3\pi}{16K_0}\cdot\frac{\mu D_k}{p}+\frac{b}{4}(\frac{3\pi}{16K_0}\cdot\frac{\mu D_k}{p})^2]$$
$$\cdot\frac{dp}{dr}\cdot 2\pi r h \cdot \rho_g \quad (20.17)$$

By separation of variables and integration, volumetric flux under the condition of steady seepage is rearranged as:

$$q_{sc} = \left[\frac{p_e^2-p_w^2}{2}+\frac{3\pi\mu D_K}{16K_0}(p_e-p_w)+\frac{b}{4}(\frac{3\pi\mu D_K}{16K_0})^2\ln\frac{p_e}{p_w}\right]$$
$$\cdot\frac{2\pi K_0 h Z_{sc} T_{sc}}{P_{sc}T\mu Z\ln\frac{r_e}{r_w}} \quad (20.18)$$

where K_0 is the absolute permeability ($\times 10^{-3}$, μm^2), M is the molar mass of the gas (kg/mol), p_e is the pressure of the boundary (MPa), p_w is the pressure at the wellbore (MPa), p_a the pressure of any point (MPa), p_{sc} is the standard state pressure (MPa), r_e is the radius of the boundary (m), r_w is the radius of the wellbore (m), φ is the porosity, μ is the viscosity of gas (mPa.s), ρ_g is the density of gas (kg/m³), ρ_{gsc} is the density of gas at the standard state (kg/m³), T is the formation temperature (K), T_{sc} is the standard state temperature (K), Z = compressibility factor, and Z_{sc} is the compressibility factor at the standard state.

20.3.2 Multi-scale seepage model considering of diffusion, slippage and desorption effect

Three gas states exist in the shale gas reservoir. Dissolved gas dissolves in the bound water. Adsorbed gas exists in the rock particles and on the organic matter, and free gas exists in the pore or fracture. The content of adsorbed gas is between 20 percent and 85 percent. In this paper, based on the non-linear model considering on diffusion and slippage effect, multi-scale seepage model considering of diffusion, slippage and desorption effect taking account of Langmuir isothermal adsorption equation is built. Volumetric flux of the free gas:

$$q_n = \left[\frac{p_e^2 - p_w^2}{2} + \frac{3\pi\mu D_K}{16K_0}(p_e - p_w) + \frac{b}{4}(\frac{3\pi\mu D_K}{16K_0})^2 \ln\frac{p_e}{p_w}\right] \cdot \frac{2\pi K_0 h Z_{sc} T_{sc}}{p_{sc} T \mu Z \ln\frac{r_e}{r_w}} \quad (20.19)$$

Volumetric flux of the absorbed gas:

$$q_d = \pi(r_e^2 - r_w^2)hp_c\left(V_m \frac{p_e}{p_L + p_e} - V_m \frac{\overline{p}}{p_L + \overline{p}}\right) \\ - \pi(r_e^2 - r_w^2)h\phi_m \quad (20.20)$$

So volumetric flux of the total gas:

$$q = q_n + q_d \quad (20.21)$$

where, V_m is Langmuir isotherm adsorption constant in cm³/g, p_L is Langmuir pressure constant.

20.4 Productivity prediction method of fracturing well

20.4.1 Productivity prediction method of vertical fracturing well

The basic assumptions are:
1. The fracture is vertical and symmetrically located at two sides of gas well.2. Section of the fracture is a rectangle, whose height is equal to the effective thickness of shale gas reservoir.3. Fracture conductivity is limited.4. The fluid in shale gas reservoir and fracture is single phase, which follows the rule of linear Darcy formula. 5. Without consideration of vertical flow in reservoir, the seepage is steady.

20.4.1.1 Outer flow field

Coordinate system x-y is built in plane Z and segment AB represents the fracture (see Figure 20.4).
Consideration of gas diffusion, slippage effect and symmetry, the expression of mass flow rate of gas well in outer flow field when the seepage of single phase is steady is:

$$q_1 = \frac{2\pi K_0 h Z_{sc} T_{sc}}{p_{sc} T \bar{\mu} \bar{Z} \ln \frac{2r_e}{L_f}} \left[\frac{p_e^2 - p_m^2}{2} + \frac{3\pi \mu D_K}{16 K_0}(p_e - p_m) + \frac{b}{4}(\frac{3\pi \mu D_K}{16 K_0})^2 \ln \frac{p_e}{p_m} \right] \quad (20.22)$$

where x_f is half length of the fracture, effective thickness of the reservoir, h, absolute permeability of the reservoir, K_0, slip factor, b, supply radius, r_e, supply boundary pressure, p_e, and well bottom pressure, p_w.
Volume flow rate of gas desorption is expressed as

$$q_d = \pi \left(r_e^2 - r_w^2\right) h p_c \left(V_m \frac{p_e}{p_L + p_e} - V_m \frac{\bar{p}}{p_L + \bar{p}} \right)$$
$$- \pi \left(r_e^2 - r_w^2\right) h \phi_m \quad (20.23)$$

20.4.1.2 Outer flow field
Percolation resistance in the fracture of vertical fractured well is expressed as

$$R_2 = \frac{1}{K_f w_f h_f \lambda} \cdot \frac{e^{\pi \lambda} + 1}{e^{\pi \lambda} - 1}$$

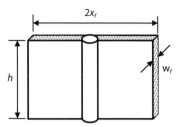

Figure 20.4: Longitudinal joints for fractured vertical well.

$$\lambda = \sqrt{2000K_m(K_f w_f \ln\frac{2R_e}{x_f})}$$

So, the expression of mass flow rate of gas well in inner flow field when the seepage of single phase is steady is:

$$q_2 = \frac{p_e^2 - p_m^2}{R2} \qquad (20.24)$$

According to equivalent flow resistance method, the relationship of inner and outer flow field is series-wound in supplying oil, so $q_1 + q_d = q_2 = q$. Considering that pressures in interface are equal and eliminating p_m, the productivity formula of multi-scale vertical fracturing well in consideration of diffusion, slip and desorption absorption is solved by programming.

Where x_f is half length of the fracture, width of the fracture, p_f, pressure of anywhere in reservoir, p, effective permeability in the horizontal direction of layer within fracture radius.

20.4.2 Productivity method of horizontal well with multi transverse cracks

There are two situations when the horizontal well fractures are multi transverse cracks (Figure 20.5):

a. If multi cracks discharge simultaneously and they don't disturb each other:

Total discharged rate is the sum of discharged rate of all cracks. According to equivalent flow resistance method and above analysis, the productivity equation of fractured horizontal well with multi transverse cracks in shale gas reservoir is:

$$Q = \sum_{i=1}^{n} q_i$$

where q_i is solved by combination of (20.34) and (20.36).

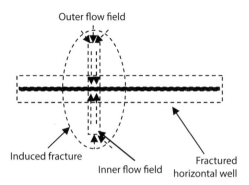

Figure 20.5 Fractured horizontal well seepage field sketch map.

b. If multi cracks discharge simultaneously and they do disturb each other:

According to equivalent percolation resistance method, when two elliptic flow discharged field are intersected, percolation resistance of this field decreases and the loss of starting pressure is influenced, but the flow resistance in fracture is not influenced. The contribution of each intersected transverse crack for gas well productivity in flow discharged elliptic field of fracture is expressed as:

$$q = \left(1 - \frac{S_i}{\pi a_i b_i}\right) q_i \tag{20.25}$$

So, when discharged elliptic flow of all transverse cracks disturb each other and in the condition of considering the starting pressure of low permeability reservoir, the productivity equation of fractured horizontal well with multi transverse cracks is expressed as:

$$Q = \sum_i^n q_i \tag{20.26}$$

If not all transverse cracks of fractured horizontal well disturb others, the productivity equation of horizontal well is the combination of above two circumstances.

In (20.25),

$$S_i = 2 \cdot \left(\frac{1}{4}\pi a_i b_i - \frac{1}{2} a_i b_i \arccos \frac{y_i}{b_i} \right) - \frac{W_i}{2} \cdot y_i$$

$$+ 2 \cdot \left(\frac{1}{4}\pi a_{i+1} b_{i+1} - \frac{1}{2} a_{i+1} b_{i+1} \arccos \frac{y_i}{b_{i+1}} \right) - \frac{W_i}{2} \cdot y_i$$

where y_i is from elliptic equation, $\dfrac{x^2}{b^2}+\dfrac{y^2}{a^2}=1$, $x_i=\dfrac{W_i}{2}$, and

$y_i = \sqrt{\left[1-\left(\dfrac{W_i}{2a}\right)^2\right]\cdot b^2}$, $i=1,2,\cdots,n-1$. If the disturb don't exit, $S_i=0$.

20.5 Production Forecasting

The data presented in Table 20.2 from a single well of shale gas reservoir in China is considered for computing a production forecast. According to the productivity model of the multi-scale vertical and horizontal fracturing well, numerical simulation results were gained by matlab programming. We analyzed the influence of fracture half length, fracture conductivity and fracture penetration ratio on the gas productivity through the calculation results, and the results showed that contribution of the free gas and the desorbed gas to total gas production.

Figure 20.6(a) shows a graph of relationship between total gas production and different production pressure drop under different half-length of hydraulic fractures. The productivity of the fracturing vertical well increases with the fracture width, but the increase extent decreases gradually.

Figure 20.6(b) shows relationship between total gas production and different fracture conductivity under different fracture penetration ratio. With the increase of the fracture conductivity, gas production increases quickly. While the fracture conductivity add to 2 $\mu m^2 \cdot cm$, the gas production increases slowly. On the other hand, the greater fracture penetration ratio, the more the gas production.

Table 20.2 Input parameters for forecasting analysis

porosity φ	0.07	compression factor Z	0.89
permeability K	0.0005md	gas viscosity μ	0.027 $mPa \cdot s$
formation tempreture T	366.15K	radius of the wellborer$_w$	0.1m
formation pressure p_e	24.13MPa	flowing bottomhole pressure p_w	1.25 MPa
pressure reliefradiir$_e$	400m	Formation thickness h	30.5m
rock densityρ_c	2.9g/cm³	fracture width w_f	3mm

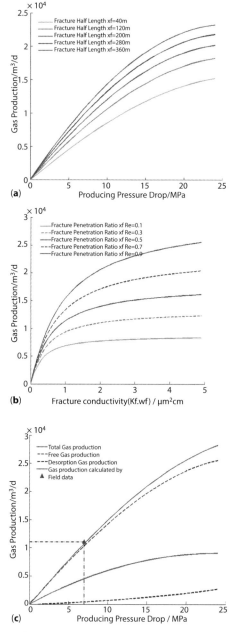

Figure 20.6 (a) Relationship between total gas production and different production pressure drop under different half-length of hydraulic fractures (b) Relationship between total gas production and different fracture conductivity under different fracture penetration ratio (c) Comparison of desorbed gas, free gas and total gas production for fractured vertical well

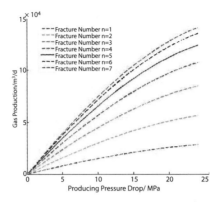

Figure 20.7 Relationship between total gas production and different production pressure drop under different number of hydraulic fractures

Figure 20.6(c) shows comparison of desorbed gas, free gas and total gas production for fractured vertical well. The free gas and desorbed gas production increase with the production pressure drop. The free gas production contributes most to the total gas production obviously. The results show that, Darcy formula has a great difference in our multi-scale seepage model. Known from the field data of fracturing vertical well, when the production pressure drop keeps 7 MPa, and the fracture half-length is 200m, the gas production is $1.2\times10^4 m^3$. Compared with the Darcy flow model, the calculation results by using multi-scale seepage model are in accordance with the field production data.

Figure 20.7 shows relationship between total gas production and different production pressure drop under different number of hydraulic fractures. Gas production increases with the number of hydraulic fractures. Due to the increase of the number of hydraulic fractures, the spacing between the fracture decreases, causing the interference among the fractures. So when the number of hydraulic fractures adds to a certain extent, gas production increases slowly.

20.6 Conclusions

The flow regime was described by combining the mechanics of continuous media and molecular kinematics method for the shale gas reservoir with Nano-micro scale pore. The fluid flow state was judged by Knudsen number. The different regional flow mechanism and flow state characteristics were illustrated. Then the flow state chart was drawn.

Under the formation pressure for the shale gas reservoir with nano-micro scale pore, Knudsen number is greater than 0.1, and the Knudsen adjustment factor shows bigger deviation between Darcy formula and the multi-scale seepage model. A new model was found based on Beskok and Karniadakis equation in consideration of different slip coefficient b. The result shows that, Darcy formula can't be applied in shale gas reservoir with nano pore.

The multi-scale seepage model considering of diffusion, slippage and desorption effect was established. By use of conformal transformation and equivalent flow resistance method, the productivity formula of vertical and horizontal fracturing well in consideration of diffusion, slip and desorption absorption was obtained.

According to the productivity model of the multi-scale vertical and horizontal fracturing well, results were gained by numerical simulation calculation. Compared with Darcy flow model, the calculation results by using multi-scale seepage model are more in accordance with the field production data, which verify the effectiveness of our model. With the increase of the fracture conductivity, gas production increases quickly. While the fracture conductivity add to $2 \ \mu m^2 \cdot cm$, the gas production increases slowly. On the other hand, the greater fracture penetration ratio, the more the gas production. The free gas and desorbed gas production increase with the production pressure drop. The free gas production contributes most to the total gas production obviously.

20.7 Acknowledgements

This work was supported by National Program on Key Basic Research Project (973 Program) (Grant No. 2013CB228002) through the effective development Research of the Southern Marine Shale Gas Reservoirs in China.

References

1. Javadpour, F., D. Fisher, M. Unsworth, Alberta Research Council. Nanoscale Gas Flow in Shale Gas sediments SPE 071006, 2007.
2. Freeman, C.M., Texas A&M University. A Numerical Study of Microscale Flow Behavior in Tight Gas and Shale Gas. SPE 141125, 2010.

3. Michel, G.G., Parametric Investigation of Shale Gas Production Considering Nano-Scale Pore Size Distribution, Formation Factor, and Non-Darcy Flow Mechanisms. SPE 147438, 2011.
4. Beskok, A., and Karniadakis, G. E., A Model for Flows in Channels,Pipes,and Ducts at Micro and Nano Scales, Microscale Thermophysical Engineering, 1999,3:43
5. Guggenheim, E. A. (1960) Elements of the Kinetic Theory of Gases. Pergamon Press. Oxford. DOI: 10.1016/0017-9310(61)90081-3.
6. Liepmann, H.W. 1961. Gas kinetics and gas dynamics of orifice flow. Journal of Fluid Mechanics, 10, p. 65.
7. Stahl, D.E. 1971. Transition range flow through microporous Vycor. Ph.D. thesis, Chemical Engineering Department, The University of Iowa.
8. Kaviany, M. 1991. Principles of Heat Transfer in Porous Media. Springer - Verlag New York Inc. New York.
9. Liu, X., Civan, F., and Evans, R.D. 1995. Correlation of the non - Darcy flow coefficient. Journal of Canadian Petroleum Technology, 34(10), pp. 50–54.
10. Chen Shangbin, Zhu Yanming, Wang Hongyan,et al.Characteristics and significance of mineral compositions of Lower Silurian Longmaxi Formation shale gas reservoir in the southern margin of Sichuan Basin[J].Acta Petrolei Sinica,2011,32(5):775–782.

21

Methane recovery from natural gas hydrate in porous sediment using gaseous CO_2, liquid CO_2, and CO_2 emulsion

Sheng-li Li, Xiao-Hui Wang, Chang-Yu Sun, Qing-Yuan and Guang-Jin Chen

State Key Laboratory of Heavy Oil Processing, China University of Petroleum, Beijing China

Abstract

Natural gas hydrate is expected to be a future energy resource. However, the decomposition of natural gas hydrate with traditional methods could lead to the weak of the ocean floor. The swapping CO_2 for CH_4 in gas hydrates is a favorable way as a long-term storage of CO_2 and enables the ocean floor to remain stable when CH_4 recovered from its hydrate. In this work, a three-dimensional apparatus was built to investigate the gas production from methane hydrate in porous media with gaseous CO_2, liquid CO_2, and CO_2 emulsion under different types of hydrate reservoir. The effects of CO_2 phase type, water saturation, hydrate saturation, temperature–pressure condition, and underlying gas room on the replacement reaction were examined. The experimental results show that the replacement rate and amount of CH_4 increase with the increase of hydrate saturation in the sediments. The replacement percent of CH_4 hydrate decreases with the increase of hydrate saturation, but increases with the increase of water saturation. Compared with injecting gaseous CO_2 method, liquid CO_2 injection is also benefit for the recovery of CH_4 from hydrate reservoir with much free water or that without underlying gas room to the extent that the injection of liquid CO_2 is kept by high gas saturation. An emulsifying agent for forming CO_2 emulsion was developed and evaluated. The rate and efficiency of replacement reaction by CO_2 emulsion is improved fundamentally from increasing the permeability of CO_2 and CH_4 in the loose hydrate layer.

21.1 Introduction

Energy and environment are key issues that confront the developed as well as developing nations in its quest for clean, affordable and reliable energy.

Ying Wu, John J. Carroll and Qi Li (eds.) *Gas Injection for Disposal and Enhanced Recovery*, (357–370) 2014 © Scrivener Publishing LLC

The exploitation of natural gas hydrate (NGH) has attracted much attention because of its enormous clean energy potential. However, the NGH plays an important role in stabilizing the stratum where it exists[1]. The exploitation of NGH with traditional methods may make the stratum unstable, and lead to geological disasters such as earthquakes, submarine landslides, and so on[2]. Another important issue is the rising atmospheric emissions as a result of fossil fuel consumption, which plays a major role in the greenhouse effect and global climate change[3]. To solve the above two issues, Hirohama et al.[4] and Ohgaki et al.[5] suggested a new method for hydrate exploitation by use of CO_2, which combines CO_2 storage and NGH exploitation together. In addition, the method of CO_2 replacement can maintain the stiffness in the granular medium[6]. With careful design of the operating conditions, it is possible to replace methane from methane hydrates with CO_2 in the solid phase whilst maintaining the geological stability[7].

In recent years, several researchers experimentally studied the replacement process in CH_4 hydrate with pressurized gaseous CO_2 using laboratory scale devices[8-12]. The microscopic mechanism of CH_4-CO_2 replacement was also investigated from molecular dynamics simulation[7, 13-15]. It was found that CH_4-CO_2 gas replacement reaction is slow due to the possible reason that new CO_2 hydrate formed on the surface of the CH_4 hydrate will wrap the CH_4 hydrate and prevent replacement reaction[16]. The effect of liquid CO_2 instead of gaseous CO_2 on the efficiency of CH_4-CO_2 replacement was also investigated. Ota et al.[11, 17] investigated the replacement of CH_4 in hydrate by liquid CO_2 using a cell with internal volume of 129.8 mL. It was found that CH_4-CO_2 replacement with liquid CO_2 (273.2 K and above 3.60 MPa) proceeds faster than that with gaseous CO_2 (273.2 K and 3.26 MPa). Zhou et al.[18] also investigated the replacement of CH_4 from its hydrate in quartz sand with CO_2-in-water emulsions and liquid CO_2 with a cell of ϕ 36 × 200 mm.

It can be seen that the current CH_4-CO_2 replacement experiments were usually performed with a laboratory scale device whatever for gaseous CO_2, liquid CO_2 or CO_2-in-water emulsion. Only the effect of temperature-pressure conditions or the hydrate cages on the replacement rate and efficiency can be examined. The obtained results and conclusions may not be suitable for the real three-dimensional behavior of CH_4-CO_2 replacement for natural gas hydrate undersea. The adaptability of the CO_2-CH_4 replacement method in different types of hydrate reservoirs can also not be simulated using the devices of laboratory scale. In this work, a three-dimensional middle-size reactor[16, 19, 20] is also used to simulate CH_4 recovery using gaseous CO_2, liquid CO_2, and CO_2 emulsion from different type of hydrate-bearing reservoir. The effects of CO_2 phase type, water saturation, hydrate saturation, temperature-pressure condition, and underlying gas room on the replacement reaction were examined.

21.2 Experiments

21.2.1 Apparatus and materials

The schematic diagram of the experimental apparatus used in this work is shown in Figure 21.1, which has been used to study the gas production from methane hydrate bearing in our previous work[16, 19, 20]. The main part of this apparatus is a reactor with effective height of 100 mm and inner diameter of 300 mm. When simulating hydrate reservoir with underlying free gas, a porous stainless steel sheet is used to separate the reactor into two parts: the upper sediments and the underlying free gas room. The maximum operating pressure of the reactor is up to 16 MPa. The reactor is immerged into a water bath containing a certain concentration of ethylene glycol to maintain constant temperature within the range of 253 K to 353 K. Sixteen thermocouples, with an accuracy of ± 0.1 K, are inserted into the reactor to detect the temperature distribution during hydrate formation and hydrate replacement process, which are divided into four groups. Each group includes four thermometers distributed at the same depth but different radius of 132 mm, 99 mm, 66 mm, and 33 mm from the center of

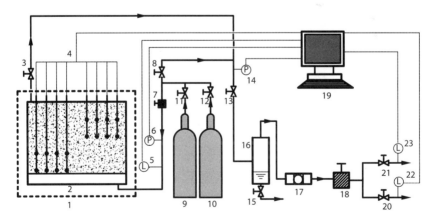

Figure 21.1: Experimental apparatus for replacing CH_4 from hydrate using CO_2. 1. water bath; 2, reactor; 3, 7, 8, 11, 12, 13, 15, 20, 21, valve; 4, thermocouple temperature transducer; 5, 22, 23, gas mass flow transducer; 6, 14, pressure transducer; 9, methane cylinder; 10, carbon dioxide cylinder; 16, gas-water separator; 17, filter; 18, back-pressure regulator; 19, computer

the reactor, respectively. The depths of four groups of thermometers are 82 mm, 58 mm, 34 mm, and 10 mm, respectively. The distribution of the thermocouples (named as T_1 to T_{16}) in the reactor is shown in Figure 21.2. The pressures of the gas room and hydrate-bearing sediments are monitored by pressure transducers with the accuracy of ± 0.02 MPa. A Monitor and Control Generated System (MCGS) are used to collect and record data of temperature, pressure, and flow rate during the experiments.

Methane and carbon dioxide with a mole fraction of 0.999 are supplied by the Beijing Beifen Gas Industry Corporation. The aqueous brine with salinity of 3.35 wt % is prepared in our laboratory. The sediment is formed by 20~40 mesh quartz sands with the porosity of 38.7%.

21.2.2 Procedure

The replacement experiments are conducted according to the following procedures, which are divided into two parts: preparation of CH_4 hydrate samples and replacement of CH_4 with CO_2.

21.2.2.1 Hydrate samples preparation for the replacement experiments.

First, a known amount of 3.35 wt % Na_2SO_4 aqueous brine was cooled to 273.2 K and quartz sands were frozen to 267.2 K and kept for 24 h. Then the brine was added into the sands, meanwhile the sands and aqueous brine were mixed and stirred immediately and adequately. The brine would become fine ice particles distributed homogeneously among the sands.

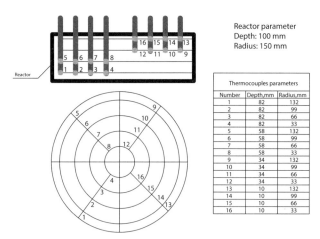

Figure 21.2: Distribution of the thermocouples in the reactor.

Thereafter, the homogeneous mixture was charged into the whole reactor or the upper space of the reactor isolated by the porous stainless steel sheet according to the different hydrate reservoir type simulated. The temperature of water bath was stabilized at 272.7 K, ensuring that water in the sand exists in ice after the mixture was loaded into the reactor, which can prevent water migration during the formation of hydrate. After that, the reactor was sealed, vacuumed for about 20 min, and purged with methane four to five times to ensure the absence of air. Then methane gas was charged into the reactor from the bottom of the reactor to the desired experimental pressure and the MCGS started to work. Hydrate would nucleate and form among the sediment gradually. Hydrate formation was considered to be finished if there were no changes of pressure in the system for 24 h. The representative hydrate sample was therefore prepared.

21.2.2.2 Replacement of CH_4 with CO_2

After the preparation of hydrate sample, the temperature of water bath was adjusted to the replacement reaction temperature of 274.7 K and kept constant. After the back-pressure regulator 18 in Figure 21.1 was set to 3.0 MPa, the outlet valves 3, 13, and 20 (or 21) were opened and the system pressure decreased. When the system pressure approached to 3.0 MPa, the valves 7 and 12 were opened, and the CO_2 was charged into the reactor from the bottom and the gaseous CH_4 was discharged at the same time from the top. During the CO_2-CH_4 exchange process in the free gas phase, the released gas from the reactor was collected into a cylinder. The composition of the released gas was measured by a gas chromatogram (HP6890). In general, when the concentration of CO_2 in the released gas was higher than 98%, it was assumed that the CO_2-CH_4 exchange process in the gas phase was completed and valves 7 and 12 were closed and CO_2 injection was stopped. At the same time, valves 3, 13, 20, and 21 were closed to stop the discharge of gas. Afterward, the CO_2-CH_4 replacement reaction proceeded in the sealed reactor and the gas samples from the reactor were analyzed by the gas chromatogram once every 24 h. The replacement of CH_4-CO_2 was assumed to be finished if the composition of CH_4 in the gas phase did not change in 24 h.

21.3 Results and Discussion

According to the definition of Moridis et al.[21], there are three commonly cited classes of hydrate reservoirs in the permafrost and marine sediment:

class 1 – hydrate layer underlain by two-phase zone of mobile gas and water, class 2 – hydrate layer underlain by one-phase zone of mobile water, and class 3 – hydrate layer with absence of underlain zones of mobile fluids. In order to investigate the favorable conditions of methane recovery from natural gas hydrate reservoir with CO_2 under different class of reservoir, different groups of hydrate sediment samples are prepared for replacing CH_4 from its hydrate with CO_2.

In order to investigate the rate of CH_4 recovery with liquid CO_2 under different hydrate samples, we defined the replacement rate of CH_4 as follows:

$$\eta_{Re} = \frac{n_{CH_4} - n_{CH_4}^0}{n_{CH_4,H}^0} \times 100\% \tag{21.1}$$

where $n_{CH_4}^0$ is the number of moles of CH_4 in gas or liquid phase at the beginning of the replacement reaction, $n_{CH_4,H}^0$ is the number of moles of CH_4 in hydrate at the beginning of the replacement reaction.

21.3.1 The replacement percent of CH_4 with gaseous CO_2

In order to investigate the favorable conditions of methane recovery from natural gas hydrate reservoir with gaseous CO_2 under different class of reservoir, three groups of hydrate sediment samples are prepared for replacing CH_4 from its hydrate with gaseous CO_2. The mass balances for the three experiments are shown in Table 21.1. In Table 21.1, nw is the total number of moles of water in the hydrate-bearing sediment samples; $n_{w,free}$ is the number of moles of free water in the hydrate-bearing sediment samples; xw is the percent of the water converted to hydrate. S_{CH_4}, S_{H_2O} and SH are the saturation of gaseous CH_4, water, and CH_4 hydrate in the hydrate sample, respectively. The definition of $n_{CH_4,Total}$, $n_{CH_4,G}$, and $n_{CH_4,H}^0$ are similar to those mentioned above, which are the total mole of CH_4 in the reactor

Table 21.1: Mass balance of the hydrate-bearing sediment samples

run	$n_{CH_4,Total}$ (mol)	$n_{CH_4,G}$ (mol)	$n_{CH_4,H}^0$ (mol)	nw (mol)	$n_{w,free}$ (mol)	xw (%)	S_{CH_4} (%)	S_{H_2O} (%)	S_H (%)
1	8.47	3.51	4.96	65.3	29.8	54.4	51.5	23.7	24.8
2	5.00	2.43	2.57	65.3	49.9	23.6	61.7	27.6	10.7
3	14.85	12.7	2.15	15	2.1	86.0	91.0	1.1	8.9

before the CH_4 hydrate formation, the mole of CH_4 in the gas phase after the CH_4 hydrate formation, and the mole of CH_4 in the hydrate phase, respectively. The initial conditions of CH_4-CO_2 replacement in hydrate are shown in Table 21.2. In Table 21.2, P is the pressure of gas phase after the CO_2 injection; T is the temperature of the reactor; $n_{CO_2,Total}$ is the total mole of CO_2 gas in the reactor; $n_{CO_2,Total}/n^0_{CH_4,H}$ is the ratio of the total mole of CO_2 to that of initial methane hydrate.

The variations of replacement percent of CH_4 versus time for three experimental runs are shown in Figure 21.3. It can be seen that run 3 has the largest replacement percent, and run 2 has the least replacement percent. As a result, it is inapplicable for replacing CH_4 from its hydrate with gaseous CO_2 for hydrate reservoir without underlying gas room (class 3 hydrate reservoir) and hydrate reservoir with underlying gas room (class 1 hydrate reservoir) with large amount of free water. On the other hand, the class 1 hydrate reservoir with high water saturation may make highly contributions to sequestrate CO_2. To sum up, the favorable conditions for replacing CH_4 from its hydrate with gaseous CO_2 are: 1) class 1 hydrate reservoir with a lower saturation of water, 2) higher ratio of the molar quantity

Table 21.2: Initial conditions of CH_4-CO_2 replacement in hydrate

run	P (MPa)	T (K)	$n_{w,free}$ (mol)	$n_{CO_2,Total}$ (mol)	$n_{CO_2,Total}/n^0_{CH_4,H}$
1	2.99	274.7	29.8	2.42	0.5
2	2.96	274.7	49.9	4.57	1.8
3	3.18	274.7	2.1	6.58	3.1

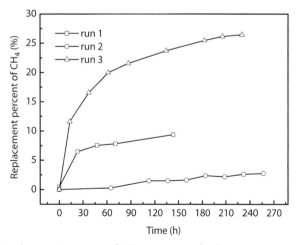

Figure 21.3: Replacement percents of CH_4 versus time for three runs.

of CO_2 injected into reactor to the molar quantity of CH_4 in hydrate, and 3) hydrate partially saturated in sediments. G

21.3.2 The replacement percent of CH_4 with liquid CO_2

In order to investigate the effects of hydrate properties on the replacement process, five different hydrate samples were formed. For experimental runs 4 to 7, the hydrate reservoirs with underlying gas room were formed; while for run 8, the hydrate reservoir without underlying gas room was formed. The experimental conditions of five runs for CH_4 hydrate formation are shown in Table 21.3 and those for CH_4-CO_2 replacement reactions are shown in Table 21.4.

The calculated replacement rates for the runs 4, 5, 7, and 8 are shown in Figure 21.4. It can be seen that all of the replacement rates decrease with time. Run 8 has the highest replacement rate in the initial stage since run 8 has the highest hydrate saturation and the largest CO_2-hydrate interface. The replacement rate in run 8 decreases very quickly because the diffusion of gas in the hydrate layer becomes the control step due to the increasing CO_2 hydrate layer around methane hydrate in the later stage of replacement reaction. For run 4 and run 7 with the similar replacement condition but different replacement temperature, it can be seen that the replacement rate in run 7 is higher than that in run 4, which also resulting from that the replacement conditions of run 7 are away from the stable zone of methane hydrate. For run 8, the replacement rate of the hydrate reservoir without underlying gas is also as good as that with underlying gas, which is also different with the replacement with gaseous CO_2[16]. This phenomenon is consistent with that observed in the CH_4-CO_2 replacement in the bulk environment by Ota et al.[11]. By comparing run 4 and run 8 in Figure 4, the replacement rate at the beginning 100 h is similar; afterward, the rate in run 8 is larger than that in run 4. This may attribute to that water

Table 21.3: Experimental condition for CH_4 hydrate formation

Runs	Temperature /K	Pressure/ MPa	Mass of sand/g	Mass of initial water/g
4	275.2	9.50	8700	500
5	275.2	9.51	8700	600
6	275.2	9.51	8700	755
7	275.2	9.50	8700	500
8	275.2	9.51	10360	600

Table 21.4: Experimental condition for CH_4-CO_2 replacement process

Runs	Temperature /K	Pressure of CO_2 / MPa	Void volume/L	Water consumed /g	CH_4 consumed /mol	Initial CO_2/mol	Water saturation/%	Hydrate saturation/%
4	275.2	4.21	3.26	372	3.44	52.46	5.42	19.7
5	275.2	4.20	3.26	576	5.33	50.88	0.17	30.5
6	275.2	4.21	3.26	367	3.40	47.62	16.44	19.4
7	280.2	4.20	3.26	366	3.38	52.45	5.68	19.4
8	275.2	4.19	2.71	321	2.97	42.43	11.82	17.0

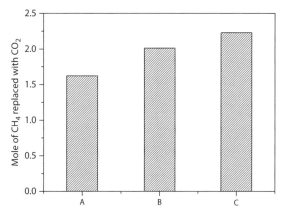

Figure 21.4: Variation of the replacement rate with time for runs 4, 5, 7, and 8

saturation in run 8 is larger than in run 4, and the hydrate formed in run 8 is not as compact as that in run 4, resulting in more water vacancy in run 8. Demurov et al.[22] considered that a water vacancy is necessary for diffusion of CO_2 molecule in hydrate layer.

21.3.3 The replacement percent of CH_4 with CO_2-in-water emulsion

Three groups of experiments, named as runs A, B, and C, were conducted in this work to investigate the performance of CO_2-CH_4 replacement with CO_2 emulsion. The effects of emulsion properties, hydrate reservoir properties, and salt in the emulsion on the replacement reaction were investigated. The experimental conditions of CH_4 hydrate formation are shown in Table 21.5. The experimental conditions of CO_2 emulsion formation for three runs are shown in Table 21.6. The experimental conditions of CH_4-CO_2 replacement reaction are shown in Table 21.7.

The moles of CH_4 replaced from hydrate in three runs are shown in Figure 21.5. As shown in Figure 5, the moles of CH_4 replaced from hydrate in run C is the largest, while that in run A is the lowest. The reason for the poor effect of run A is that it is chiefly the thermal dissociation of hydrate and only part of CH_4 is replaced by CO_2. In addition, the hydrate saturation in run A is the lowest. The salt in emulsion in run C will promote CH_4 hydrate dissociation, and decrease the hydrate particles aggregation, resulting in the best replacement effect in run C.

Figure 21.6 shows the replacement efficiencies with CO_2 emulsion of run C (obtained in this work), liquid CO_2[20], and gaseous CO_2[16]. It can be seen that the replacement efficiency of CO_2 emulsion can reach 47.8%,

Table 21.5: Experimental conditions for CH_4 hydrate formation

Run	Temperature /K	CH_4 initial pressure /MPa	CH4 pressure after hydrate formation /MPa	Mass of sand /g	Mass of water /g
A	273.2	9.4	7.0	10360	601
B	273.2	10.3	6.6	10360	601
C	273.2	10.1	6.9	10360	600

Table 21.6: Experimental conditions for CO_2-in-water emulsion formation

Run	Distilled water/ vol%	SDS/ wt%	Tween 80/ wt%	Na_2SO_4/ wt%	Temperature/ K	Pressure /MPa	Rotation speed /r/min
A	19.6	0.5	5.0	0	293.2	15	1200
B	31.1	0.5	5.0	0	290.2	15	1200
C	32.2	0.5	5.0	3.35	290.2	15	1200

Table 21.7: Experimental conditions for CH_4-CO_2 replacement reaction

Run	Temperature /K	mass of water consumed /g	mole of CH_4 consumed / mol	water saturation %	hydrate saturation /%
A	281.2	327	3.03	10.08	15.09
B	275.7	528	4.88	2.66	24.36
C	275.7	487	4.60	3.82	23.00

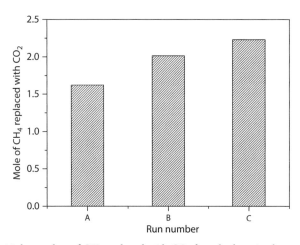

Figure 21.5: Mole number of CH_4 replaced with CO_2 from hydrate in three runs.

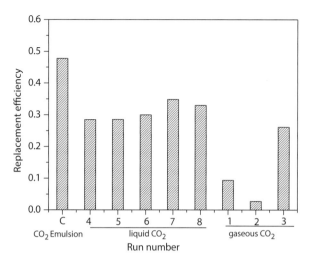

Figure 21.6: Comparison of CH_4 hydrate replacement efficiency with gaseous CO_2, liquid CO_2, and CO_2 emulsion.

which is much higher than that of gaseous CO_2 and liquid CO_2. The reason of good effect in run C in this work is chiefly due to three aspects: first, the diffusion of CO_2 emulsion is better than liquid CO_2, so the CO_2 emulsion can reach larger area and induce larger interface between CO_2 and CH_4 hydrate; second, Na_2SO_4 in emulsion can inhibit the hydrate formation and promote CH_4 hydrate dissociation, and prevent massive hydrate formation; third, the SDS in emulsion can make the newly formed hydrate layer looser, which will increase the permeability of CO_2 and CH_4 in hydrate layer.

21.4 Conclusion

A three-dimensional middle-size reactor was built to investigate the favorable conditions for methane recovery from different types of hydrate reservoir with gaseous CO_2, liquid CO_2, and CO_2 emulsion. The effects of CO_2 phase type, water saturation, hydrate saturation, temperature-pressure condition, and underlying gas room on the replacement reaction were examined. The experimental results indicated that hydrate reservoir with underlying free gas, high saturation of free gas and low saturation of water is appropriate for recovering CH_4 with gaseous CO_2. Compared with injecting gaseous CO_2 method, liquid CO_2 injection is also benefit for the recovery of CH_4 from hydrate reservoir with much free water or that without underlying gas room to the extent that the injection of liquid

CO_2 is kept by high gas saturation. An emulsifying agent for forming CO_2 emulsion was developed and evaluated. Compared with injecting liquid CO_2 and gaseous CO_2, the replacement efficiency with CO_2 emulsion is highest.

21.5 Acknowledgements

The financial support received from National 973 Project of China (No. 2012CB215004), National Natural Science Foundation of China (Nos. 20925623, U1162205, 21076225), and the Research Funds of China University of Petroleum, Beijing (01JB0171) are gratefully acknowledged.

References

1. Zhou, X., S. Fan, D. Liang. Advancement in research on replacement of CH_4 from hydrate with CO_2. **Chem. Indust. Eng. Prog**, 2006, 25: 524–527.
2. Maslin, M., N. Mikkelsen, C. Vilela, B. Haq. Sea-level –and gas-hydrate–controlled catastrophic sediment failures of the Amazon Fan. **Geology**, 1998, 26(12): 1107–1110.
3. Hatzikiriakos, S.G., P. Englezos. The relationship between global warming and methane gas hydrates in the earth. **Chem. Eng. Sci.**, 1993, 48(23): 3963–3969.
4. Hirohama, S., Y. Shimoyama, A. Wakabayashi, S. Tatsuta, N. Nishida. Conversion of CH_4 hydrate to CO_2 hydrate in liquid CO_2. **J. Chem. Eng. Jpn.**, 1996, 29(6): 1014–1020.
5. Ohgaki, K., K. Takano, H. Sangawa, T. Matsubara, S. Nakano. Methane exploitation by carbon dioxide from gas hydrates—Phase equilibria for CO_2-CH_4 mixed hydrate system. **J. Chem. Eng. Jpn.**, 1996, 29(3): 478–483.
6. Espinoza, D.N., J.C. Santamarina. P-wave monitoring of hydrate-bearing sand during CH_4–CO_2 replacement. **Int. J. Greenh. Gas. Con.**, 2011, 5(4): 1031–1038.
7. Tung, Y.-T., L.-J. Chen, Y.-P. Chen, S.-T. Lin. In situ methane recovery and carbon dioxide sequestration in methane hydrates: a molecular dynamics simulation study. **J. Phys. Chem. B**, 2011, 115(51): 15295–15302.
8. Lee, H., Y. Seo, Y.-T. Seo, I.L. Moudrakovski, J.A. Ripmeester. Recovering methane from solid methane hydrate with carbon dioxide. **Angew. Chem. Int. Ed.**, 2003, 42(41): 5048–5051.
9. Yoon, J.-H., T. Kawamura, Y. Yamamoto, T. Komai. Transformation of methane hydrate to carbon dioxide hydrate: in situ raman spectroscopic observations. **J. Phys. Chem. A**, 2004, 108(23): 5057–5059.

10. Ota, M., Y. Abe, M. Watanabe, R.L. Smith Jr, H. Inomata. Methane recovery from methane hydrate using pressurized CO_2. **Fluid Phase Equilib.**, 2005, 228–229: 553–559.
11. Ota, M., T. Saito, T. Aida, M. Watanabe, Y. Sato, R.L. Smith, H. Inomata. Macro and microscopic CH_4–CO_2 replacement in CH_4 hydrate under pressurized CO_2. **AIChE J.**, 2007, 53(10): 2715–2721.
12. Schicks, J.M., M. Luzi, B. Beeskow-Strauch. The conversion process of hydrocarbon hydrates into CO_2 hydrates and vice versa: thermodynamic considerations. **J. Phys. Chem. A**, 2011, 115(46): 13324–13331.
13. Geng, C.-Y., H. Wen, H. Zhou. Molecular simulation of the potential of methane reoccupation during the replacement of methane hydrate by CO_2. **J. Phys. Chem. A**, 2009, 113(18): 5463–5469.
14. Qi, Y., M. Ota, H. Zhang. Molecular dynamics simulation of replacement of CH_4 in hydrate with CO_2. **Energy Convers. Manage.**, 2011, 52(7): 2682–2687.
15. Bai, D., X. Zhang, G. Chen, W. Wang. Replacement mechanism of methane hydrate with carbon dioxide from microsecond molecular dynamics simulations. **Energ. Environ. Sci.**, 2012, 5(5): 7033–7041.
16. Yuan, Q., C.-Y. Sun, X. Yang, P.-C. Ma, Z.-W. Ma, B. Liu, Q.-L. Ma, L.-Y. Yang, G.-J. Chen. Recovery of methane from hydrate reservoir with gaseous carbon dioxide using a three-dimensional middle-size reactor. **Energy**, 2012, 40(1): 47–58.
17. Ota, M., K. Morohashi, Y. Abe, M. Watanabe, Smith, Jr., H. Inomata. Replacement of CH_4 in the hydrate by use of liquid CO_2. **Energy Convers. Manage.**, 2005, 46(11–12): 1680–1691.
18. Zhou, X., S. Fan, D. Liang, J. Du. Replacement of methane from quartz sand-bearing hydrate with carbon dioxide-in-water emulsion. **Energy Fuels**, 2008, 22(3): 1759–1764.
19. Yang, X., C.-Y. Sun, K.-H. Su, Q. Yuan, Q.-P. Li, G.-J. Chen. A three-dimensional study on the formation and dissociation of methane hydrate in porous sediment by depressurization. **Energy Convers. Manage.**, 2012, 56: 1–7.
20. 20. Yuan, Q., C.-Y. Sun, B. Liu, X. Wang, Z.-W. Ma, Q.-L. Ma, L.-Y. Yang, G.-J. Chen, Q.-P. Li, S. Li, K. Zhang. Methane recovery from natural gas hydrate in porous sediment using pressurized liquid CO_2. **Energy Convers. Manage.**, 2013, 67: 257–264.
21. 21. Moridis, G.J., T.S. Collett, S.R. Dallimore, T. Satoh, S. Hancock, B. Weatherill. Numerical studies of gas production from several CH_4 hydrate zones at the Mallik site, Mackenzie Delta, Canada. **J Petrol. Sci. Eng.**, 2004, 43(3–4): 219–238.
22. 22. Demurov, A., R. Radhakrishnan, B.L. Trout. Computations of diffusivities in ice and CO_2 clathrate hydrates via molecular dynamics and Monte Carlo simulations. **J. Chem. Phys.**, 2002, 116(2): 702–709.

22

On the Role of Ice-Solution Interface in Heterogeneous Nucleation of Methane Clathrate Hydrates

PaymanPirzadeh and Peter G. Kusalik

Department of Chemistry, University of Calgary, Calgary, AB, Canada

Abstract

Clathrate hydrates are specific cage-like structures formed by water molecules around a guest molecule. Despite many studies performed on clathrate hydrates, the actual molecular mechanism of both their homogeneous and heterogeneous nucleation has yet to be fully clarified. Here, by means of molecular simulations, we demonstrate how the interface of hexagonal ice facilitates the heterogeneous nucleation of methane clathrate hydrate from an aqueous methane solution. Our results indicate an initial accumulation of methane molecules, which promote induction of defective structures, particularly coupled 5–8 ring defects, at the ice surface. Structural fluctuations originating from these defective motifs, mainly 5-member rings, support hydrate cage formation next to an ice interface. The cage-like structures that are formed act as a sink of methane molecules in the solution and enhance the stability and growth of the amorphous nucleus, which converts into a proper crystal upon annealing. The insights provided by the present results further our understanding of the molecular mechanisms involved in, for instance, gas hydrate clogs formed in oil transport pipelines that may result in environmental hazards or delays in energy supply.

22.1 Introduction

Every year oil and gas companies spend considerable financial resources on keeping their pipelines clear of deposits to ensure flow of energy products to their markets [1-12]. A major threat to hydrocarbon flow as a blocking factor is formation of clathrate hydrates, also known as gas hydrates [1–12]. These solid species have an appearance similar to water ice as water is a major structural component in the hydrate structure. High pressures and low temperatures increase the formation probability of these solids [1–12]. Hydrate formation is becoming more of a challenging issue as the industry is moving towards exploration in deeper and colder waters. Therefore, having a better knowledge on formation of these species would provide insights in designing safe and secure operation technologies for uninterrupted delivery of hydrocarbons to consumers. In particular, solid surfaces such as the surfaces of (even) tiny mineral rocks, ice particles and pipe walls can facilitate and trigger formation of gas hydrates that may clog valves and wellheads [13–22].

Clatharte hydrates are inclusion compounds enjoying a specific topology of the hydrogen bond network of water molecules, comprising the structural framework of the clathrate, and guest molecules which facilitate formation and stabilize the specific network topology[1–12]. In fact, one may find a hierarchy of water structural motifs within a clathrate crystal. The hydrogen bonding ability of water molecules enables them to form intermediately ranged structural motifs known as rings [23]. It is well established through both experiments and theoretical studies that such rings are randomly formed and dissociated in pure liquid water, as they lack long-range order to stabilize them [23]. Studies have shown that one may dominantly find 4, 5, 6 and 7-member rings in pure liquid water [23]. On the other hand in hexagonal ice, the common solid form of water, molecules form stable 6-member rings in a puckered conformation, with a long-range order among them resulting in a crystalline configuration [23].

In clathrate hydrates, however, water molecules form flat 5 and 6-member ring motifs giving rise to cage structures of 5^{12} (common to all hydrate structures), $5^{12}6^2$(specific to type I hydrates) and $5^{12}6^4$(specific to type II hydrates) [1–12]. As this implies, the hydrate structural motifs do not match with those present in ice. But experimental evidence has shown that an ice surface can facilitate nucleation of clathrate hydrates. In a recent molecular simulation study of methane hydrates, it is demonstrated how an ice surface can play a facilitating role in heterogeneous nucleation of methane hydrate from a super saturated solution [24]. Induction-promotion-nucleation (IPN) has been suggested as a three-stage mechanism for the heterogeneous nucleation of methane hydrate on the surface of hexagonal ice [24]. Key to this facilitation is the dynamics of water molecules within the interfacial region between the ice and bulk solution, which aids in the induction of surface defects on the surface of ice and promoting stability of hydrate-like structural motifs [24, 25]. Here we mainly focus on the promotion stage of heterogeneous nucleation of methane clathrate hydrate next to a hexagonal ice surface.

22.2 Method Summary

Molecular dynamics has been used to reveal the molecular aspects of heterogeneous nucleation of methane hydrate near the surface of hexagonal ice. A six-site model and a united atom model represented the water and methane molecules, respectively [26, 27]. Hexagonal ice and a supersaturated methane solution comprised the simulation setup. Hexagonal ice was oriented in a way that its prism face of (10-10) was positioned next to a 10 mol% methane solution. Because of period boundary condition in all directions, two solid-liquid interfaces were introduced in the direction of heterogeneity, identified as the z-axis, in the system. The temperature was kept at 265 K in all parts of the system except at one interface [24]. The pressure was kept at 1000 bars to ensure nucleating conditions for methane molecules with the employed models and to prevent possible phase separations (i.e. bubble formation) in the system.

A custom post-analysis code was used to analyze ring structures present in system configuration [23]. The criteria applied to identify hydrogen bonded water molecules included: (i) H---O distance should be less than 2.2 Å, (ii) the angle between O-H and O-O vectors should be less than 30°, and (iii) assuming OG is the bisector of H-O-H angle in a water molecule, then the G-O-O angle (angle between GO and oxygen of another water molecule) should cover a range of 110-180°. Once a list of H-bonded neighbours is established for each molecule, the algorithm searches first for smaller rings (to avoid over-counting them when searching for larger rings). The translational mean square displacement was also calculated for each particle as $\Delta r_i^2(t) = [r_i(t) - r_i(0)]^2$ for a period of 100ps. Further technical details have been provided in a recent work [24].

22.3 Results and Discussion

Nucleation of clathrate hydrate requires an increased local density of the guest molecules within a region in the system [5–8, 24]. In the presence of a solid surface such as ice or silica, next to a supersaturated solution, guest molecules tend to accumulate within the interfacial region of the liquid and solid [24, 28–30]. In the present work, looking at the time evolution of the density of methane molecules during the simulation reveals a similar trend in Figure 22.1. As pointed out elsewhere, migration of methane molecules towards the ice-water interface results in a decrease in the free energy of the system; thus accumulation of methane molecules next to ice is a thermodynamically favorable event [24, 28–30]. One can also justify this observation from a molecular structure point of view. It has been shown that the ice-water interface is a region where water molecules tend to form more stable 5 and 7-member rings relative to the bulk liquid

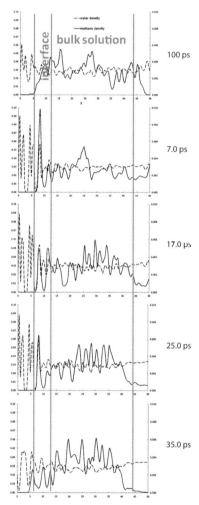

Figure 22.1: Density profile of methane guest solutes and water along the z-axis during nucleation of methane hydrate on the (10-10) hexagonal ice-water interface. Density profile of water (dashed black line) and methane (solid red line) are presented on the left and right axes, respectively. Displacement of methane molecules across the box can be seen from the profiles. Empty cages are highlighted by dashed squares.

water [23]. These structural motifs are particularly attractive to methane molecules (as hydrophobic entities) and water molecules tend to form similar structures around methane molecules to maximize their ability to form hydrogen bonds [2–4, 11]. This natural avoidance (force) can push the methane molecules towards a phase boundary region to minimize their disruption upon liquid water molecules. While the interfacial region gradually gets enriched in methane, the local density of water molecules is somewhat reduced (see density profiles in Figure 22.1). This can result in local fluctuations in density appropriately to trigger hydrate nucleation. However, presence of a nearby structurally unmatched substrate can either hinder or slow down the nucleation process.

Heterogeneous nucleation on a solid substrate can happen in two ways [31]. One is an epitaxial growth where the nucleation barrier is overcome by formation of a new solid layer on a mineralogically similar substrate [31]. The second method is topotaxial nucleation where the new solid phase overcomes the nucleation barrier by forcing formation of structural fits on a mineralogically different substrate [31]. In the case of methane hydrate nucleation on ice, which does not have a crystalline match to clathrate hydrate, the motifs to help achieve structural fits had been identified as the coupled 5-8 ring defects [24]. These structural motifs had also been previously identified as motifs allowing accommodation of cubic ice (with a similar crystalline symmetry to hydrates) within the lattice of hexagonal ice [25]. These defects have coupled 5-member rings (also common in hydrate crystals) next to an 8-member ring which resembles the void within a 5^{12} cage of a clatharte crystal [24, 25]. A particular feature on coupled 5-8 ring defects is that their manifestation on the (10-10) prism face of hexagonal ice can be visualized through formation of unusually stretched 6-member rings if one looks at the hexagonal ice crystal from a different plane vector of the prism face (1-210), as highlighted by dashed white rectangles in Figure 22.2. These defects arise because the accumulated methane molecules can disrupt the positioning of water molecules at the surface of the ice and modify the pattern of the hydrogen bond network (refer to insets of Figure 22.3). This stage demonstrates the "induction" phase of the heterogeneous nucleation of methane hydrate through an IPN mechanism [24].

Figure 22.2 also provides evidence for the second stage of the heterogeneous nucleation of methane hydrate, the "promotion" stage. In Figure 22.2, water rings are probed as well as the mobility of the water molecules in the system. 4, 5 and 6-member rings are presented as magenta, green and blue sticks, respectively. As mentioned earlier, the coupled 5-8 ring defect has coupled 5-member rings as a structural motif (also refer to insets of

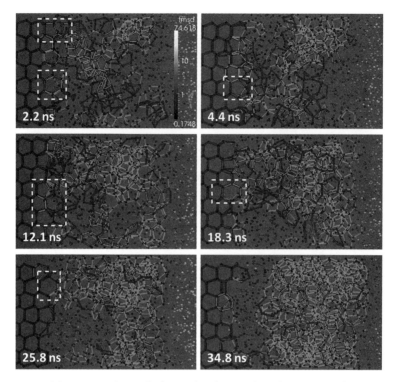

Figure 22.2: The impact of water hydrogen bond network in the presence of 5-8-member ring defects on methane hydrate nucleation. The (10-10) hexagonal ice-water system projected on the y-z plane at various times in trajectory is presented where the rings in the system are highlighted. The blue, green and magenta colors highlight the 6, 5 and 4-member water rings, respectively. Methane molecules are not shown for visualization purposes. Water molecules are represented by their oxygen atoms which are colored based on their translational mean square displacement (tmsd), see legend. As it can be observed, close to the end of the simulation, there exists a layer of disordered water molecules between the final hydrate crystal and ice interface. This layer might be present because of the structural mismatch between the ice and hydrate. The regions containing the 5-8-member rings are enclosed by dashed squares.

Figure 22.3). It can be observed that the hydrogen bond network propagating from these surface defects of ice is mainly in the form of 5-member rings. These 5-member rings can form cage-like structures that can attract methane molecules from the bulk; it has previously been shown [2-4] that the corresponding potential of mean force is attractive. As more methane molecules migrate towards the ice-water interface, the population of 5-member rings, as well as the population of hydrate cages, increases. As it is apparent from Figure 22.2, during the development of the new hydrate phase water rings are stabilized through their connections to the surface of

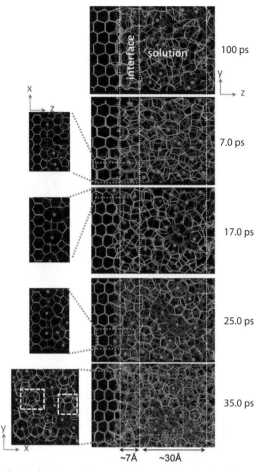

Figure 22.3: Snapshots of an ice-methane solution system during heterogeneous nucleation of methane hydrate. Configurations of the system are shown in which precursors of clathrate cages are highlighted. The green spheres are the methane molecules. Water molecules are represented by the hydrogen bonds (white sticks) they make with their neighbor molecules. The yellow, red, purple, cyan, blue, magenta and orange highlight the 5^{12}, $5^{12}6^2$, $5^{11}6^24^1$, $5^{12}6^4$, $5^{12}6^3$, $5^{10}6^24^1$ and incomplete cages, respectively. The incomplete cages may include $5^{10}6^2$, $5^{11}6^1$ and $5^96^24^1$ motifs or 7-8-member rings which might anneal later. The brown colored regions underline the 5-8-member ring defects. The insets on the left panel are cross-sections of the defected region. Empty cages are highlighted with white dashed rectangles.

ice. As the hydrate phase begins to mature, the connections between the two structurally mismatched crystals become scarcer, and eventually the two solid phases separate. In Figure 22.2, the mobility of the water molecules (based on their translational root mean square displacement) are also highlighted. Accordingly, least mobile/solid-like water molecules are colored in black while the most mobile ones are shown with white. It is observed from Figure 22.2 that early in the trajectory as one gets closer to the ice-water interface, the mobility of water molecules decreases. However, as time elapses, the population of solid-like particles within the interfacial region of the ice-solution interface also rises, and expands beyond the interfacial region into the bulk solution. It can be seen that the mobility of particles within the new solid hydrate phase is comparable to those in ice. However, after stabilization of the hydrate phase, a mobile layer of water molecules is left between the two solid phases. This mobile layer may act as a structural transition layer between the two solid phases.

Figure 22.3 provides a presentation of the variety of cages that form during the heterogeneous nucleation of methane hydrate next to the prism face of hexagonal ice. Cage motifs such as 5^{12}, $5^{12}6^2$, $5^{12}6^4$ and $5^{12}6^3$ (previously identified as [32] an early motif during homogeneous nucleation as well as a mediator between type I and II) are highlighted with yellow, red, cyan and blue, respectively. The coupled 5-8 ring defect regions are shown with brown sticks. Looking at cross-sectional slices of the system also reveals that empty hydrate cages (highlighted with white dashed squares in Figure 22.3) can form among neighboring filled cages due to methane density fluctuations. It should be emphasized that although the new hydrate phase appears to be amorphous, it is stable enough to continue to grow with the presence of further solutes. Also, this amorphous structure will anneal but requires continuing the simulation for a very long period of time, which is not feasible in the present study. Early in the trajectory (see in Figure 22.2) the presence of short-lived hydrate like structural motifs originating from the ice surface facilitates nucleation of the new hydrate phase about twice as fast as a homogeneous nucleation process under comparable conditions [10, 12]. Overall, the present results, when combined with other recent theoretical studies, suggest that most of the hydrate nucleation observed in laboratory experiments and in real scenarios arise from the presence of a surface, i.e. heterogeneous nucleation of hydrates [33].

22.4 Summary

Heterogeneous nucleation of clathrate hydrate on the surface of ice appears to be a clear example of a topotaxial nucleation explained by the IPN mechanism. Formation of coupled 5-8 ring defects, which are induced by the presence of methane molecules and have features consistent with the lattice structures and hydrogen bond pattern of both hexagonal ice and methane hydrate, seems to be the core event of this nucleation phenomenon. This set of defects can act as a platform of propagation of 5-member rings, which promote formation of cage-like structures, from the surface of ice into the bulk methane solution. The transient hydrate like motifs next to the ice surface attract further methane molecules from the bulk solution. This migration of guest molecules increases the local density of methane molecules and eventually results in rapid nucleation of a new hydrate phase. The results presented provide insights that can be useful to pipeline designers and metallurgic experts on what type of molecular-level patterns may contribute to triggering hydrate nucleation.

References

1. Sum, K.A.; Koh, C.A.; Sloan, E.D. Clathrate hydrates: From laboratory science to engineering practice. *Ind. Eng. Chem. Res.* **2009**, 48, 7457–7465.
2. Guo, G. J.; Zhang, Y. G.; Li, M.; Wu, C. H. *J. Chem. Phys.* **2008**, 128, 194504.
3. Guo, G. J.; Li, M.; Zhang, Y. G.; Wu, C. H. *Phys. Chem. Chem. Phys.* **2009**, 11, 10427–10437.
4. Guo, G-J.; Zhang, Y-G.; Liu, H. *J. Phys. Chem. C* **2007**, 111, 2595–2606.
5. Jacobson, L.C.; Molinero V. *J. Am. Chem. Soc.* **2011**, 133, 6458–6463.
6. Jacobson, L.C.; Hujo, W.; Molinero, V. *J. Am. Chem. Soc.* **2010**, 132, 11806–11811.
7. Jacobson, L.C.; Hujo, W.; Molinero, V. *J. Phys. Chem. B* **2009**, 113, 10298–10307.
8. Jacobson, L.C.; Hujo, W.; Molinero, V. *J. Phys. Chem. B* **2010**, 114, 13796–13807.
9. Walsh, M.R.; Koh, C.A.; Sloan, E.D.; Sum, A.K.; Wu, D.T. *Science* **2009**, 326, 1095–1098 (2009)
10. Walsh, M.R.; Rainey, J.D.; Lafond, P.G.; Park, D.H.; Beckham, G.T.; Jones, M.D.; Lee, K.H.; Koh, C.A.; Sloan, E.D.; Wu, D.T.; Sum, A.K. *Phys. Chem. Chem. Phys.* **2011**, 13, 19951–19959
11. Guo, G.J.; Zhang, Y.G.; Liu, C.J.; Li, K.H. *Phys. Chem. Chem. Phys.* **2011**, 13, 12048–12057.
12. Walsh, M.R.; Beckham, G.T.; Koh, C.A.; Sloan, E.D.; Wu, D.T.; Sum, A.K. *J. Phys. Chem. C* **2011**, 115, 21241–21248.
13. Bai, D.; Chen, G.; Zhang, X.; Wang, W. *Langmuir* **2011**, 27, 5961–5967
14. Liang, S.; Rozmanov, D.; Kusalik, P.G. *Phys. Chem. Chem. Phys.* **2011**, 13, 19856–19864
15. Pietrass, T.; Gaede, H. C.; Bifone, A.; Pines, A.; Ripmeester, J. A. *J. Am. Chem. Soc.* **1995**, 117, 7520–7525.
16. Moudrakovski, I. L.; Ratcliffe, C. I.; McLaurin, G. E.; Simard, B.; Ripmeester, J. A. *J. Phys. Chem. A* **1999**, 103, 4969–4972.
17. Chan, J.; Forrest, J. A.; Torrie, B. H. *J. Appl. Phys.* **2004**, 96, 2980–2984.
18. Kuhs, W.F.; Staykova, D.K.; Salamatin, A.N. *J. Phys. Chem. B* **2006**, 110, 13283–13295.

19. Staykova, D.K.; Kuhs, W.F.; Salamatin, A.N.; Hansen, T. *J. Phys. Chem. B* **2003**, 107, 10299–10311.
20. Falenty, A.; Genov, G.; Hansen, T.C.; Kuhs, W.F.; Salamatin, A,N. *J. Phys. Chem. C* **2011**, 115, 4022–4032.
21. Zhang, Y.; Debenedetti, P.G.; Prud'homme, R.K.; Pethica, B.A. *J. Phys. Chem. B* **2004**, 108, 16717–16722 (2004)
22. Davies, S.R.; Hester, K.C.; Lachance, J.W.; Koh, C.A.; Sloan, E.D. *Chem. Eng. Sci.* **2009**, 64, 370–375
23. Pirzadeh, P.; Beaudoin, E. N.; Kusalik, P.G. *Chem. Phys. Lett.* **2011**, 517, 117–125
24. Pirzadeh, P.; Kusalik, P.G. *J. Am. Chem. Soc.* **2013**, 135, 7278–7287
25. Pirzadeh, P.; Kusalik, P.G. *J. Am. Chem. Soc.* **2011**, 133, 704–707
26. Nada, H.; van der Eerden, J.P.J.M. *J. Chem. Phys.* **2003**, 118, 7404–7413
27. Jorgensen, W.L.; Madura, J.D.; Swenson, C.J. *J. Am. Chem. Soc.* **1984**, 106, 6638–6646
28. Boewer, L.; Nase, J.; Paulus, M.; Lehmkühler, F.; Tiemeyer, S.; Holz, S.; Pontoni, D.; Tolan, M. *J. Phys. Chem. C* **2012**, 116, 8548–8553
29. Partay, L. B.; Jedlovszky, P.; Hoang, P. N. M.; Picaud, S.; Mezei, M. *J. Phys. Chem C* **2007**, 111, 9407–9416
30. Bagherzadeh, S.A.; Englezos, P.; Alavi, S.; Ripmeester, J.A. *J. Phys. Chem. C* **2012**, 116, 24907–24915
31. Worden, R.H.; Smalley, P.C.; Cross, M.M. *J. Sediment. Res.* **2000**, 70, 1210–1221
32. Vatamanu, J.; Kusalik, P.G. *Phys. Chem. Chem. Phys.* **2010**, 12, 15065–15072
33. Knott, B.C.; Molinero, V.; Doherty, M.F.; Peters, B. *J. Am. Chem. Soc.* 2012, 134, 19544 19547

23
Evaluating and Testing of Gas Hydrate Anti-Agglomerants in (Natural Gas + Diesel Oil + Water) Dispersed System

Chang-Yu Sun, Jun Chen, Ke-Le Yan, Sheng-Li Li, Bao-ZiPeng, and Guang-Jin Chen

State Key Laboratory of Heavy Oil Processing, China University of Petroleum, Beijing, China

Abstract

The anti-agglomeration performance of single or compounded commercial chemical additives with/without the addition of alcohol as co-surfactants was evaluated using a sapphire cell and a focused beam reflectance measurement (FBRM) device. The experimental results showed that AEO-3 combined with some commercial chemical additives, especially Span 20, exhibits good anti-agglomeration performance. The hydrate slurry thus formed has a high stability and will not result in agglomeration for a long period of time in an autoclave reactor. The performance of the anti-agglomerant was also evaluated using a flow loop device in which hydrate slurries were formed from (natural gas + diesel oil + water) dispersed system. The flow characteristics, shutdown/restart behavior, and morphology of hydrate slurries were investigated experimentally at different water cut. It was found that the adopted anti-agglomerant can disperse hydrate particles in fluid phase apparently. The formed hydrate slurries can safely flow and be easily restarted even though the initial water cut was up to 30.0 vol%.

23.1 INTRODUCTION

Flow assurance is becoming increasingly important with the strengthening petroleum industry interest in deep-water hydrocarbon exploration and development. The flow assurance issues in deepwater flow infrastructure are primarily due to the deposition of solid, such as gas hydrates, waxes, asphaltene, etc. Gas hydrate is considered to be the largest problem by an order of magnitude relative to the others.

Low dosage hydrate inhibitors (LDHI) such as kinetic inhibitors (KI) and anti-agglomerants (AA), have attracted much attention in recent years because of their low dosage and potential savings in operating costs. The cost saving arise because they are use in significantly smaller amounts than the tradition thermodynamic inhibits such as methanol or glycol.

In this study we examine the AA inhibitors. Their purpose is not to prevent the formation of a gas hydrate, *per se*, but to prevent the hydrate from accumulating into a flow-blocking plug.

23.2 Experimental Apparatus And Analysis

The experimental apparatus used in this work consists of a high pressure sapphire cell and a flow loop. The pictures of these devices are shown in Figure 23.1.

23.3 Results And Discussion

23.3.1 Measurement of water-droplet size in emulsion

The experimental results of water-droplet diameter measurement of six groups of emulsions are listed in Table 23.1. The experimental results show that the water droplet diameter increases with increasing initial water cut.

Measurements were made at 277.15 K (4.0°C, 39.4°F) and fro pressure up to 7.65 MPa (1110 pisa). Part of this study included an investigation of the effect of pressure on the slurtry formed.

Figure 23.1: Pictures of the Experiential Apparatus Used in This Study, Sapphire Cell on the Left and Flow Loop on the Right.

Table 23.1: Information Regarding the Preparation of the Emulsions Used in This Study

Water cut (Qw) (vol%)	Mixing speed (rpm)	Mixing time (min)	Water droplet diameter (nm)	Standard deviation (nm)
5.0	6000	10	115.0	39.10
10.0	6000	10	348.1	119.5
15.0	6000	10	768.1	218.1
20.0	6000	10	1615.0	390.8
25.0	6000	10	1841.0	456.6
30.0	6000	10	2428.0	862.7

23.3.2 Morphology of hydrate slurry formed in emulsion

Figure 23.2 shows the morphologies of hydrate slurry formed from methane + water + diesel oil + AA dispersed system with six water cuts (from 5 to 30 vol%) at 277.15 K and 7.65 MPa. It can be easily found that the hydrate can be apparently dispersed in oil phase with the addition of anti-agglomeration.

23.3.3 Gas consumption in the hydrate formation process in emulsion

The gas consumptions in the hydrate formation process during each experiment were calculated and the results are shown in Figures 23.3 through 23.5. These figures show that the hydrate formation rate increases with the increase of water cut of emulsion, and the hydrate formation rate is much dependent on temperature but relatively less on initial pressure.

23.3.4 Flow characteristic and morphology of hydrate slurry in a flow loop apparatus

The morphology of natural gas hydrate slurry or emulsion in flow loop was observed through a visual window and shown in Figure 23.6. It can be clearly seen from these morphologies that the oil and gas can be safely transported by forming stable and flowable hydrate slurries.

The dosage of anti-agglomerant added is specified as mass percent in the water phase and its content in the aqueous phase is constant at 2.0

Figure 23.2: Morphologies of the Hydrate Slurry Formed from Methane + Water + Diesel Oil + AA Dispersed System with Different Water Cuts at 277.15 K and 7.65 MPa.

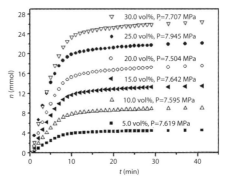

Figure 23.3: Experimental Results of Gas Consumption Moles with Time During Hydrate Formation in Methane + Water + Diesel Oil + AA Dispersed System at Different Water Cuts and 277.15 K.

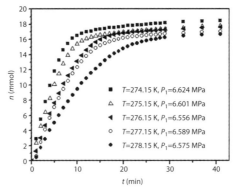

Figure 23.4: Experimental Results of Gas Consumption Moles with Time During Hydrate Formation in Methane + Water + Diesel Oil + AA Dispersed System at Different Temperatures and a 20 vol% Water-cut.

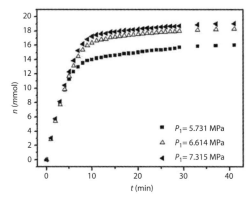

Figure 23.5: Experimental Results of Gas Consumption Moles with Time During Hydrate Formation in Methane + Water + Diesel Oil + AA Dispersed System at Different Initial Pressure and 274.15 K and a 20 vol% Water-cut.

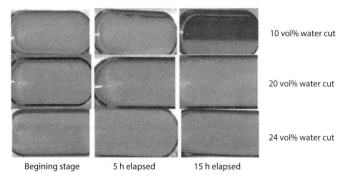

Figure 23.6: Morphologies of Emulsion and Hydrate slurry at Different Water-cut and Times for a Natural Gas + Water + Diesel Oil System

wt% in all experimental runs. In each run, the pressure and temperature of the system are set to 2.10 MPa and 274.2 K. It can be seen from Figure 23.7 that the flow rate decreases with the undergoing of hydrate formation and then becomes stable after the completion of hydrate formation. On the other hand, the pressure drop increases with the increase of hydrate quantity formed and gradually becomes stable accompanied with some fluctuation.

23.4 CONCLUSION

Experimental data on the kinetics of hydrate formation from pure methane in water-in-oil emulsions and the flow characteristics and morphology

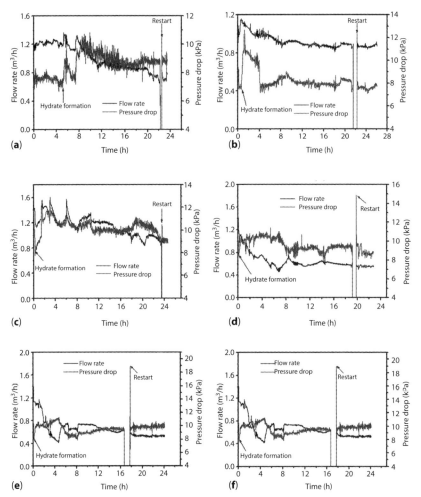

Figure 23.7: Variation of the Fluid Flow Rate and the Pressure Drop of the Hydrate Slurry with the Elapsed Time at Different Initial Water-cuts for Natural Gas + Water + Diesel Oil System (a) 5 vol%, (b) 10 vol%, (c) 15 vol%, (d) 20 vol% (e) 22 vol%, and (f) 24 vol%

of hydrate slurry have been presented in this work. The experimental results demonstrate that the absorption process of methane in the emulsion reaches equilibrium very quickly. The induction time increases with decrease of initial pressure and increase of temperature in the cell. The hydrate growth rate increases with increase of initial pressure and decrease of temperature in the cell. The water cut of emulsion influences on the hydrate growth rate very much. In addition, diesel oil and natural gas can be safely transported by forming stable and flowable hydrate slurries with the presence of AA.

24

"Is That a Bacterium in Your Trophosome, or Are You Just Happy to See Me?" - Hydrogen Sulfide, Chemosynthesis, and the Origin of Life

Neil Christopher Griffin

University of Calgary, Calgary, AB, Canada

Abstract

Life on Earth can survive in the most inhospitable environments. But one group of organisms do not just survive in inhospitable environments, they thrive. These are the extremophiles, a collection of bacterial species that colonize the most unforgiving locales on Earth. Of the extremophiles, few are more specialized than the chemosynthetic bacteria that dwell on the ocean floor, 2000 m beneath the surface. Chemosynthetic bacteria eschew normal sources of energy (the sun, or eating other animals), and instead fuel their growth and development by capturing the hydrogen sulfide spewed from deep sea vents in prodigious quantities. Using this unlikely, and toxic, energy source, chemosynthetic bacteria form the backbone of an entire ecosystem – an ecosystem that may harbor clues to the origin of life on Earth.

24.1 Introducing the extremophiles

The Miracle of Life, contrary to popular belief, is not the act of reproduction (after all, even a slime mold can do that)[1]. The real miracle of life is that it exists at all - despite low probability - and continues to exist in virtually every imaginable environment. In the last four billion years, life has thrived across Earth[2-4]. From super-heated hot springs[5], to Antarctica sea ice[6], to the barren tundra of northern Canada - "life finds a way" (to quote the inimitable film *Jurassic Park)*. But not all species are equally capable of survival in difficult places. One group rises above all others, the undisputed champions of eking a living from inhospitable environments: the extremophiles.

Extremophiles, literally 'lovers of the extreme', are organisms that not only survive, but thrive in environments that sit well outside the comfort zones of most Earthly life [7,8]. These organisms, usually prokaryotic bacteria and archaea, make a living out of colonizing the most inhospitable environments on Earth (and also possibly outer space)[9,10,11]. They're a cosmopolitan group. There are thermophiles[12], acidophiles and alkaliphiles[13], halophiles[14], and psychrophiles (which sounds as though it should be a lover of psychotropic drugs - *Hunter sthompsonii*, perhaps, but actually means cold temperatures)[7,8,15]. The more researchers search for extremophiles in places that had previously been thought to be dead zones, the more they turn up new species[8,10]. Even the Dead Sea turns out to have a pulse when halophilic (salinity-loving) archaea are considered[14,16]. But if there is a gold medal in the Extremophile Olympics, the winner is easy to decide: the polyextremophiles[7]. Like a decathlete, a polyextremophile is not content to excel at only one thing - instead it insists on surviving in environments of multiple extremes (and, like a decathlete, along the way they make all other extremophiles feel slightly inadequate). And few environments are more challenging than the bottom of the ocean.

The ocean floor is like Goldilocks' misappropriated porridge - it's either too cold[17], or (if you're near a hydrothermal vent) too hot[10]. It's dark: light barely penetrates below 200 m, and is absent completely below 1000 m[18]. And to make things more complicated, the atmospheric pressure on the sea floor can be up to 1000 times greater than at sea level[19,20]. This cavalcade of complicating conditions makes life in the deep sea difficult, even for extremophilic bacteria. Luckily, they've evolved a solution: they shield themselves from some of the worst excesses of their environment by finding sanctuary inside the body of another living organism.

24.2 Tempted by the guts of another

Meet the Giant tubeworm *(Riftiapachyptila)*, a brilliant example of creativity in scientific naming. *Riftiapachyptila* is a tubeworm, and a member of the phylum Annelida, the worms. Giant tubeworms are invertebrates that live on the sea floor by using their tail to anchor their bodies onto the substrate[12]. After anchoring, they secrete a mineral-based tube around their body, which they can use to hide inside. Picture a giant tube of lipstick: the mineral casing is the tube, and the worm's body itself is the lipstick.

Except in this case the lipstick is a 2 m long worm.

The Giant tubeworm grows in the Pacific Ocean at depths of up to 2,500 m, favoring areas near "black smokers" - hydrothermal vents that continuously spew plumes of sulfide rich smoke into the ocean. Hydrogen sulfide is poisonous, corrosive, and smells awful. In most animals, hydrogen sulfide inhibits aerobic respiration and enzyme function, and leads to abnormal blood pigments - and, subsequently, death[15,21]. Generally not a substance one wants in the environment.

It's possible that the Giant tubeworms prefer to live near black smokers because of the lower price of property (even tubeworms feel the squeeze of a recession), and given that they have no eyes, ears, or sense of smell, and only a rudimentary nervous system, they can probably handle the smell and the unsightly black smoke. But instead, they prefer black smokers because, for a Giant tubeworm, hydrogen sulfide is a perfect source of energy[12]. But it isn't easy to turn a poisonous gas into a gourmet meal, and it requires a little help.

Giant tubeworms, along with their aural, optic, and neural deficiencies, also don't have guts. That's not a slander on their courage or bravery, but a biological fact: tubeworms lack a traditional digestive system (coincidentally, this makes the insult "you gutless worm" a little bit of a tautology).

Instead of a gut, giant tubeworms have an organ called the trophosome ("you trophosome-less worm" would be a more accurate putdown, although it doesn't scan as well)[22]. The trophosome is a spongy, dark-green tissue found in the coelom of the tubeworm - a fluid-filled cavity that lies inside the body walls, beneath skin and muscle tissue. The role of the trophosome is to be an internal factory for the tubeworm, a factory filled with millions of extremophilic bacterial workers - up to one million bacterial cells per milligram of tubeworm tissue[23].

(For those who find the idea a little disconcerting, keep in mind that inside a human body the number of bacterial cells outnumbers the number of body cells by about 10:1 - we are, essentially, more bacterium than human: at least tubeworms keep it confined to the trophosome[24]).

The trophosome provides a home for the bacteria, and in return, the bacteria pay rent in the form of energy, by using chemosynthesis to convert hydrogen sulfide into carbohydrates.

24.3 Chemosynthesis 101

In the parlance of an ecologist, all of life's functions can be boiled down to four categories: feeding, fleeing, fighting, and reproduction. Feeding, the acquisition of energy, is probably the most important (if not necessarily the most fun). Without energy, no organism can perform the other three functions. The most familiar forms of energy acquisition are heterotrophy - 'feeding' in the traditional sense of consuming another organism and converting its matter into usable energy; and autotrophy - the acquisition of energy from nonorganic sources. The most common type of autotrophy, and the most well known, is photosynthesis: the process by which green plants and algae convert sunlight and carbon dioxide into carbohydrates.

Photosynthesis and heterotrophy are all well and good for most organisms on Earth. But when you live at the bottom of the ocean, and you're a microscopic bacterium, they aren't viable options. Instead, you find a third-path, a unique form of autotrophy called chemosynthesis.

Chemosynthesis is the conversion of carbon-based molecules into organic compounds (such as carbohydrates) through the oxidization of inorganic molecules (often sulfur-based). The process was first suggested by Sergei Nikolaievich Winogradsky, a Russian-Ukrainian scientist who along with proposing chemosynthesis, was also a pioneer of microbiology, a discoverer of numerous biogeochemical cycles, and a talented pianist[25,26]. (What is the Russian word for "over-achiever"?) Winogradsky's research suggested that there existed microbial life forms that could subsist entirely on inorganic matter - a groundbreaking suggestion at the time. Unfortunately for Winogradsky, technology took some time to catch up with theory, and it wasn't until 90 years later when chemosynthesis was proved conclusively†.

In the late 1970s, the manned submersible *Alvin* discovered deep-sea hydrothermal vents along the Galapagos Rift[27,28]. The first researchers to see the vents, especially the black smokers, were surprised by the complexity of the ecosystem they supported[29]. Much of the deep-sea floor is barren desert, but the area around black smokers is full of life: complex communities of tubeworms, sea fans, mussels and crabs grow and mingle around the vents like campers huddled around a fire on a cold night[27,29].

The original discoverers of the hydrothermal vents hypothesized, like Winogradsky, that chemosynthetic organisms supported the ecosystems, but the proof didn't arrive until a few years later. In 1981, the microbiologist Colleen Cavanaugh cracked the mystery. She collected samples of tissue from the giant tubeworms, and examined them under a microscope.

---

† Perhaps a useful strategy to employ when being asked for evidence to back-up an assertion: "The evidence will be provided...in approximately one-hundred years. But may I have the grant money now please?"

Trapped inside the trophosome tissue, she found evidence of crystallized sulfur - and using electron microscopy she discovered bacteria living within the trophosome[12]. Careful experimentation revealed that the bacteria were capable of generating metabolic energy by oxidizing the hydrogen sulfide spewed from the black smokers (24.1)[12,30].

$$12H_2S + 6CO_2 \rightarrow C_6H_{12}O_6 + 6H_2O + 12S + ATP \qquad (24.1)$$

The formula for hydrogen sulfide chemosynthesis.

For the first time, chemosynthesis had been proved as a sole source of energy, capable of supporting an entire ecosystem. No less a scientific authority than Bill Nye (the Science Guy) has called chemosynthesis one of the top 10 greatest discoveries of all time[31]. Winogradsky, for his part, can rightfully say, "I told you so."

24.4 Chemosynthetic bacteria and the origins of life

But, honestly, who really cares about what's going on near some hot, smelly, dark hole at the bottom of the ocean? As it turns out, an inordinately large number of people, and for very good reason: the ecosystems that have evolved around hydrothermal vents may hold clues to the origins of life on Earth.

The standard narrative of the origin of life is that the building blocks of organic life arose from a primordial ooze (also called a soup, or stew, depending on your culinary inclinations). But it wasn't long after hydrothermal vents were discovered before scientists began hypothesizing that perhaps they were the site of first life [32] (one particularly creative scientist proposed warm, alkaline hydrothermal vents as an origin *before* such hydrothermal vents were known to exist[33]).

Experimentally, the chemistry of hydrothermal vents is similar to that which is suggested to have lead to biologically life in the primordial soup [34]. At high pressures and temperature, the sulfur based exhalations from hydrothermal vents can react with carbon monoxide to form, among other things, pyruvic acid [34]. For those who never had the misfortune of memorizing metabolic pathways in college, pyruvic acid (and its conjugate base, pyruvate) is a critical compound in both anaerobic and aerobic respiration.

If laboratory experiments aren't enough, genetics can be added to the mix. Phylogenetic studies of microbial life have found that the earliest common ancestor of the three major biological domains (Eukarya,

Archaea and Bacteria) have a chemosynthetic origin, possibly traceable to hydrothermal vents [35].

Life on Earth may ultimately owe itself to the relationship between a notoriously poisonous gas, a giant lipstick-esque worm, and the bacterium that lives inside it. Miracle of life, indeed.

References

1. McGuinness, M. D. & Haskins, E. F. Genetic Analysis of the Reproductive System of the True Slime Mold Comatricha lurida. *Mycologia***77,** 646–653 (1985).
2. Lineweaver, C. H. & Chopra, A. The Habitability of Our Earth and Other Earths: Astrophysical, Geochemical, Geophysical, and Biological Limits on Planet Habitability. *Annu. Rev. Earth Planet. Sci.***40,** 597–623 (2012).
3. Battistuzzi, F. U., Feijao, A. & Hedges, S. B. A genomic timescale of prokaryote evolution: insights into the origin of methanogenesis, phototrophy, and the colonization of land. *BMC Evol Biol***4,** 44 (2004).
4. Sleep, N. H. & Bird, D. K. Evolutionary ecology during the rise of dioxygen in the Earth's atmosphere. *Philosophical Transactions of the Royal Society B: Biological Sciences***363,** 2651–2664 (2008).
5. Lau, M., Aitchison, J. C. & Pointing, S. B. Bacterial community composition in thermophilic microbial mats from five hot springs in central Tibet - Springer. *Extremophiles***13,** 139–149 (2009).
6. Thomas, D. N. Antarctic Sea Ice--a Habitat for Extremophiles. *Science***295,** 641–644 (2002).
7. Rothschild, L. J. & Mancinelli, R. L. Life in extreme environments. *Nature***409,** 1092–1101 (2001).
8. Pikuta, E. V., Hoover, R. B. & Tang, J. Microbial Extremophiles at the Limits of Life. *Critical Reviews in Microbiology***33,** 189–209 (2007).
9. Cavicchioli, R. Extremophiles and the Search for Extraterrestrial Life. *Astrobiology***2,** 281–292 (2002).
10. Canganella, F. & Wiegel, J. Extremophiles: from abyssal to terrestrial ecosystems and possibly beyond. *Naturwissenschaften***98,** 253–279 (2011).
11. de Vera, J. P. P. et al.The adaptation potential of extremophiles to Martian surface conditions and its implication for the habitability of Mars.*EGU General Assembly Conference Abstracts***14,** 2113 (2012).
12. Cavanaugh, C. M., Gardiner, S. L., Jones, M. L., Jannasch, H. W. & Waterbury, J. B. Prokaryotic Cells in the Hydrothermal Vent Tube Worm Riftia pachyptila Jones: Possible Chemoautotrophic Symbionts. *Science, New Series***213,** 340–342 (1981).
13. Bräuer, S. L., Cadillo-Quiroz, H., Yashiro, E., Yavitt, J. B. & Zinder, S. H. Isolation of a novel acidiphilic methanogen from an acidic peat bog. *Nature***442,** 192–194 (2006).

14. Oren, A. Molecular ecology of extremely halophilic Archaea and Bacteria. *FEMS Microbiology Ecology* **39**, 1–7 (2002).
15. Grieshaber, M. K. & Völkel, S. Animal Adaptations for Tolerance and Exploitation of Poisonous Sulfide. *Annu.Rev. Physiol.* **60**, 33–53 (1998).
16. Timpson, L. M. et al. Characterization of alcohol dehydrogenase (ADH12) from Haloarcula marismortui, an extreme halophile from the Dead Sea. *Extremophiles* **16**, 57–66 (2011).
17. Ridgway, N. M. Temperature and salinity of sea water at the ocean floor in the New Zealand region. *New Zealand Journal of Marine and Freshwater Research* **3**, 57–72 (1969).
18. Clarke, G. L. & Wertheim, G. K. Measurements of illumination at great depths and at night in the Atlantic Ocean by means of a new bathyphotometer. *Deep Sea Research (1953)* **3**, 189–205 (1956).
19. Nogi, Y. & Kato, C. Taxonomic studies of extremely barophilic bacteria isolated from the Mariana Trench and description of Moritella yayanosii sp. nov., a new barophilic bacterial isolate. *Extremophiles* **3**, 71–77 (1999).
20. Pathom-aree, W. et al. Diversity of actinomycetes isolated from Challenger Deep sediment (10,898 m) from the Mariana Trench. *Extremophiles* **10**, 181–189 (2006).
21. Völkel, S. & Grieshaber, M. K. in … *of Systemic Regulation: Acid—Base Regulation …* (Heisler, N.) **22**, 233–257 (Springer Berlin Heidelberg, 1995).
22. Katz, S., Klepal, W. & Bright, M. The Osedax Trophosome: Organization and Ultrastructure. *Biological Bulletin* **220**, 128–139 (2011).
23. Somero, G. N. Symbiotic Exploitation of Hydrogen Sulfide. *Physiology* **2**, 3–6 (1987).
24. Backhed, F. Host-Bacterial Mutualism in the Human Intestine. *Science* **307**, 1915–1920 (2005).
25. Dworkin, M. Sergei Winogradsky: a founder of modern microbiology and the first microbial ecologist. *FEMS Microbiology Reviews* **36**, 364–379 (2011).
26. Waksman, S. A. Sergei Nikolaevitch Winogradsky: 1856-1953. *Science* **118**, 36–37 (1953).
27. Corliss, J. B., Dymond, J., Gordon, L. I. & Edmond, J. M. Submarine thermal springs on the Galapagos Rift. *Science* **203**, 1073–1083 (1979).
28. Crane, K. & Ballard, R. D. The Galapagos Rift at 86° W: 4. Structure and morphology of hydrothermal fields and their relationship to the volcanic and tectonic processes of the Rift Valley. *J. Geophys. Res.* **85**, 1443 (1980).
29. Lutz, R. A. & Kennish, M. J. Ecology of deep-sea hydrothermal vent communities: A review. *Rev. Geophys.* **31**, 211 (1993).
30. Cavanaugh, C. M. Symbiotic chemoautotrophic bacteria in marine invertebrates from sulfide-rich habitats. *Nature* **302**, 58–61 (1983).
31. *100 Greatest Discoveries.* (Discovery Science Channel, 2004).
32. Martin, W. & Russell, M. J. On the Origin of Biochemistry at an Alkaline Hydrothermal Vent.*Philosophical Transactions: Biological Sciences* **362**, 1887–1925 (2007).

33. Russell, M. J., Daniel, R. M., Hall, A. J. & Sherringham, J. A. A hydrothermally precipitated catalytic iron sulfide membrane as a first step toward life. *J Mol Evol* **39,** 231–243 (1994).
34. Cody, G. D. *et al.* Primordial Carbonylated Iron-Sulfur Compounds and the Synthesis of Pyruvate. *Science, New Series* **289,** 1337–1340 (2000).
35. Pace, N. R. A Molecular View of Microbial Diversity and the Biosphere. *Science* **276,** 734–740 (1997).

Index

absorber column 155, 157, 159–176, 188, 194, 196–199, 201–205, 208–214, 234, 240
acid gas 99–101, 114, 129–130, 133, 148, 150–151, 155–156, 175, 178, 195, 202, 229
acid gas injection 1, 29, 99, 243–246, 248–253, 255–256, 275–284, 286
adsorption 179, 228, 230, 293, 295–302, 347
alkanolamine 153–157, 162–166, 168–172, 174, 177, 179–191, 195–201, 214, 218, 220, 228, 231–235, 240, 258, 262,
amine (see alkanolamine)
ammonia 229
AQUAlibrium 29, 30, 54–60, 61–63, 130, 133, 141, 146, 151

Benedict-Webb-Rubin 29–30, 37–43
Benzene (see BTEX)
BTEX 129
bubble point 101, 104, 279
butane 146,
BWR (see Benedict-Webb-Rubin)

caprock 244–245, 275–278, 280–286, 311
capital cost 150, 205, 231–232, 241,
carbon capture 1, 153–176, 177–191, 193–226, 295, 302, 305–310, 311–329
carbon dioxide

carbon sequestration
carbonate 249, 258, 264, 267–271, 275, 278, 280, 281, 312
casing 244, 246, 251, 271, 277, 279
cement 277–280,
clathrate (see gas hydrate)
CO_2 EOR 129, 131–132, 142, 146, 195, 294
CO_2 storage (see carbon storage)
coal 293–302
coal-fired 153, 155, 162, 178–179, 194, 196, 205, 220.
compressibility factor 1–27, 29–63, 279, 345, 347, 351
compression 2, 112, 129, 131–133, 135, 137, 141, 142, 145, 150, 193, 196, 202, 213, 220, 224, 255, 256, 258, 259–264, 270, 272, 279,
corrosion 130, 133, 135–136, 140–142, 150, 199, 205, 209–211, 213, 218, 228, 268, 270, 280

DEA (see amine)
dehydration 100, 111–112, 129–151, 201, 246, 255
density 1–27, 29–63, 70, 88, 112–115, 117–122, 137, 156, 232, 239, 246, 257–259, 279, 281, 297, 301, 316, 335, 346–347, 351, 373–375, 378
dew point 84–85, 87, 93, 104, 136, 279
DexPro™ 129–151
dolomite 249, 281,

395

elemental sulfur (see sulfur)
enhanced recovery (see CO_2 EOR)
EOS (see specific equations of state)
equation of state (see specific equations of state)

foam 177, 227–228, 234, 240

gas compressibility (see compressibility factor)
gas hydrate 99–101, 129–131, 135–137, 141, 148–151, 255–273, 305–310, 311–329, 371–379, 381–385
glycol 111–125, 129, 140, 201, 230, 359, 382

H_2S (see hydrogen sulfide)
hydrate (see gas hydrate)
hydrogen sulfide 99–101, 105, 111–116, 122, 146, 245, 250, 257, 260–261, 267, 273, 278, 281, 387–392

interfacial tension (see surface tension)

Lee-Kesler 29, 31–33, 62–63, 65–67, 69–70, 73, 76
limestone 249

mass transfer 153–174, 177–191, 203, 219, 314,
MDEA (see amine)
MEA (see amine)
mercaptan 277
methane 1–27, 29–62, 111, 113–116, 118, 122, 146, 294, 312–315, 319–322, 327–328, 357–369, 371–379, 383–386
methanol 99–108, 129–151, 229, 256, 262–263, 266, 269, 271–272, 382

nitrogen 1–27, 29–62, 146, 178–179, 203, 297,

PC-SAFT 111–125
Peng-Robinson 29–30, 43–48, 60–63, 69, 102, 130, 151
permeability 79, 244–245, 279–281, 285, 314–315, 317, 332, 335–336, 339–340, 343–345, 347–351, 357, 368
phase behavior 112–113, 121, 275, 277–278, 286
pipeline 100, 129–133, 135–136, 140–142, 244, 246, 249–250, 261–262, 264, 268–269, 281, 371, 379,
porosity 244–245, 280–281, 285–286, 294–296, 302, 314–315, 317, 325, 335, 340, 347, 351, 360
PR (see Peng-Robinson)
Prausnitz-Gunn 19–27
ProMax 130, 135, 151, 205
propane 131, 146
pumparound 189–191

regenerator (also see stripper) 157, 168, 171–172, 196, 198–202, 212–213

safety 266, 268, 280, 281, 283, 312, 315
safety factor 133, 136, 150
SAFT (see PC-SAFT)
sandstone 249, 264
shale gas 79–96, 178, 227, 339–354
Soave-Redlich-Kwong 29–30, 49–54, 87, 102, 119
solubility 99, 111, 130, 182–184, 189, 233, 279, 312, 314
sour gas 129, 186, 228–229, 244, 256
SRK (see Soave-Redlich-Kwong)
stripper (also see regenerator) 155, 195–196, 198, 203–205, 208–209, 213–214, 218–220
sulfur 178, 264, 267, 275, 281, 391
sulfur dioxide 177–179
surface tension 111–112, 118–120, 122, 156, 186, 282

Index 397

TAG (see acid gas)
TEG (see glycol)
thiol (see mercaptan)
toluene (see BTEX)
triethylene glycol (see glycol)

viscometer 112
viscosity 111–113, 117–118, 120, 122, 156, 203, 279, 285, 335, 344, 345, 347
VMGSim 29–30, 69, 130, 133–137, 141, 145–146

water coning 331–336
water content 100, 129–130, 133, 135–142, 144, 147–150, 277, 279
well 1, 79, 80, 86, 92, 100, 244–246, 250, 251, 255, 256, 258–261, 263–270, 272–273, 275–277, 279–281, 283, 286, 313–316, 318–319, 323–324, 328, 331–332, 335, 339–340, 347–354

z-factor (see compressibility factor)

Also of Interest

Check out these forthcoming related titles coming soon from Scrivener Publishing

Now Available from the Same Author:
Acid Gas Injection and Carbon Dioxide Sequestration, by John J. Carroll, ISBN 9780470625934. Provides a complete overview and guide on the hot topics of acid gas injection and CO2 sequestration. *NOW AVAILABLE!*

Carbon Dioxide Thermodynamic Properties Handbook, by Sara Anwar and John J. Carroll, ISBN 9781118012987. The most comprehensive collection of carbon dioxide (CO2) data ever compiled. *NOW AVAILABLE!*

Advances in Natural Gas Engineering Series, Volumes 1, 2, and 3:
Sour Gas and Related Technologies, edited by Ying Wu, John J. Carroll, and Weiyao Zhu, ISBN 9780470948149. Written by a group of the most well-known and knowledgeable authors on the subject in the world, volume three focuses on one of the hottest topics in natural gas today, sour gas. This is a must for any engineer working in natural gas, the energy field, or process engineering. *NOW AVAILABLE!*

Acid Gas Injection and Related Technologies, edited by Ying Wu and John J. Carroll, ISBN 9781118016640. Focusing on the engineering of natural gas and its advancement as an increasingly important energy resource, this volume is a must-have for any engineer working in this field. *NOW AVAILABLE!*

Carbon Dioxide Sequestration and Related Technologies, edited by Ying Wu and John J. Carroll, ISBN 9780470938768. volume two focuses on one of the hottest topics in any field of engineering, carbon dioxide sequestration. *NOW AVAILABLE!*

Other books in Energy and Chemical Engineering:

Zero-Waste Engineering, by Rafiqul Islam, February 2011, ISBN 9780470626047. In this controvercial new volume, the author explores the question of zero-waste engineering and how it can be done, efficiently and profitably.

Advanced Petroleum Reservoir Simulation, by M.R. Islam, S.H. Mousavizadegan, Shabbir Mustafiz, and Jamal H. Abou-Kassem, ISBN 9780470625811. The state of the art in petroleum reservoir simulation. *NOW AVAILABLE!*

Energy Storage: A New Approach, by Ralph Zito, ISBN 9780470625910. Exploring the potential of reversible concentrations cells, the author of this groundbreaking volume reveals new technologies to solve the global crisis of energy storage. *NOW AVAILABLE!*

Ethics in Engineering, by James Speight and Russell Foote, December 2010, ISBN 9780470626023. Covers the most thought-provoking ethical questions in engineering.